PARAMETERIZATION SCHEMES: KEYS TO UNDERSTANDING NUMERICAL WEATHER PREDICTION MODELS

Numerical weather prediction models play an increasingly important role in meteorology, both in short- and medium-range forecasting and global climate change studies. Arguably, the most important components of any numerical weather prediction model are the subgrid-scale parameterization schemes. These parameterization schemes determine the amount of energy that reaches the Earth's surface; determine the evolution of the planetary boundary layer; decide when subgrid-scale clouds and convection develop and produce rainfall; and determine the influence of subgrid-scale orography on the atmosphere. The analysis and understanding of parameterization schemes is a key aspect of numerical weather prediction.

This is the first book to provide in-depth explorations of the most commonly used types of parameterization schemes that influence both short-range weather forecasts and global climate models. Each chapter covers a different type of parameterization scheme, starting with an overview explaining why each scheme is needed, and then reviewing the basic theory behind it. Several parameterizations are summarized and compared, followed by a discussion of their limitations. Review questions at the end of each chapter enable readers to monitor their understanding of the topics covered, and solutions are available at www.cambridge.org/9780521865401

Parameterization Schemes: Keys to Understanding Numerical Weather Prediction Models will be an essential reference for academic researchers, meteorologists, weather forecasters, and graduate students interested in numerical weather prediction and its use in weather forecasting.

DAVID J. STENSRUD is a research meteorologist at the National Severe Storms Laboratory, National Oceanic and Atmospheric Administration, Norman, Oklahoma. He is also an adjunct professor at the University of Oklahoma and was Editor or Joint Chief Editor for *Weather and Forecasting*, a professional journal of the American Meteorological Society, from 1999 to 2006.

PARAMETERIZATION SCHEMES

Keys to Understanding Numerical Weather Prediction Models

DAVID J. STENSRUD

National Severe Storms Laboratory
National Oceanic and Atmospheric Administration
Norman, Oklahoma

CAMBRIDGE UNIVERSITY PRESS
Cambridge, New York, Melbourne, Madrid, Cape Town, Singapore,
São Paulo, Delhi, Dubai, Tokyo

Cambridge University Press
The Edinburgh Building, Cambridge CB2 8RU, UK

Published in the United States of America by Cambridge University Press, New York

www.cambridge.org
Information on this title: www.cambridge.org/9780521126762

© D. Stensrud 2007

This publication is in copyright. Subject to statutory exception
and to the provisions of relevant collective licensing agreements,
no reproduction of any part may take place without the written
permission of Cambridge University Press.

First published 2007
This digitally printed version 2009

A catalogue record for this publication is available from the British Library

ISBN 978-0-521-86540-1 Hardback
ISBN 978-0-521-12676-2 Paperback

Cambridge University Press has no responsibility for the persistence or
accuracy of URLs for external or third-party internet websites referred to in
this publication, and does not guarantee that any content on such websites is,
or will remain, accurate or appropriate.

For Audrey, Matthew, and Caitlin,
the three great joys of my life,
and my parents, Stanton and Betty, who set a wonderful example.

"Make a joyful noise to the Lord, all ye lands." (*Psalms 100:1*)

Contents

Preface

Numerical weather prediction models are playing an ever increasing role in meteorology. Not only are numerical models the foundation of short- and medium-range forecasting efforts, they are key components in studies of global climate change. Gaining insight into the physical processes behind many atmospheric phenomena often rests upon studying the output from numerical models, since observations of sufficient density or quality are not available. This ubiquitous use of numerical models suggests that all aspects of numerical models need to be understood very well by those who use the models or examine their output. However, it is my belief that the meteorological community has stressed an understanding of computational fluid dynamics above an understanding of the subgrid-scale parameterization schemes that play a significant role in determining model behavior. While some may argue in defense that parameterization schemes are fluid and changing all the time, implying that studying them is fruitless, closer inspection reveals that many basic assumptions used in the parameterization of specific processes have changed little over the past decades. The study of parameterization schemes also opens a window that allows one to examine our most fundamental ideas about how these important physical processes function and explore how they behave. Thus, the study of parameterization is a vital and necessary component in the study of numerical weather prediction and deserves far greater attention.

There are two main goals of this book. First, to provide the reader with sufficient background to be able to read and scrutinize the literature on the major model physical process parameterization schemes. While your personal list of the major parameterization schemes may differ from those in the text, the schemes selected for examination are used in a large majority of the numerical models available today. The second goal is to develop a deeper understanding of the various physical processes that are parameterized in numerical models and why their parameterization is necessary.

Each chapter begins with an overview that summarizes why this particular physical process is important to represent in a numerical model. Most chapters also contain a review of the theory behind the physical process being parameterized or highlight the observations upon which the parameterization is built. Several different parameterization schemes for this physical process are then presented and compared. Unfortunately, it is impossible to present all the parameterization schemes available and so choices had to be made. The parameterization schemes presented are intended to represent the spectrum of schemes available in commonly used numerical models, but are in no respect an exhaustive list. Each chapter also contains a discussion section in which concerns that have been raised regarding the parameterization are highlighted and future areas of research outlined. Questions are available at the end of each chapter that allow one to delve deeper into the various schemes and to gain an appreciation of their sensitivities. I encourage everyone to answer these questions and to ponder their implications. It is my fervent hope that this book will help those who use models in their research, and those who use model output in making forecasts, to understand both these tools and the atmosphere better.

My interest in parameterization schemes arose during my Ph.D. studies at The Pennsylvania State University when I first began using a numerical weather prediction model (The Pennsylvania State University – National Center for Atmospheric Research Mesoscale Model). While exploring the simulation of a mesoscale convective system in a weakly forced large-scale environment, sensitivities to small changes in the convective parameterization were seen. As the model matured over the following years and began to include a greater number of parameterization scheme options, the sensitivities of a given simulation to changes in the parameterizations selected became readily apparent in nearly all the simulations I studied. These sensitivities may not be seen in all models, due to the limited variety of parameterization schemes available for use in some models, but they are significant and raise a number of important issues regarding the way we view and use numerical weather prediction models. Thus, a graduate-level course in parameterization schemes was born.

This book began as a set of handwritten notes for a graduate course I taught within the School of Meteorology at the University of Oklahoma (OU) in 1994. The class has since been offered every 2 years, allowing every graduate student at OU the opportunity to take it if they so choose. Over the years the students asked for copies of my notes and in response I slowly put them into an electronic format. Once the notes were finalized, the transition to a book format was manageable if not easy. However, I could not have completed this book without the support of Jeff Kimpel and David Rust of the National Severe Storms Laboratory (NSSL) and Fred Carr of OU. Jeff and Dave have supported both my regular teaching stints at OU and the time it has taken to complete this book. Fred has graciously allowed me to teach this rather unusual course within the School of Meteorology all these years as an adjunct faculty member, entrusting many of the School's students to my care each time I teach. Other faculty members have encouraged their students to take my class, and I am thankful for their support and encouragement. Since teaching and dealing with university regulations is not my regular job, I greatly appreciate the support and kindness of everyone in the School of Meteorology office. I also want to recognize and thank Bob Maddox for first inspiring me to teach this course at OU and for supporting my teaching during his tenure at NSSL.

A number of other people also have my sincere thanks and heartfelt gratitude. The students who were brave enough to take this course over the past 12 years contributed greatly to the evolution of this book by the questions they asked and their enthusiasm for meteorology. I have enjoyed teaching all of you! I also have a long list of chapter reviewers who helped me to improve the material presented. Thus, my deepest thanks go to Jeff Anderson, Jian-Wen Bao, Stan Benjamin, Toby Carlson, Fei Chen, Brad Ferrier, Matt Gilmore, Jack Kain, Young-Joon Kim, Brian Mapes, Bob Rabin, Chris Snyder, Roland Stull, Steve Weiss, Lou Wicker, Bob Zamora, and Conrad Ziegler for providing very helpful and professional reviews. The roots for Chapter 2 are in my notes from a course in bioclimatology taught by Toby Carlson at Penn State, and I am grateful to Toby for his wonderful teaching ability and his encouragement. I also want to recognize Bob Davies-Jones, John Locatelli, Chris Godfrey, and Tadashi Fujita for their assistance with specific aspects of the book. Joan O'Bannon drafted several of the more complex figures and her assistance is greatly appreciated. Working with Matt Lloyd, Lindsay Barnes, Louise Staples and Dawn Preston at Cambridge University Press was very easy and I appreciate their support and assistance.

As with all human endeavors, mistakes are guaranteed. I hope that my mistakes will not hamper your reading and understanding of the material. Choosing the notation for many parts of this book was particularly challenging, and so several variables have different meanings in different chapters. While this is not the best situation, the only other alternative was to depart from the notation commonly used in the literature for certain parameterizations. I decided to stay as close as possible to the conventional notation and accept the consequences. If you find any mistakes, I would be grateful if you would let me know via email to David.Stensrud@noaa.gov so that I may correct them. Finally, my admiration goes out to all those scientists who develop the parameterization schemes that help make numerical weather prediction models the success they are today.

David J. Stensrud

List of principal symbols and abbreviations

a	Albedo (0 to 1), various constants
b	Cloud cover fraction (0 to 1), various constants
c_g	Soil heat capacity
c_p	Specific heat at constant pressure ($1004\,\mathrm{J\,K^{-1}\,kg^{-1}}$)
d	Distance from sun to the Earth, displacement depth for the log wind profile
$\bar{d}$	Mean distance from sun to the Earth ($1.50 \times 10^{11}\,\mathrm{m}$)
e	Vapor pressure
$\bar{e}$	Turbulent kinetic energy
$e_s(T)$	Saturated vapor pressure at temperature T
f	Coriolis parameter
g	Acceleration due to gravity at the surface of the Earth ($9.81\,\mathrm{m\,s^{-2}}$)
h	Local hour of the sun
k	von Karman's constant (~ 0.4)
k_e	Entrainment coefficient (<1)
k_g	Thermal molecular conductivity
k_v	Plant resistance factor
k_ν	Absorption coefficient for radiation
k_w	Hydraulic conductivity
l	Mixing-length
l^2	Scorer parameter
n_o	Intercept parameter for drop distributions
p	Pressure
q	Specific humidity
q_c	Cloud water mixing ratio
q_i	Cloud ice mixing ratio
q_r	Rain water mixing ratio

q_s	Snow mixing ratio
q_v	Water vapor mixing ratio
$q_s(T)$	Saturation specific humidity at temperature T
r	Drop radius (cloud droplets, raindrops)
r_a	Resistance in atmospheric surface layer
r_b	Resistance of the interfacial sublayer to sensible heat flux
r_{bv}	Resistance of the interfacial sublayer to latent heat flux
r_c	Canopy resistance
r_H	Resistance to sensible heat flux
r_M	Resistance to momentum flux
r_V	Resistance to latent heat flux
u	East–west component of velocity
u_*	Friction velocity
v	North–south component of velocity
w	Vertical component of velocity
w^*	Free convection scaling velocity
w_p	Total column precipitable water
z	Height
z_{eff}	Effective height
z_0	Roughness length for momentum
z_{0h}	Roughness length for sensible heat flux
z_{0v}	Roughness length for latent heat flux
A_c	Cloud fractional area
API	Antecedent precipitation index
B_ν	Planck function
$CAPE$	Convective available potential energy
CIN	Convective inhibition
CWP	Cloud water path
D_v	Water vapor diffusivity
D_w	Soil water diffusivity
E	Collection efficiency (0 to 1)
E_p	Potential evaporation
EL	Equilibrium level
F_D	Downward directed radiative flux
F_U	Upward directed radiative flux
Fr	Inverse Froude number
I	Radiance, or intensity
K	Mixing coefficient used in K-theory
K_e	Kersten number

K_m	Eddy viscosity
L	Monin–Obukhov length, empirical length scale
LAI	Leaf area index
L_f	Latent heat of fusion ($3.34 \times 10^5\,\mathrm{J\,kg^{-1}}$ at 273.16 K)
L_v	Latent heat of vaporization ($2.500 \times 10^6\,\mathrm{J\,kg^{-1}}$ at 273.16 K)
LCL	Lifting condensation level
LFC	Level of free convection
M	Moisture availability (0 to 1)
M_c	Convective mass flux
N	Brunt-Väisälä frequency
$NDVI$	Normalized difference vegetation index
PAR	Photosynthetically active radiation
Q_1	Apparent heat source
Q_2	Apparent moisture source
Q_S	Incoming solar radiation
Q_{Lu}	Upwelling longwave radiation
Q_{Ld}	Downwelling longwave radiation
Q_H	Sensible heat flux
Q_E	Latent heat flux
Q_{EB}	Bare soil evaporation
Q_{EV}	Vegetation transpiration
Q_{EW}	Canopy evaporation
Q_G	Ground heat flux
Q_R	Heating rate due to radiation
R_d	Individual gas constant for dry air ($287\,\mathrm{J\,K^{-1}\,kg^{-1}}$)
R_{NET}	Total radiation from both shortwave and longwave components
R_v	Individual gas constant for water vapor ($461\,\mathrm{J\,K^{-1}\,kg^{-1}}$)
R_r	Roughness Reynolds number
Re	Reynolds number
Re^*	Roughness Reynolds number
Ri	Richardson number
RH	Relative humidity
S	Solar irradiance, supersaturation, maximum intercepted canopy water
T	Temperature
T_c	Temperature at cloud base
T_g	Ground temperature
T_{SST}	Sea surface temperature
V_r	Fall speed of precipitation particles
W_c	Intercepted canopy liquid water content

α_c	Charnock's constant
δ	Solar declination angle
δ_i	Unit vector
δ_{mn}	Kronecker delta
ε_a	Atmospheric emissivity
ε_{ijk}	Alternating unit tensor
ε_g	Ground surface emissivity
ζ	Solar zenith angle
θ	Potential temperature
θ_v	Virtual potential temperature
κ	Thermal diffusivity of the air
κ_m	Molecular thermal diffusivity of air ($0.18 \times 10^{-4}\,\mathrm{m^2\,s^{-1}}$)
κ_{soil}	Soil hydraulic conductivity
κ_w	Soil water thermal conductivity
λ	Entrainment coefficient, slope parameter
ν	Kinematic viscosity
ν_g	Soil thermal diffusivity
π	Exner function
ρ, ρ_a	Air density
ρ_w	Water density ($1000\,\mathrm{kg\,m^{-3}}$ at $273.16\,\mathrm{K}$)
σ	Stefan–Boltzmann constant ($5.6767 \times 10^{-8}\,\mathrm{W\,m^{-2}\,K^{-4}}$)
σ_f	Vegetation fraction (0 to 1)
τ	Momentum flux or Reynolds stress, optical thickness
τ_s	Transmissivity
τ_ν	Transmission function
φ	Latitude
ψ_g	Gravitational potential
ψ_m	Correction to log wind profile for non-neutral conditions
ψ_m	Matric potential
ψ_o	Osmotic potential
ψ_p	Plant potential
ψ_s	Saturation matric potential
ψ_{soil}	Soil potential
ω	Vertical velocity in pressure coordinates
$\overline{\Delta}$	Change in saturation vapor pressure with respect to temperature
Θ	Volumetric water content
Θ_{fc}	Volumetric water content at field capacity
Θ_S	Soil porosity
Θ_w	Wilting point

1

Why study parameterization schemes?

1.1 Introduction

The weather forecasts depicted in brilliant colors on television, in the newspapers, and on the Internet are providing ever greater details about how the atmosphere is going to evolve over the coming days and even the coming seasons. Both these details and the length of these predictions are due in large part to the increasing processing power of computers and the improving numerical weather prediction models that run on them. Numerical weather prediction models are computer software programs based upon the mathematical equations of motion describing the flow of fluids. Given the present state of the atmosphere, as estimated from weather observations across the globe, these models are able to move the atmosphere forward in time using a sequence of small steps and thereby predict a future state. Not only are these models a critical component in making weather forecasts for the coming week and season, but versions of these models are used to examine how increasing greenhouse gases influence future global climate. Thus, numerical models are important in making decisions not only about daily human activities but about how to be good stewards of planet Earth.

The initial models used for numerical weather prediction (NWP) were simplified versions of the complete equations of motion and were applied over relatively small portions of the globe. In 1949, Charney, Fjørtoft, and von Neumann produced the first one-day weather forecast using a one-layer barotropic model (Charney *et al.* 1950). Another 5 years passed before a barotropic model was used to produce routine forecasts of the 500 hPa flow patterns out to 3 days in an operational forecast center (Shuman 1989; Kalnay 2003; Persson 2005a, b, c). A number of problems still had to be overcome before a multi-vertical-layer quasi-geostrophic model became operational in 1962, with the acquisition of a faster computer system being an important enabling factor (Shuman 1989). A six-vertical-layer primitive equation model became operational in 1966 (Shuman and Hovermale 1968). Each of these models had to be able to compete with or improve upon the manual human-generated forecasts available at the time and each new model needed to be more skillful than its predecessor. Under these constraints, over a short time span of 17 years numerical weather prediction evolved from a simplified model forecast of a single atmospheric layer to a multi-layer primitive equation model capable of predicting cyclone development.

Numerical models also were being developed and used by the research community for a variety of purposes. Non-hydrostatic cloud-scale numerical models to study thunderstorms, hydrostatic mesoscale models to examine the details of cyclogenesis, and hydrostatic general circulation models to study climate all appeared in the late 1970s and slowly started to develop their own user communities. While the simulations often required days of computer time on supercomputers at national research centers, these models were developed to provide an improved understanding of atmospheric processes and so the computer time required to complete the simulation was not a significant factor. As more and more researchers began using these models, the model developers responded by making the models more user friendly, more computer efficient, and began providing documentation on how to use the models.

While both the models used in operations and in research continued to advance throughout the ensuing years, arguably the next major advancement in operational weather prediction was the arrival in 1980 of models that made predictions for the entire globe (Sela 1980). In the early 1990s local computer resources became sufficient for numerical model forecasts to be produced at universities and smaller research laboratories (Warner and Seaman 1990; Cotton *et al.* 1994). These local modeling activities have only increased over the years and have provided many unique opportunities for public education and public service (Mass and Kuo 1998; Mass *et al.* 2003). Improved computer capabilities also led to the operational implementation of the first mesoscale model in the United States in 1995 (Black 1994). Since the 1990s, numerical models have continued to develop along with even greater improvements and affordability in the needed computational resources. Thus, the complexity and sophistication of numerical weather prediction models has increased tremendously since the first forecast in 1949, thanks to both model improvements and the continued availability of ever larger and faster computers.

The evolution towards increasing numerical model complexity has both positive and negative aspects. The positive aspects of this evolution are that the model forecasts are more accurate, more skillful, and produce features that often very closely resemble what actually occurs in the atmosphere. The improvement in model skill has contributed greatly to the improved forecasts that are delivered to the public and influence numerous weather-sensitive industries across the world. The negative aspects of this evolution are that the model behaviors are more difficult to understand, and that errors in the model can be much more difficult to find and correct.

In tandem with the improvement in numerical models came the realization that the atmosphere is sensitive to slight changes in initial conditions (Lorenz 1963). Even the tiniest of errors in the atmospheric initial state is capable of growing quickly and eventually overwhelming the numerical forecast. By introducing very small differences into a model's initial condition, and comparing the forecasts generated from this perturbed initial condition to the forecast from the original unperturbed initial condition, one finds that the differences grow with time. After about two weeks the differences are large enough that the two forecasts are as different as two forecasts started from initial conditions on the same day but from different years (Lorenz 1969). Thus, instead of producing a single (deterministic) forecast, Epstein (1969) and Leith (1974) suggested producing an ensemble of forecasts in order to provide information on forecast uncertainty.

An ensemble is simply a group of forecasts valid over the same time period. A common method of generating an ensemble is to create a number of different model initial conditions, that all lie within the range of analysis uncertainty, and to use each of these initial conditions to produce a separate model forecast. More recently the use of different models in the ensemble has been shown to provide additional value. Ensembles have been used in medium-range forecasting since the early 1990s (see Kalnay 2003) and for short-range forecasting since 2001. The use of an ensemble alters the way in which model guidance is used from a deterministic perspective to a probabilistic perspective that provides richer information for the end users of weather information. The computational cost of ensembles is high, however, since each additional member requires a separate forecast to be produced. While ensembles have been found to be very beneficial, the basic tool – the numerical model – is the same. Only the way in which this numerical tool is applied has been changed when using ensembles instead of single model forecasts.

1.2 Model improvements

Most of the improvements in numerical models that occurred over the past 50 years can be categorized as either improved numerical techniques, improved model resolution, or improved model physical process parameterization schemes. In addition to model improvements, data assimilation methods that judiciously incorporate the wide variety of available observations into numerical model initial conditions also have been very important in increasing forecast skill and their importance should not be underestimated. Daley (1991) discusses many of the advancements and remaining challenges in data assimilation.

Numerical techniques are the methods by which the equations of motion are stepped forward in time. The equations of motion that govern fluid flow are called partial differential equations and are common in mathematical models of many different physical, chemical, and biological phenomena. The partial differential equations that govern fluid flow cannot be solved analytically and so numerical techniques are needed to convert the continuous partial differential equations into a set of algebraic equations that can be solved using a computer (Durran 1999). The numerical techniques differ in the strategies used for representing the original continuous equations by a finite data set that can be stored on a computer and in the computation of derivatives. The basic approaches used are grid point methods, series expansions, and finite-element methods (Haltiner and Williams 1980; Durran 1999; Kalnay 2003).

Model resolution refers to the horizontal and vertical scales that can be resolved or reproduced by the numerical model. Since each algebraic operation used to step the numerical model forward in time requires computer processor time, memory, and disk storage, the number of computations that can be completed in a given time period or for a given memory and disk size is limited even for the fastest computers. This means that the atmosphere cannot be represented perfectly by the numerical model and instead is approximated by a finite data set. Regardless of the type of model, at some point in the computation the atmosphere is represented by a three-dimensional set of points, called a grid, that covers the region of interest (e.g., country, continent, or globe). These points often are regularly spaced horizontally and represent the state of the atmosphere at the point in question (Fig. 1.1). It is easily seen that the number of discrete grid points determines how well the atmospheric structures are represented by the model. As the number of discrete grid points increases, the structures in the atmosphere are represented (or resolved) with increasing accuracy (Fig. 1.1b). However, for a given number of grid points there are always structures small enough that they cannot be captured. When looking at the horizontal distribution of grid points, the grid created often resembles a chessboard when viewed from above with the value of the variables at each grid point representing the conditions within the surrounding grid cell, or the rectangular area for which the grid point is the center point (Fig. 1.2).

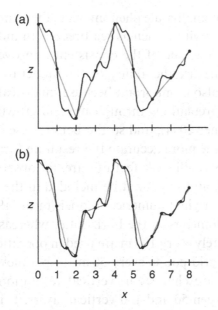

Figure 1.1. Idealized grid point approximation (gray) of a function z (black) plotted on the interval $[0, 8]$. Grid points every 1 in (a) and every 0.5 in (b) are indicated by black circles. Note how the idealized grid point approximation follows the function more closely in (b), although some wave structures are still not captured as clearly seen between the interval $[2.5, 3.5]$.

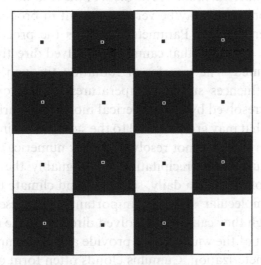

Figure 1.2. Each white or black square represents a grid cell or grid box, whose properties are represented by the grid point value in the middle of the box. In a numerical model, these grid cells are approximately cube-like in shape with an additional dimension in the vertical direction.

Since there are many small-scale phenomena in the atmosphere that are very important to human activities, such as sea breezes, thunderstorms, and snow bands to name just a few, one of the consistent improvements to numerical models has been to increase the ability of the model to resolve smaller and smaller features. This also is important because small-scale features can influence the larger-scale circulations through upscale growth (Thompson 1957; Tribbia and Baumhefner 2004), and so the better these observed small-scale features are captured the more accurate the resulting forecast. The difficulty, of course, is that many small-scale features are not observed with the present observational network and so cannot be included in the model initial conditions. Regardless, the original numerical models in the 1950s had grid points every few hundred kilometers in the horizontal, whereas today models with grid points approximately every 10 km are used in operations and models used in research may have grid points every 50 m. In parallel with the changes in horizontal resolution are changes in vertical resolution, such that models today often have between 50 and 100 vertical layers. It is important to note that a factor of 2 decrease in model grid spacing requires eight times as many grid points on a three-dimensional grid, and the time step generally also must be reduced by a factor of 2, thereby requiring 16 times more computer time! This simple analysis further highlights the important role that increases in computer processor speed play in numerical modeling.

There are always physical processes and scales of motion that cannot be represented by a numerical model, regardless of the resolution. Unfortunately, these unresolved processes may be very important in producing an accurate and useful weather forecast. Parameterization is the process by which the important physical processes that cannot be resolved directly by a numerical model are represented. The transfer of radiation through the atmosphere, which strongly influences surface temperatures, occurs on the molecular scale and so is not resolved by any numerical model. Similarly, the formation of cloud droplets, that may grow and fall to the surface as raindrops, occurs on the molecular scale and so is not resolved by any numerical model. Yet high and low temperatures and precipitation are arguably the most important forecast concerns of people on daily, seasonal, and climate timescales. Thus, processes at the molecular scale are important to represent in numerical models, even though they cannot be resolved directly by the model.

A quick look out of the window may provide another simple illustration of the need for parameterization. Cumulus clouds often form on a sunny afternoon (Fig. 1.3). These clouds are a few hundred meters across, and likely have a similar depth. Most operational forecast models in use in 2005 have horizontal grid spacings of 10 km or larger, roughly 10 to 20 times the size of the

Figure 1.3. A small cumulus cloud developing on a sunny afternoon.

cumulus clouds. When examined from above, it becomes clear that these cumulus clouds cannot be represented explicitly even using 5 km horizontal grid spacing (Fig. 1.4). Since it takes eight or more grid points to represent to some (undefined) level of accuracy any wave-like feature (Haltiner and Williams 1980; Walters 2000), a grid spacing of perhaps 25 m is needed to resolve small cumulus clouds well. Since the grid box is larger than the feature of interest – the cumulus cloud – cloud development and evolution is therefore a subgrid process. If the effects of cumulus clouds are to be included in the evolution of the model variables, then a parameterization that represents the effects of cumulus clouds is needed in the numerical model. If the effects of cumulus clouds are not parameterized, then the model never knows that a cloud formed.

1.3 Motivation

Parameterization schemes are important because they strongly influence model forecasts and interact with each other indirectly through their changes to the model variables. A wet ground surface can lead to strong latent heat flux during the daytime and the development of a shallow and moist planetary boundary layer. The amount of radiation that reaches the ground, the latent heat flux from the ground, and the boundary layer evolution are all determined by parameterization schemes. Furthermore, at the top of the boundary layer cumulus clouds may form and a few of these could eventually grow into

Figure 1.4. Satellite view of a field of cumulus clouds at 1 km resolution forming over Louisiana and Mississippi. Open white polygons represent different sizes of grid cell, varying from 160 km on a side down to 5 km. Even within the 5 km grid cell there are cumulus clouds smaller than the grid cell. Satellite image courtesy of the National Oceanic and Atmospheric Administration.

thunderstorms that produce rain. The development and evolution of these cloud processes also are determined by a parameterization scheme. When clouds are present in the atmosphere, they alter the amount of radiation that reaches the ground. Once the rain reaches the ground surface, it moistens the ground and feeds back to influence the partitioning of the sensible and latent heat fluxes. All of these physical processes from radiation to surface heat fluxes and from the boundary layer to cloud processes are parameterized in most numerical weather prediction models and are critical to the predicted evolution of the atmosphere. So many interactions occur within and between the parameterization schemes and the numerics of models that it is increasingly challenging to discern cause and effect.

Parameterizations almost always focus on the effects of the subgrid physical processes within the vertical column of each individual model grid cell, and only rarely examine what is happening at neighboring grid cells (Fig. 1.5). The vertical orientation of parameterization schemes is chosen since many of the physical processes naturally rearrange energy in this direction. Since

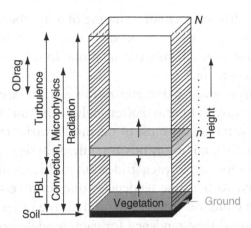

Figure 1.5. Idealized vertical column of model grid cell, with N vertical layers and vertical layer n highlighted. Soil layers are denoted by black shading. Vertical layers where radiation, convection, microphysics, planetary boundary layer (PBL), turbulence, orographic drag (ODrag), and vegetation (and/or bare soil and/or water) parameterizations are likely to affect the model variables are indicated. Vertical arrows emphasize that the parameterizations affect the vertical column only.

parameterizations represent subgrid physical processes for which the model has no direct information, parameterization schemes must relate the subgrid processes to known model variables. For example, the amount of solar radiation that passes through the atmosphere can be related to the cloud cover and water vapor in the vertical column. The formation of cloud droplets can be related to the relative humidity within the grid cell, and the formation of raindrops then can be related to the number of cloud droplets. The specified relationships between the subgrid processes and the known model variables define the parameterization scheme closure. While parameterizations are simplified and idealized representations of complex physical processes, it is common and useful for parameterizations to retain the essential behavior of the process they represent. It also is important to recognize that without parameterization model forecasts are not very interesting or helpful, since the parameterized processes are the most important factors in the forecasts of sensible weather that concern people. Thus, parameterization is a requirement if a model is to provide forecasts for use in weather prediction on any scale of motion.

Parameterization schemes by necessity distill only the essential aspects of the physical processes they represent. Only a limited amount of complexity is possible within a parameterization since it is difficult enough to correctly reproduce the basic behaviors of the physical process for a variety of environmental conditions. Thus, parameterization schemes are an idealized window through which one can gain an understanding of a number of complex atmospheric physical processes when reduced to their most fundamental form. Studying parameterization schemes can help one to understand weather and how weather affects climate.

The outputs from a given parameterization scheme are used to step the numerical model forward in time and often include the time tendencies ($\partial X/\partial t$, where X is a model variable) for most of the model variables, such as temperature, specific humidity, mixing ratios for microphysical particles, and the horizontal wind components at each grid point and vertical level. These time tendencies are added to the time tendencies due to advection, all the other forcing terms in the equations of motion, and all the other parameterization schemes to yield a total time tendency for each model variable. Some parameterization schemes are called at every time step in the model integration, while others may be called less frequently. The individual parameterization scheme time tendencies typically are held constant until the scheme is called again and the values of the time tendencies updated. The total time tendency calculated by the numerical model defines how the model variables change with time and so is very important to the evolution of the model forecasts. Parameterization schemes play a large role in determining this time tendency, further underscoring their importance to numerical weather prediction.

Finally, there are many challenges that increasingly better model forecasts provide to forecasters (Bosart 2003). With every model improvement it seems to become harder and harder for human forecasters to produce forecasts that disagree with the model, even when forecasters are faced with evidence that the model forecast may be incorrect. Bosart suggests that part of this inability to discard model forecasts may be due to the human forecasters losing their analysis skills in this age of enhanced automation, and part of it may be forecaster apathy. However, another contributing factor may be a lack of understanding of how models function, and in particular of how the physical process parameterization schemes behave. These schemes are the dominant players in deciding how the model develops features that are important to daily activities, such as high and low temperatures, rainfall, cloudiness, and winds. These schemes also are crucial to seasonal forecasts and climate assessments. Increased knowledge of model physical process schemes may provide enough understanding and confidence that forecasters are enabled to disagree with the models when presented with good evidence.

In the following chapters the main types of parameterization schemes for the land surface, vegetation, planetary boundary layer and turbulence, convection, microphysics, radiation in clear skies, cloud cover and radiation in cloudy skies, and orographic drag are explored. Many more parameterization schemes exist than can be realistically discussed in a single book. Thus, the intent is not to conduct a complete survey of the literature, but instead to provide the reader with sufficient background information and some reasonable appreciation for the breadth of the literature. It is hoped that by reading the material one can gain enough perspective and sufficient understanding of the parameterization process to read the available literature easily.

1.4 Question

1. Do an Internet search looking for information on operational and/or research numerical weather prediction models. Read over the outlines of what is available with regard to these systems. Based upon the information available on these web pages, list the positive and any negative aspects of each modeling system. Which one seems most complete and user friendly?

2

Land surface–atmosphere parameterizations

2.1 Introduction

It is always enjoyable to walk outside on a cool, clear, crisp day and feel the sun warming your clothes and skin, taking the chill away. The sun's radiation also plays a similar role for the Earth's surface. As sunlight reaches the surface, some of it is reflected back to space, but much of it is absorbed and acts to warm the Earth's surface. In addition, the atmosphere emits longwave (infrared) radiation that acts to warm the surface. The energy provided by these two radiation sources must either be stored in the ground or transferred to the atmosphere via sensible and latent heat flux and longwave radiative flux. The partitioning of this energy into the storage and sensible and latent heat components plays a large role in determining the temperature and humidity near the ground surface, making a day strolling in the park either pleasant or uncomfortable.

The variation in the sensible and latent heat fluxes from location to location is significant (Fig. 2.1). Some of the variations in total flux are due to location influencing the amount of radiation reaching the ground surface. However, much of the horizontal variation in the relative amounts of sensible and latent heat flux is due to the ground surface conditions. The soil type, vegetation type and health, and soil moisture all strongly influence how much energy is partitioned into sensible heat flux (warming) and how much energy is partitioned into latent heat flux (moistening). One can take the same location on the same day and artificially alter the ground conditions in a model such that the curves of sensible and latent heat flux are reversed from what is seen in Fig. 2.1, further illustrating the wide variety of responses that can occur due to surface conditions. This partitioning of sensible and latent heating has a huge influence on the near-surface variables, the depth and structure of the boundary layer, and the potential for precipitation.

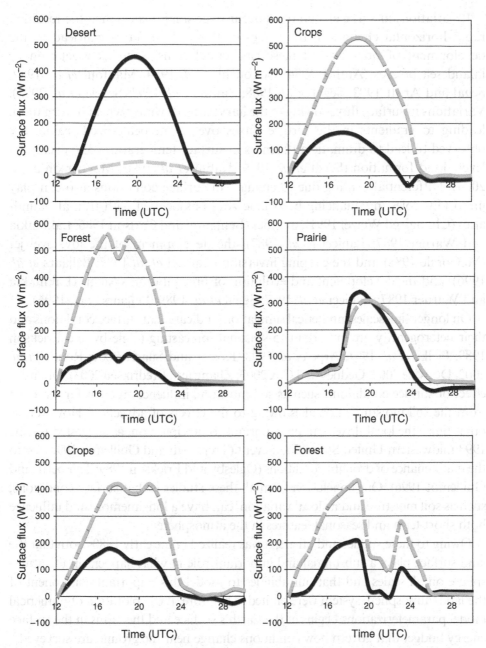

Figure 2.1. Numerical model predictions of sensible (black) and latent (gray) heat fluxes versus time (UTC in hours) for different locations over North America. The general vegetation present at the ground surface is indicated. Note that maximum of the sum of the sensible and latent heat flux is roughly 600 W m^{-2} in many of the curves, yet the partitioning between the sensible and latent heat fluxes is quite different. Thus, the underlying ground surface and vegetation are important to predicting the near-surface variables in NWP models.

If variations in surface conditions occur over small scales, then the resulting rapid horizontal changes in the values of the surface fluxes can lead to the development of non-classical mesoscale circulations, such as vegetation or inland sea breezes (Anthes 1984; Ookouchi *et al.* 1984; Mahfouf *et al.* 1987; Segal and Arritt 1992; Segal *et al.* 1988; Smith *et al.* 1994; Segele *et al.* 2005). Variations in surface fluxes owing to the harvesting of winter wheat or irrigation, leading to gradients in the surface fluxes over a mesocale-sized region, are observed to lead to dramatic differences in surface temperatures and even low-level cloud formation (Segal *et al.* 1989; Rabin *et al.* 1990; McPherson *et al.* 2004). Differential heating due to changes in surface conditions also can play important roles in enhancing baroclinic zones associated with frontal boundaries (Chang and Wetzel 1991), drylines (Benjamin and Carlson 1986; Lakhtakia and Warner 1987; Lanicci *et al.* 1987), the development of the low-level jet (McCorcle 1988) and the capping inversion (Lanicci *et al.* 1987; Beljaars *et al.* 1996), and the development and evolution of precipitation systems (Lakhtakia and Warner 1987; Lanicci *et al.* 1987; Segal *et al.* 1989; Beljaars *et al.* 1996).

On longer timescales, numerical simulations indicate that surface conditions and their heterogeneity are important to seasonal forecasting (Oglesby and Erickson 1989; Pielke *et al.* 1991; Arpe *et al.* 1998; Koster and Suarez 2001; Ronda *et al.* 2002; Douville 2003; Gedney and Cox 2003; Zhang and Fredriksen 2003). Often the effect of surface conditions, such as soil moisture, influences regions far removed from the soil moisture anomalies owing to the effects of advection. However, at other times the local fluxes can be important to precipitation, as suggested for the 1993 midwestern United States flood event (Trenberth and Guillemot 1996), or to the maintenance of drought conditions (Oglesby and Erickson 1989; Trenberth and Guillemot 1996). One conclusion from all these studies is that surface conditions, such as soil moisture and its local variation, can have a long memory and influence both short-term and seasonal features in the atmosphere.

Owing to these studies and others, it was realized in the early 1980s both that the land surface plays a substantial and important role in what happens in the atmosphere on all scales and that our abilities to model this important component of the land–atmosphere system were limited. Therefore, the exploration of numerical model parameterizations begins at the Earth's surface and the terms in the surface energy budget that govern how conditions change near the ground are surveyed.

2.2 Overview of the surface energy budget

The surface energy budget that defines the energy balance at the infinitesimally thin Earth–atmosphere boundary is composed of four main terms: net radiation, sensible heat flux, latent heat flux, and ground heat flux. The net radiation

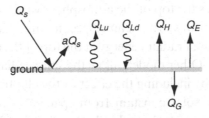

Figure 2.2. Schematic of the fluxes associated with the surface energy budget during a typical daytime situation when incoming solar radiation is present. The arrows indicate the direction of flux relative to the ground surface. Terms are defined in the text.

term typically is separated into parts representing the incoming shortwave radiation from the sun, the reflected shortwave radiation, longwave downwelling radiation, and longwave upwelling radiation. These terms can be sketched on a surface energy budget diagram as in Fig. 2.2, where the direction of the arrow designates the direction of energy transfer.

In the figure, Q_S is the incoming solar radiation, aQ_S is the reflected solar radiation (where a is the albedo), Q_{Lu} is the upwelling longwave radiation from the surface, Q_{Ld} is the downwelling longwave radiation from the atmosphere, Q_H is the sensible heat flux, Q_E is the latent heat flux, and Q_G is the ground heat flux. Unfortunately, the sign conventions for the flux values are inconsistent in the literature. The plus sign is sometimes used to mean that the fluxes are upward into the atmosphere, but sometimes the minus sign is used. The best way to approach this problem is to pay attention to the direction of the flux arrows, i.e. determine if the flux is acting to warm or cool the surface, and then apply the sign that is needed. The surface energy budget simply specifies that the amount of energy going into the surface equals the amount of energy leaving the surface. There is no storage of energy since the Earth–atmosphere boundary is assumed to have zero thickness and no mass.

What processes affect these surface energy budget terms, and thus need to be included in any numerical weather prediction model? Let us take a look term by term.

2.2.1 Incoming solar radiation (Q_S)

The incoming solar radiation that reaches the Earth's surface depends upon a number of factors. The first factor is the amount of radiant energy from the sun that reaches the top of the atmosphere. This amount is related to the solar constant, the amount of radiant energy flux per unit area passing through a plane normal to the solar beam at the Earth's mean distance from the sun. Thus, the radiant energy from the sun that reaches the top of the atmosphere varies with the distance from the sun, and since the Earth's orbit is an ellipse, it varies throughout the year. In some models, the amount of radiant energy at the top of the atmosphere represents the yearly average value. Other models allow the solar radiant energy at the top of the atmosphere to vary by including the effect of the elliptical orbit.

Measurements of the solar constant from space also indicate variability on timescales much longer than the orbital period (Fig. 2.3). Sunspot activity appears to influence the solar constant. In addition, different instruments provide different values for the constant. It appears that there is still more to learn about the solar constant, although some value must be used. Typical values range from 1365 to 1374 W m^{-2}, with the generally accepted value being 1368 W m^{-2}.

Incoming solar radiation may be viewed as a continuous electromagnetic spectrum of waves with wavelengths of approximately 0.15–4.0 µm and with a maximum emission near 0.475 µm corresponding to blue light. Approximately 40% of this spectrum lies in wavelengths within the visible region of 0.4–0.7 µm, 50% in wavelengths longer than visible (infrared), and 10% in wavelengths shorter than visible (ultraviolet). The solar radiation that reaches the Earth's surface is influenced by the path length – the total distance through the atmosphere that the radiant energy passes. When the sun is directly overhead, the path

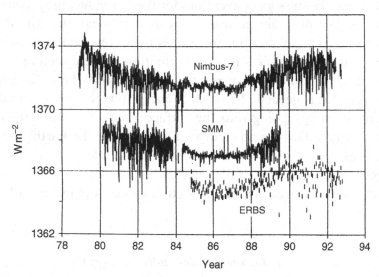

Figure 2.3. Measurements of the solar constant over a more than ten-year period from three different satellites in orbit around the Earth. Note both the longer period cycles, that correspond to sunspot activity, and the different average values produced from the three sensors. The solar constant is difficult to measure accurately, owing to uncertainties in the absolute calibrations, leading to uncertainty in its actual value. From Foukal (1994).

length is defined to be 1. As the sun moves toward the horizon, the path length increases. This is important to the intensity of incoming solar radiation at the surface because as the path length increases more of the radiant energy is scattered by molecules in the atmosphere. In addition, the angle between the Earth's surface and the beam of incoming solar radiation, called the angle of incidence, directly affects the intensity of the radiation that reaches the surface. Just as longer shadows occur early and late in the day, the incoming solar radiation is spread over a larger area when the beam of light from the sun is not perpendicular to the surface, thereby decreasing its intensity over a fixed area.

Cloudiness has a large influence on solar radiation. Cloud droplets act to both scatter and absorb solar radiation. When a photon enters a cloud, it is scattered back and forth by particles of water or ice until it either escapes or is absorbed. Clouds typically appear white because all the colors within sunlight scatter out of the cloud equally. Individually, small clouds may not have a huge influence on the amount of solar radiation that reaches the surface, although the combined effects of many small clouds can be considerable. However, when light enters a large storm cloud, very little of it passes through to the ground. In these situations, blue or green colors can sometimes be seen as one looks up into the cloud. Particles from pollution and aerosols also influence the incoming solar radiation in ways very similar to those of clouds.

Clouds arguably represent the greatest uncertainty in climate change simulations that examine the potential consequences to continued global increases in greenhouse gases owing to the burning of fossil fuels. In addition to both absorbing and scattering solar radiation, clouds absorb and emit longwave radiation, making them very important contributors to the radiation balance of the atmosphere.

Finally, the slope of the ground surface changes the angle of incidence. In regions of steep terrain, some slopes may not receive any direct solar radiation, while in other areas the slopes may receive direct radiation only starting several hours after sunrise (Fig. 2.4). Other parts of the steep terrain may be oriented nearly perpendicular to the direct beam of the solar radiation during certain parts of the year, and thereby receive more solar radiation than expected from a flat surface at these times. Thus, gradients in the incoming solar radiation can be quite large when the terrain is complex, and can feed back and influence the vegetation that survives (Fig. 2.5).

2.2.2 Albedo (a)

Albedo is the fraction (0 to 1) of the incoming solar radiation that is reflected upward from the Earth's surface. For bare surfaces, the albedo is influenced by the soil type. Sand reflects more solar radiation than black dirt, which explains

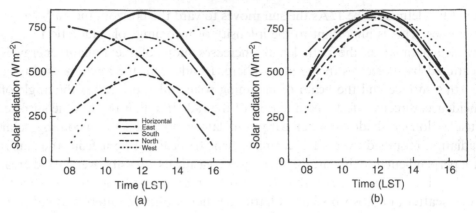

Figure 2.4. Diurnal variation of solar global radiation received on a horizontal surface and on uniformly sloping terrain facing north, south, east, and west on 15 July at a latitude of 37° N. Uniform slope inclinations are (a) 45° and (b) 15°. From Avissar and Pielke (1989).

Figure 2.5. A mountain ridge near Fort Collins, Colorado, illustrating the vegetation contrast on a south-facing slope (left side of ridge) and a north-facing slope (right side of ridge). Also see Avissar and Pielke (1989).

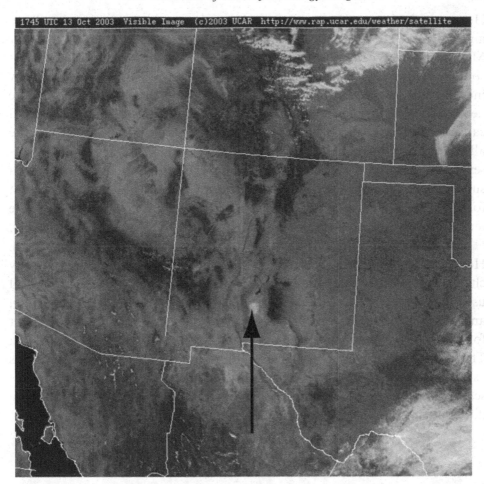

Figure 2.6. Visible satellite image from late morning local time from 13 October 2003 over the southwestern USA. Note the white spot in south-central New Mexico, as indicated by the arrow, which shows the location of the white sands that cover the ground over a relatively large area. © 2006 University Corporation for Atmospheric Research, used with permission.

why White Sands, New Mexico, is so easy to see in visible satellite images (Fig. 2.6). Albedo also is influenced by the vegetation type and coverage, and changes throughout the vegetation life cycle (Moore *et al.* 1996). The angle of incidence also can have a large effect on the albedo, especially for water surfaces. Anyone who enjoys boating or water sports knows that a lot of sunlight is reflected by the water when the sun is low on the horizon, and very little when the sun is directly overhead. Waves on the water also affect the albedo.

Snow cover can change the albedo of a surface dramatically in a very short period of time. New snow is very bright and has an albedo close to 1.0. As snow ages, it typically has a lower albedo as dirt gets blown on top of the snow.

However, the interactions between snow and vegetation are complex. Betts and Ball (1997) find an 850 hPa temperature error in the European Center for Medium-range Weather Forecasting (ECMWF) medium-range forecasts over southwestern Canada during the winter. They hypothesize that when snow initially falls on the trees in the coniferous forests, it accumulates on the tree needles. It remains on the needles until the wind begins to blow. The wind causes the snow to fall off the trees and onto the ground below. Thus, the albedo changes from one representing a snow-covered surface to one representing a dark green surface of pine needles over a very short time period. Tests suggest that when the albedo in southwestern Canada is modified to account for the decrease in albedo associated with the falling of snow off the trees, the temperature errors in the forecasts are reduced.

Finally, human beings influence the albedo in a wide variety of ways. Homes, office buildings, parking lots and roads are built that dramatically change the reflective properties of the ground surface. The vegetation and land use of large regions are altered through farming, irrigation, logging, burning, and watering. Chase *et al.* (2001) propose that changes in land use can account for most of the observed surface warming of the past several hundred years. Thus, one could argue that parameterizations are needed for human activities as a component of NWP, at least when examining seasonal or longer time-scales. This is already accounted for in global climate models by examining different emission scenarios for greenhouse gases.

2.2.3 *Longwave upwelling radiation (Q_{Lu})*

Longwave upwelling radiation is the amount of longwave radiation emitted from the Earth's surface. It may be viewed as a continuous electromagnetic spectrum of waves with wavelengths from 4 to 100 μm. Therefore, longwave radiation has very little overlap with the wavelength spectrum associated with incoming solar radiation. Since the intensity of longwave radiation is explained by the Stefan–Boltzmann law, it is influenced most strongly by the temperature of the ground surface and by the value of surface emissivity (the power emitted by a body at temperature T to the power emitted by a black-body at temperature T). Of course, the ground surface temperature is influenced by incoming solar radiation, albedo, downwelling longwave radiation, soil type, soil moisture, and the temperature of the soil below the ground surface. One can already begin to see the numerous feedbacks in this complex Earth system that sustains life. Vegetation also influences the longwave upwelling radiation, as thick forest canopies can act to retard radiation emitted from the ground via multiple reflectances. The upwelling longwave radiation from thick forests is determined largely by the temperature and emissivity of the vegetation canopy.

2.2.4 Longwave downwelling radiation (Q_{Ld})

Longwave downwelling radiation is the amount of longwave radiation emitted from the atmosphere that reaches the ground surface. It is influenced by the atmospheric temperature, especially from the lower levels of the atmosphere. But many other factors influence downwelling longwave radiation. Clouds play a large role, acting to absorb and emit longwave radiation very efficiently. Greenhouse gases, such as carbon dioxide, ozone, and methane also contribute significantly, as does water vapor. Together, the effects of increasing concentrations of greenhouse gases and the feedback due to water vapor constitute the major unresolved questions regarding global warming. Pollution and aerosols also influence the downwelling radiation, although they vary in concentration and in their characteristics from location to location.

2.2.5 Sensible heat flux (Q_H)

Sensible heat flux is the rate of heat transfer per unit area from the ground to the atmosphere. During a typical cloudless summer day, the sensible heat flux increases during the morning hours, reaching a maximum in the early afternoon before decreasing back towards zero after the sun sets (Fig. 2.7). It is most strongly affected by the difference in temperature between the ground (or water) surface and the overlying atmosphere immediately above the ground. However, observations indicate that the conditions throughout the lower atmosphere also play a role, as the heat flux is modified by wind speed, wind shear, and the vertical temperature gradient within the planetary boundary layer. The presence and type of vegetation also influences the sensible heat flux, owing to the height, coverage, and structure of the vegetation affecting the ground and vegetation canopy temperatures, and the low-level wind shear. In very dense canopies, almost all of the sensible heat flux comes from the vegetation layer.

2.2.6 Latent heat flux (Q_E)

Latent heat flux is the rate of moisture transfer per unit area from the ground (or water) surface to the atmosphere. Similarly to sensible heat flux, it is affected by the surface temperature and the temperature, wind speed, wind shear, and stability of the lower levels of the atmosphere. Vegetation plays an even larger

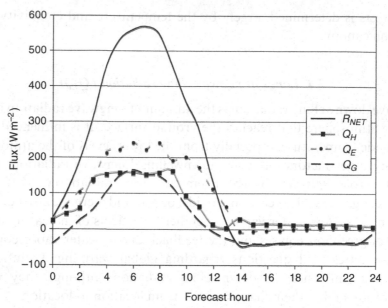

Figure 2.7. Monthly mean diurnal cycle of observed components of the surface energy budget at Norman, Oklahoma, for July 1997. Forecast hour 0 corresponds to 1200 UTC, or 7 a.m. local time. Hourly values of net radiation (R_{NET} is the sum of all four radiation terms), sensible heat flux (Q_H), latent heat flux (Q_E), and ground heat flux (Q_G) are shown. After Marshall *et al.* (2003), where data are from the Oklahoma Mesonet.

role in determining the latent heat flux, as vegetation can transfer moisture from deep soil layers to the atmosphere. Soil type also is important, as water retention differs among the various soil types. And of course, the soil moisture content and the presence of water (lakes, rivers, oceans) are important factors.

A useful parameter for comparing the relative amounts of sensible and latent heat flux is the Bowen ratio (β), defined as the ratio of the sensible to the latent heat fluxes at the surface (Bowen 1926). Over semi-arid regions, the Bowen ratio can exceed a value of 5. Over grasslands and forests, values of 0.5 are typical. Bowen ratios drop to 0.2 over irrigated orchards or grass, and to 0.1 over water. Values of Bowen ratios greater than one correspond to surfaces with more sensible heat flux than latent heat flux, and vice versa when the Bowen ratio is less than one.

As mentioned previously, differences in the relative amounts of sensible and latent heat flux can lead to very different boundary layer depths (Chang and Wetzel 1991; Segele *et al.* 2005), setting limits on whether or not precipitation is likely to occur. In addition, horizontal gradients of sensible and latent heating can enhance the low-level baroclinicity and produce secondary circulations that influence boundary layer structure and convective development

(Lakhtakia and Warner 1987; Chang and Wetzel 1991; Segele *et al.* 2005). These differences can further feed back to the atmosphere and influence rainfall in more distant locations.

2.2.7 Ground heat flux (Q_G)

Ground heat flux is the rate of heat transfer from the ground surface into the deeper soil levels. As with the other fluxes from the surface, it is influenced by surface temperature, soil type, soil moisture, and vegetation. Observations suggest that the upper 25 cm of soil has the largest diurnal changes in temperature, with temperature changes in the deeper soil levels occurring much more slowly. However, on yearly timescales, soil temperatures vary at depths below 1.5 m. While the ground heat flux is often smaller in magnitude than the sensible and latent heat fluxes, as shown in Fig. 2.7, it can be comparable in magnitude and is rarely insignificant.

While there are even more processes that affect these terms in the surface energy budget, this discussion provides some perspective on the complexity of the atmospheric system. Many of the processes listed are interrelated, and have complicated interactions, making their prediction very challenging. As a thought experiment, vary one of the variables and see how many terms in the surface energy budget are influenced by changes to just this one variable. It is very likely that most of the terms in the energy budget are influenced by this chosen variable but in different ways. This complexity and interrelatedness is at the heart of the difficulty that weather prediction faces every day. A brief look at how each term of the energy budget is described mathematically follows, building the foundation for a more detailed look at the soil–vegetation–atmosphere parameterization schemes that occurs in Chapter 3 and water–atmosphere parameterizaton schemes that occurs in Chapter 4.

2.3 Net radiation

2.3.1 Incoming solar radiation

Incoming solar radiation is often calculated, in its simplest form, using

$$Q_s = S\left(\frac{\bar{d}}{d}\right)^2 (1 - a)\cos(\zeta)\tau_s, \tag{2.1}$$

where ζ is the solar zenith angle, S is the solar irradiance (1365–1374 W m^{-2}), d is the distance from the sun to the Earth, $\bar{d}$ is the mean distance from the sun to the Earth, a is the albedo, and τ_s is the transmissivity. The $(\bar{d}/d)^2$ term accounts for the elliptical orbit of the Earth around the sun and departs by no more than 3.5% from unity (Liou 1980). The zenith angle is determined from

$$\cos(\zeta) = \sin(\varphi)\sin(\delta) + \cos(\varphi)\cos(\delta)\cos(h), \qquad (2.2)$$

where φ is the latitude, δ is the solar declination angle, and h is the local hour of the sun. Note that $h=0$ at local solar noon, when the sun is directly overhead. The zenith and declination angles are defined and shown graphically in Figs. 2.8 and 2.9, respectively. The declination angle can be calculated as

$$\delta = 23.45° \cos\left[\frac{2\pi(d - d_{solstice})}{d_y}\right], \qquad (2.3)$$

where d is the Julian day, $d_{solstice}$ is 173, or the day of the summer solstice, and d_y is 365.25, the average number of days in a year. Finally, the local hour of the sun (h) is defined as

$$h = \frac{(t_{UTC} - 12)\pi}{12} + \frac{\lambda\pi}{180}, \qquad (2.4)$$

where t_{UTC} is the coordinated universal time (UTC) in hours (0 to 24), and λ is the longitude (plus for east, minus for west) in degrees. Note that when $\cos(\zeta) < 0$, then $Q_S = 0$ for the sun is below the horizon and there is no solar radiation reaching the ground surface.

The albedo is the fraction (0 to 1) of incoming radiation that is reflected back to space. Values of albedo can range from 0 to near 1 depending upon the land

Figure 2.8. Illustration of the solar zenith angle (ζ), defined as the angle between the direction of a direct beam of radiation from the sun and a line normal to the surface of the Earth.

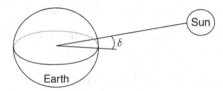

Figure 2.9. Illustration of the declination angle (δ), defined to be the angular distance of the sun north (positive) or south (negative) of the equator when a line is connected between the center of the Earth and the center of the sun.

Table 2.1. *Typical values for albedo for different surface conditions.*

Surface	Albedo
Dark, wet soils	0.05
Coniferous forests	0.10
Grass and agricultural lands	0.20
Light colored soils	0.40
Snow	0.95
Water when sun is directly overhead	0.05
Water when sun elevation angle is low	1.0

surface, and vary most strongly over water. Typical values of albedo are shown in Table 2.1.

The transmissivity term (τ_s) plays a large role in determining the amount of incoming solar radiation. This term incorporates the influences of aerosols, water vapor, and clouds, and as such can be quite complex. A much more complete discussion of it is found in Chapters 8 and 9.

2.3.2 Upwelling longwave radiation

Upwelling longwave radiation is calculated as

$$Q_{Lu} = \varepsilon_g \sigma T_g^4, \qquad (2.5)$$

where ε_g is the emissivity, σ is the Stefan–Boltzmann constant ($5.67 \times 10^{-8}\,\mathrm{W\,m^{-2}\,K^{-4}}$), and T_g is the temperature (K) of the ground surface. Emissivity values for most surfaces on the Earth range from 0.9 to 0.99, where emissivity is defined as the ratio of the emittance to the emittance expected from a blackbody.

2.3.3 Downwelling longwave radiation

The downwelling longwave radiation from the atmosphere to the Earth's surface is much more complicated to calculate and estimate. It is often calculated using complex radiative transfer models as discussed in Chapters 8 and 9. However, a number of approaches have been developed based upon empirical relationships between the downwelling radiation and various meteorological parameters that are routinely measured. While these approaches are not as accurate as the radiative transfer models, and generally should not be used in

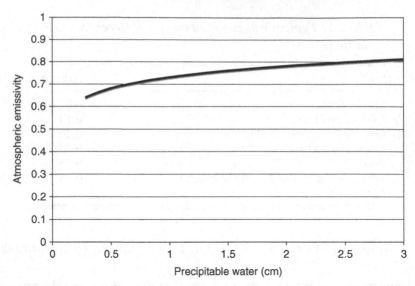

Figure 2.10. Atmospheric emissivity (ε_a) versus total column precipitable water (cm) using the formula from Monteith (1961).

numerical weather prediction models today, they provide a good summary of what is important in the calculation of downwelling radiation. For example, under clear skies Monteith (1961) finds that

$$Q_{Ld} = \varepsilon_g \varepsilon_a \sigma T_a^4, \tag{2.6}$$

where T_a is the air temperature (K) at approximately 40 hPa above ground level, w_p is the total column precipitable water (cm), and ε_a is the atmospheric emissivity, defined as

$$\varepsilon_a = 0.725 + 0.17 \log_{10}(w_p). \tag{2.7}$$

The empirical relationship expressed by (2.6) and (2.7) only considers the temperature 40 hPa above the ground because, in general, more than half the radiant flux received at the ground from the atmosphere comes from gases in the lowest 100 m and roughly 90% from gases within the lowest 1 km. Thus, the emissivity of the air increases in tandem with increasing total column precipitable water as shown in Fig. 2.10.

Typically, clouds dense enough to cast a shadow on the ground emit as a blackbody at the temperature of the water droplets or ice crystals from which they are formed. Clouds strongly affect Q_{Ld} since they emit in the 8–13 μm waveband, called the atmospheric water vapor window, since other absorbers are generally not present in this band. Since most of the atmospheric radiation reaching the surface is emitted from below the cloud base, and the cloud base is

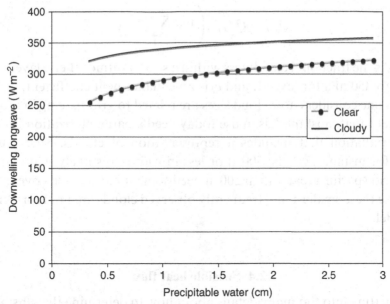

Figure 2.11. Calculated downwelling longwave radiation ($\mathrm{W\,m^{-2}}$) versus the total column precipitable water (cm) for clear (gray line) and 60% cloudy skies (black line) using the formula in (2.8). The air temperature is assumed to be 290 K and cloud base temperature is assumed to be 270 K.

located most often above 1 km, it is possible to treat the gaseous component of downward radiation separately from the cloud component. One approach that is used in simple radiation schemes is

$$Q_{Ld} = \varepsilon_g \varepsilon_a \sigma T_a^4 + b \varepsilon_g (1 - \varepsilon_a) \sigma T_c^4, \qquad (2.8)$$

where b is the cloud fractional area (0 to 1), ε_a is the atmospheric emissivity, and T_c is the temperature of the base of the cloud. The first term on the right-hand side represents the clear sky contribution, with the second term acting to increase the downwelling radiation due to clouds. Assuming an air temperature of 290 K and a cloud temperature of 270 K with 60% cloud cover, the downwelling longwave radiation reaching the surface is enhanced by 11–26% due to the clouds (Fig. 2.11). Cloud cover clearly plays a greater role in the amount of downwelling radiation when the atmospheric precipitable water is low, although clouds do not form unless there is sufficient moisture to reach saturation. Other simple models use a slightly different approach, such as

$$Q_{Ld} = Q_{Ld_{clear}} (1 - b) + b Q_{Ld_{overcast}}, \qquad (2.9)$$

where $Q_{Ld_{clear}}$ is defined similarly to (2.6). Alternatively, one can add enhancements to $Q_{Ld_{clear}}$ by dividing the clouds into layers, such as

$$Q_{Ld} = Q_{Ld_{clear}} \left(1 + \sum_{i=1}^{3} c_i n_i \right),\qquad(2.10)$$

where n_i is the fractional cloudiness within a specific atmospheric layer, say the surface to 850 hPa for layer 1, and c_i is an enhancement coefficient.

When one considers how cloud sizes are related to grid spacing in models, it is apparent that most models in use today need a parameterization for downwelling radiation that includes a representation of clouds. Since cumulus clouds, for instance, can be 500 m or less in width, it is likely that all models with a grid spacing greater than 200 m need some subgrid cloud cover scheme; otherwise these smallest and frequently observed clouds are not represented in the model.

2.4 Sensible heat flux

Before getting into too many details about how to determine the sensible heat flux, it is wise to review the definition of flux. Flux is a rate of flow, or the transfer of a quantity per unit area per unit time. Thus, flux F can be written as

$$F = -u_T(\bar{X}_{top} - \bar{X}_{bottom}),\qquad(2.11)$$

where u_T is a transport velocity or conductivity. Thus, flux depends upon the change in the quantity of interest (X) across the zone over which the flux is calculated (top – bottom) and the speed of the transport (u_T).

In fields outside of meteorology, fluxes are often viewed from a slightly different framework. To understand this framework one must recall Ohm's law from an introductory physics class. This law states that for a given conductor, at a given temperature, the current (I) is directly proportional to the difference of potential (voltage) between the ends of the conductor, such that

$$\text{current} = \frac{\text{voltage}}{\text{resistance}}.\qquad(2.12)$$

Thus, the larger the resistance (R) becomes, the smaller the current that results from a given voltage. As is seen later, this basic resistance framework for defining current can be used to define surface sensible and latent heat fluxes as well. Note that any combination of resistances can always be replaced by a single, equivalent value of resistance.

Resistors in a parallel circuit are illustrated in Fig. 2.12, and lead to the relationship

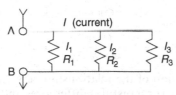

Figure 2.12. Illustration of a parallel circuit with three resistors (R_1, R_2, and R_3) and three different currents (I_1, I_2, and I_3).

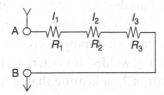

Figure 2.13. Illustration of a series circuit with three resistors (R_1, R_2, and R_3) and the same current across each resistor (I_1, I_2, and I_3).

$$\frac{1}{R} = \frac{1}{R_1} + \frac{1}{R_2} + \frac{1}{R_3}, \tag{2.13}$$

where R is the total resistance and R_1, R_2, and R_3 are the resistances of the individual resistors. Here the voltage is the same across each resistor, which leads to the three currents being defined as

$$I_1 = \frac{V_{AB}}{R_1}, \qquad I_2 = \frac{V_{AB}}{R_2}, \qquad I_3 = \frac{V_{AB}}{R_3}, \tag{2.14}$$

and the total current I equals

$$I = I_1 + I_2 + I_3 = V_{AB}\left(\frac{1}{R_1} + \frac{1}{R_2} + \frac{1}{R_3}\right). \tag{2.15}$$

In contrast, for resistors in a series circuit we have the situation illustrated in Fig. 2.13, where the total resistance is the sum of the individual resistances, such that

$$R = R_1 + R_2 + R_3. \tag{2.16}$$

In a series circuit, the current is the same across each resistor, such that

$$V_{AB} = V_1 + V_2 + V_3 = IR_1 + IR_2 + IR_3 = I(R_1 + R_2 + R_3). \tag{2.17}$$

A few simple examples can be used to highlight the factors that influence the value of the resistance. It is well-known that all metals have a specific

Voltage	=	Current	*	Resistance
$\Delta T = T_2 - T_1$	=	Fluxes $Q_H \ Q_E \ Q_G$	*	??

Figure 2.14. An illustration of the relationship between electric voltage and current, and temperature (or moisture) differences and fluxes.

resistance, so if one takes a wire of a given length l and doubles the length to $2l$, then what happens to the resistance? Since the length of the wire is doubled, it is as though two wires of length l are added together in series. Thus, the resistance doubles, indicating that $R \propto$ length. If instead the cross-sectional area of the wire is doubled, then what happens to the resistance? In this case it is like two identical wires being welded together, and the wires are in parallel. Thus, R is reduced by a factor of 2, indicating that $R \propto 1/$area. Combining these two relations yields $R \propto$ length/area. Since numerical weather prediction models typically define heat flux per square meter, the resistance is only a factor of length – the distance from the ground surface to the first model level. This distance can play a large role in how well the model responds to surface forcing.

It may not yet be obvious how this relates to meteorology and the calculation of the sensible heat flux. One finds that there are analogous relationships between the temperature or moisture difference between the ground surface and the atmosphere, and the electric voltage between two points on a circuit, and also between the electric current and fluxes (Fig. 2.14). However, these analogies lack the precise information needed to calculate the sensible heat flux. Instead, a well-known physical process is used as the basis for deriving an expression for the sensible heat flux.

We know that for conduction, the transport of energy occurs solely as a consequence of the random motion of individual molecules, so the heat flux Q_H in the vertical direction is given by

$$Q_H = -k_a \frac{\partial T}{\partial z}, \tag{2.18}$$

where k_a is the thermal conductivity of air (W m^{-1} K^{-1}). By analogy, one can assume that the vertical heat flux due to turbulent eddies behaves similarly. So, we define a thermal diffusivity (κ), which is equivalent to the thermal conductivity in (2.18) but with the heat transfer being due to turbulence instead of conduction, such that

$$\kappa = \frac{k_a}{\rho c_p}, \tag{2.19}$$

where ρ is the air density and c_p is the specific heat at constant pressure. Note that diffusivity can simply be viewed as another name for a transfer coefficient, such that

$$\overline{w'T'} = -\kappa \frac{\partial \bar{T}}{\partial z}, \tag{2.20}$$

where κ has units of $m^2 s^{-1}$. This expression defines a flux-gradient relationship. If one assumes that temperature varies only in the vertical direction, (2.18) can be rewritten for heat conduction due to turbulent eddies as

$$dT = -\frac{Q_H}{k_a} \, dz = -\frac{Q_H}{\rho c_p} \frac{dz}{\kappa}. \tag{2.21}$$

Recall that resistance is proportional to length divided by area, and that resistance must always be defined over a specified length. If Q_H is assumed to be constant in the surface layer over which the flux is calculated, which is a reasonable assumption, then

$$\int_{T_1}^{T_2} dT = -\int_{z_1}^{z_2} \frac{Q_H \, dz}{\rho c_p \, \kappa} = -\frac{Q_H}{\rho c_p} \int_{z_1}^{z_2} \frac{dz}{\kappa} = -\frac{Q_H}{\rho c_p} r_H, \tag{2.22}$$

where r_H is defined as the resistance to heat flux. Resistance has units that are the inverse of velocity ($s\,m^{-1}$). Resistance depends upon the separation distance over which the flux is determined, so that the model vertical grid spacing becomes very important in determining the resistance over the lowest model layers. In fact, all other things being equal, the thicker the lowest model layer is the larger the resistance, and the slower the response of the model to surface heating. This is one very important way in which the model vertical grid influences directly the physical process parameterization schemes. Rearranging (2.22), the expression for sensible heat flux written in resistance form is

$$Q_H = \frac{\rho c_p \Delta T}{r_H}, \tag{2.23}$$

where $\Delta T = T(z_1) - T(z_2)$ is the temperature difference across a layer of the atmosphere, $r_H > 0$ is the resistance across this same layer, and $z_2 > z_1$. Typically, for the sensible heat flux calculations in numerical models, the layer over which the heat flux is calculated is from the ground surface to the lowest model level so that $z_1 = 0$ and z_2 is some positive value determined by the model vertical grid spacing. When the ground temperature is warmer than the air temperature, then $\Delta T > 0$ and $Q_H > 0$. Note that some investigators prefer to use conductance g instead of resistance where $g \equiv 1/r_H$ (e.g., Cox *et al.* 1998).

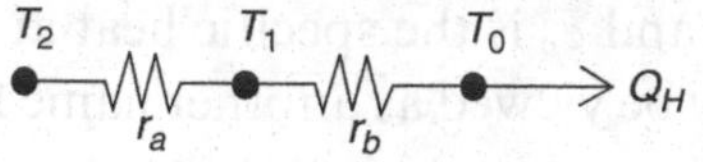

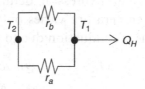

Figure 2.15. Illustration of three different temperatures ($T_2 > T_1 > T_0$), separated by air with resistances r_a and r_b, and with flux Q_H.

Figure 2.16. Illustration of two different temperatures ($T_2 > T_1$), separated by air or vegetation with resistances r_a and r_b, and with flux Q_H.

Now for a few examples of how sensible heat flux is calculated for various situations. First, if the resistances are in series and the flux is constant, i.e. no losses or gains occur along the path, then the situation is as shown in Fig. 2.15, indicating that

$$Q_H = \frac{\rho c_p (T_2 - T_1)}{r_a} = \frac{\rho c_p (T_1 - T_0)}{r_b} = \frac{\rho c_p (T_2 - T_0)}{r_a + r_b} = \frac{\rho c_p (T_2 - T_0)}{r_{total}}. \quad (2.24)$$

Resistances in a series configuration occur, for example, when the sensible heat flux from the ground surface interacts first with a vegetation layer and then with the atmosphere. If the resistances are in parallel, then the situation is as shown in Fig. 2.16, indicating that

$$Q_H = \frac{\rho c_p (T_2 - T_1)}{r_a} + \frac{\rho c_p (T_2 - T_1)}{r_b} = \rho c_p (T_2 - T_1) \left(\frac{1}{r_a} + \frac{1}{r_b} \right)$$
$$= \frac{\rho c_p (T_2 - T_1)}{\bar{r}}, \quad (2.25)$$

where

$$\bar{r} = \left(\frac{1}{r_a} + \frac{1}{r_b} \right)^{-1}. \quad (2.26)$$

Resistances in parallel occur, for example, when both sides of a leaf interact with the atmosphere to transfer water from the plant to the atmosphere. Putting these two scenarios together, then if both parallel and series resistors are found (Fig. 2.17), the heat flux equals

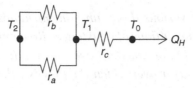

Figure 2.17. Illustration of three different temperatures $(T_2 > T_1 > T_0)$, separated by air or vegetation with resistances r_a, r_b and r_c, and with flux Q_H.

$$Q_H = \frac{\rho c_p (T_2 - T_0)}{\bar{r} + r_c},$$ (2.27)

where the parallel resistances r_a and r_b are replaced by a single resistance $\bar{r}$ as shown earlier.

Resistors in series occur in the equations for both sensible and latent heat flux, as the resistances are calculated from the ground surface to the atmospheric layer within the vegetation canopy, and then from this canopy layer to the atmospheric layer above the canopy. Resistors in parallel also occur in the sensible and latent heat flux equations when there are flux contributions from both bare soil and vegetated surfaces within the same model grid cell. Thus, the scenario shown in Fig. 2.17, and the expression for heat flux in (2.27), are representative of how heat fluxes are defined in a resistance formulation within many models.

Similar to the derivation of the resistance form of the sensible heat flux equation, as shown in (2.18)–(2.23), one can derive resistance equations for the other surface flux terms in the surface energy budget as well. Thus, there are diffusivities for momentum (eddy viscosity), heat (eddy conductivity), and water vapor (eddy diffusivity), and for other variables that are influenced by turbulent motions, such as carbon dioxide, ozone, and methane. The various diffusivities are shown in Table 2.2.

While the resistance to heat flux r_H has been defined, it has not yet been expressed mathematically. In earlier meteorological models, heat flux is typically calculated in terms of bulk transfer coefficients that are multiplied by the wind speed. Even from everyday experiences of living and breathing within the atmospheric surface layer, the lowest 10–40 m above ground level, it is clear that wind gusts act to remove the hotter air that develops near the ground. Thus, intuition suggests that the resistances used in the calculations of sensible and latent heat fluxes should be related to the wind speed. But how are these resistances calculated? The process begins by calculating the momentum flux, from which everything else follows.

The goal is to calculate the resistance to momentum flux r_M within the atmospheric surface layer, since this defines the momentum flux τ and sets

Table 2.2. *Definitions of various fluxes and the variables used to define the diffusivities used in the calculations of resistance. Remember that while the values of diffusivity are assumed to be constant over a given location, the values of resistance change as the layer over which the flux is calculated changes.*

Variable	Flux	Diffusivity
Temperature (T)	$Q_H = \dfrac{\rho c_p \Delta T}{r_H}$	κ
Specific humidity (q)	$Q_E = \dfrac{\rho L_v \Delta q}{r_V}$	D_v
Wind speed (V)	$\tau = \dfrac{\rho \Delta V}{r_M}$	ν

the stage for the other resistance calculations. Close to the ground we know from the kinetic theory of gases that for laminar flow

$$\tau = \nu \frac{\partial}{\partial z}(\rho \bar{u}), \qquad (2.28)$$

which states that the stress is directly proportional to the wind speed gradient. This is valid in the lowest few millimeters above z_0, the roughness height, where the mean wind speed goes to zero and molecular processes are dominant.

It is assumed that the same equation can be generalized to turbulent flow, such that

$$\tau = K_m \frac{\partial}{\partial z}(\rho \bar{u}), \qquad (2.29)$$

where K_m is an eddy viscosity ($\sim 1 \ \mathrm{m^2 \, s^{-1}}$). This expression again defines a flux-gradient relationship. One important difference between these two equations for momentum flux, (2.28) and (2.29), is that while ν is constant for a given p and T, K_m varies depending upon the surface roughness, buoyancy forces, and the height above the ground.

At this point, some insight into the behavior of the atmosphere is needed to decide how best to proceed. What is known observationally that could help to determine the momentum flux or stress? Observations and scaling analyses show that in neutral planetary boundary layers, where the potential temperature is constant with height and the sensible heat flux is zero, one finds the relationship

$$\frac{\bar{u}(z)}{u_*} = \frac{1}{k} \ln\left(\frac{z}{z_0}\right). \qquad (2.30)$$

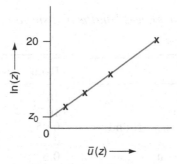

Figure 2.18. Illustration of the log wind profile, where the crosses are observed mean wind speeds.

This famous expression is called the log wind profile. Here k is von Karman's constant (~ 0.4), u_* is the friction velocity ($\mathrm{m\,s}^{-1}$), and z_0 is the roughness length (m). The log wind profile is a characteristic change of wind speed with height that is observed under neutral stability conditions across the globe. When the wind speed is plotted versus the natural logarithm of the height of the observation, a straight wind speed profile is found (Fig. 2.18). One of the key parameters for this wind profile is the roughness length, z_0, the height above ground at which the mean wind speed goes to zero. By taking the derivative of (2.30) with respect to z, the log wind profile specifies that

$$\frac{\partial \bar{u}}{\partial z} = \frac{u_*}{kz}. \tag{2.31}$$

Thus, the change in wind speed with height is inversely proportional to the height above ground level of the observation, and the wind speed changes most rapidly just above the ground and more slowly as the height increases.

Numerous observational studies have evaluated the log wind profile and calculated the roughness length over various surfaces. Results indicate that over ice or smooth water the roughness length can be as small as 10^{-4} m, while over forests or cities the roughness length can be as large as several meters. A crude estimate for the roughness length is obtained using $z_0 \sim h/8$, where h is the height of the vegetation. Table 2.3 summarizes some of these results to illustrate the variations in roughness length that are observed.

Returning to (2.29), the equation for momentum flux that started the hunt for an expression for r_M, and assuming that ρ is a constant throughout the layer in which the log wind profile is observed, one finds that the momentum flux can be written as

$$\tau = \rho K_m \frac{\partial \bar{u}}{\partial z}. \tag{2.32}$$

Table 2.3. *Roughness lengths measured over various surface types.*

Surface	Roughness length (m)
Ice	10^{-4}
Grass (mown)	10^{-2}
Long grass, rocky ground	0.05
Pasture	0.20
Suburban housing	0.6
Forests, cities	1–5

Knowing that $u_*^2 = (\tau/\rho)$ and assuming that u_* is a constant, one can rearrange a few terms to get an expression for the eddy viscosity,

$$K_m = \frac{(\tau/\rho)}{(\partial \bar{u}/\partial z)} = \frac{u_*^2}{(u_*/kz)} = u_* kz. \qquad (2.33)$$

Therefore, using the equation for momentum stress in resistance form from Table 2.2, one can arrive at an equation for the momentum resistance r_M, such that

$$r_M = \frac{\rho \Delta u}{\tau} = \frac{\rho \Delta u}{\rho K_m(\partial \bar{u}/\partial z)} = \frac{\Delta \bar{u}}{u_* kz(\Delta \bar{u}/\Delta z)} = \frac{\Delta z}{u_* kz}. \qquad (2.34)$$

When integrated from the ground to the top of the surface layer in question (and in which the log wind profile generally is valid), this becomes

$$r_M = \int_{z_0}^{z} \frac{dz}{u_* kz} = \frac{1}{u_* k} \ln\left(\frac{z}{z_0}\right) \equiv \frac{1}{u_* k} \ln\left(\frac{z_{eff}}{z_0}\right), \qquad (2.35)$$

where z_{eff} is the effective height. Finally, then, we have a mathematical expression for the resistance to momentum transfer!

Before moving on, there are a few things that need to be mentioned. First, recall that resistance is always defined over a particular depth as indicated in (2.35). As the thickness of model layers changes, the resistance also changes. Second, the use of the variable z_{eff} in (2.35) is a relatively standard form of the equation. The effective height is due to the observation that some roughness elements when packed close together, such as forest canopies, act like a displaced ground surface. This typically only occurs over a fairly uniform forest canopy or in urban areas with relatively high-density housing. In this case, the flow behaves as though the ground surface has been raised upward to near the height of the tree or house tops. Although the log wind profile is still observed, its apparent origin is moved upward to a height d above the ground surface. The effective height is defined as $z_{eff} = z - d$ to account for this displacement when needed. The exact height of d for a given canopy depends upon how the drag force is distributed through the foliage, and the structure of the mean wind and turbulence within the canopy. In general, a value of d representing a height that is 75% of the average canopy height seems to work well (Kaimal and Finnigan 1994). For most applications, $d = 0$ is used in the calculations.

In completing this derivation of the resistance, several assumptions are used that are true within the atmospheric surface layer. Remember that the surface layer is a "thin resistor" where fluxes of sensible heat, latent heat, and momentum change by less than 10%. Typically the surface layer is less than 50 m deep during the daytime and less than 20 m deep at night. The assumptions made are that the mean motion is one-dimensional (no horizontal fluxes are important), the Coriolis force can be neglected, the shearing stress and pressure gradient force can be neglected with respect to the viscous force, and the mixing length depends upon the distance of the fluid from the boundary (see Munn 1966).

Complications arise because the atmosphere is not often neutral within the surface layer. Thus, the standard form of the log wind profile,

$$\bar{u}(z) = \frac{u_*}{k}\ln\left(\frac{z}{z_0}\right) = \frac{u_*}{k}\ln\left(\frac{z_{eff}}{z_0}\right) = \frac{u_*}{k}\ln\left(\frac{z - d}{z_0}\right), \qquad (2.36)$$

is not valid very often. In cases for which the surface layer is unstable or stable, observations indicate that the actual wind profile varies from the values it takes under neutral conditions (Fig. 2.19). The wind speeds are greater than those found under neutral conditions for unstable conditions and less than those found under neutral conditions for stable conditions. Thankfully,

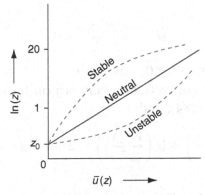

Figure 2.19. Illustration of the typical wind profiles observed under stable, unstable, and neutral conditions. The log wind profile is the wind profile for neutral conditions.

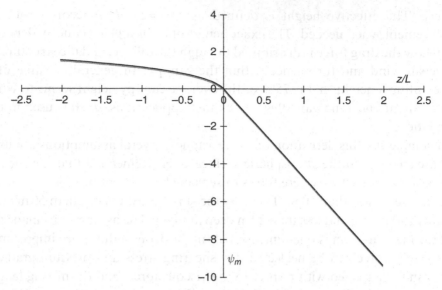

Figure 2.20. ψ_m plotted as a function of z/L, where L is the Monin–Obukhov length. Values of $z/L < 0$ are for unstable conditions, while those of $z/L > 0$ are for stable conditions.

observational results indicate that empirical corrections can be made to the log wind profile to account for departures from neutral stability and that these apply in a variety of locations (Dyer 1963, 1974; Businger *et al.* 1971). The form of this correction (Fig. 2.20) is

$$\bar{u}(z) = \frac{u_*}{k} \left[\ln\left(\frac{z_{eff}}{z_0}\right) - \psi_m\left(\frac{z_{eff}}{L}\right) \right], \tag{2.37}$$

where L is the Monin–Obukhov length. For stable conditions ($L > 0$), the ratio z_{eff}/L represents the relative suppression of mechanical turbulence by the stratification, and

$$\psi_m = -\frac{4.7 z_{eff}}{L} < 0. \tag{2.38}$$

For unstable conditions ($L < 0$), the ratio $-z_{eff}/L$ represents the relative importance of buoyant convection in comparison to mechanical turbulence (Panofsky and Dutton 1984), and

$$\psi_m = 2 \ln\left(\frac{1+x}{2}\right) + \ln\left(\frac{1+x^2}{2}\right) - 2 \tan^{-1}(x) + \frac{\pi}{2} > 0, \tag{2.39}$$

where

$$x = \left(1 - 15\frac{z_{eff}}{L}\right)^{1/4} \tag{2.40}$$

and

$$L = -\frac{\bar{\theta}_v u_*^3}{kg\overline{(w'\theta_v')}_{sfc}}.$$ (2.41)

The Monin–Obukhov length L is generally small and negative on strongly convective days ($L \sim -10$ m), is likely near -100 m on days with strong winds and some solar heating, and is positive and very large for days with mostly mechanically generated turbulence (Panofsky and Dutton 1984).

This correction to the log wind profile produces an alteration to the equation for momentum resistance, such that the resistance to momentum for both neutral and non-neutral surface layers is expressed as

$$r_M = \frac{1}{ku_*}\left[\ln\left(\frac{z_{eff}}{z_0}\right) - \psi_m\left(\frac{z_{eff}}{L}\right)\right].$$ (2.42)

A brief examination of (2.42) and Fig. 2.20 shows that the effect of turbulent flux generally is to reduce r_M during the day and to increase r_M at night. This makes physical sense, since the effects of turbulent motions from strong heating of the ground are expected to produce stronger mixing, and hence greater heat flux. In contrast, when conditions are stable and the turbulence from buoyancy is greatly reduced, the heat flux should be relatively small.

It is important to be aware that other forms of this correction to the log wind profile under non-neutral conditions are given in the literature (e.g., Swinbank 1968; Zilitinkevich and Chalikov 1968; Lobocki 1993) and generally yield similar behaviors. Yaglom (1977) suggests that the differences in these various formulations are due to instrumentation and measurement error. However, the different formulations yield different values of ψ_m that influence the low-level wind speed profiles.

Now that an expression for the resistance to momentum has been derived, which applies for neutral, unstable, and stable boundary layer environments, it can be used to help derive an equation for the resistance to heat flux. Recall that the resistance form of the sensible heat flux equation is

$$Q_H = \frac{\rho c_p (T_{z_a} - T_{z_b})}{r_{H(a,b)}},$$ (2.43)

where z_a and z_b are two specified vertical levels. When calculating the sensible heat flux from the ground surface to the atmosphere, T_{z_a} is often assumed to be equal to the ground temperature (T_g) and T_{z_b} is often assumed to be equal to the air temperature. Thus, the temperature difference in the numerator of (2.43) is the difference between the ground temperature and the temperature of the closest model level to the ground surface. This suggests, and observations also indicate, that the "origin" of the heat flux lies somewhere below the origin for momentum flux, which occurs at the height of the roughness length z_0 (e.g., Garratt and Hicks 1973; Blyth and Dolman 1995). This changes the depth over which the resistance is calculated when compared to the momentum resistance.

As discussed by Garratt (1992), the transfer of momentum very close to the ground surface is influenced by pressure fluctuations in the turbulent wakes that occur behind the roughness elements. However, for heat transfer no such dynamical mechanism exists to transfer the heat, so the heat is transferred by molecular diffusion across the layer from the surface to the height where turbulence processes dominate. This layer is called the interfacial sublayer and plays an important role in controlling the sensible and latent heat fluxes. The interfacial sublayer is located between the ground surface and the surface layer.

For neutral conditions, it is assumed that $r_M = r_H = r_V$ over the same depth, such that

$$r_{H(z,z_0)} = \frac{1}{ku_*} \ln\left(\frac{z_{eff}}{z_0}\right), \tag{2.44}$$

which is valid from height z down to z_0. However, as is shown later, this expression does not account for the resistance within the interfacial sublayer that must be accounted for in the resistance calculations for heat flux. For non-neutral conditions, observations again suggest that corrections can be made (Paulson 1970), yielding an expression very similar to the resistance to momentum, such that

$$r_{H(z,z_0)} = \frac{1}{ku_*}\left[\ln\left(\frac{z_{eff}}{z_0}\right) - \psi_h\left(\frac{z_{eff}}{L}\right)\right], \tag{2.45}$$

where

$$\psi_h = 2\ln\left[\frac{(1+x^2)}{2}\right] \quad \text{for} \quad L < 0 \tag{2.46}$$

and

$$\psi_h = -\frac{4.7z}{L} \quad \text{for} \quad L > 0, \tag{2.47}$$

where x in (2.46) is defined by (2.40). As with the equations for ψ_m, several versions of the equations for ψ_h are available in the literature and yield similar differences in the actual values of ψ_h as discussed earlier for ψ_m (see Yaglom 1977). Weidinger *et al.* (2000) indicate that the use of different functions for ψ_h can result in important differences in the surface flux calculations.

Now that a corrected equation for the resistance to heat flux has been derived for both neutral and non-neutral conditions within the surface layer, it is time to account for the additional resistance that occurs in the interfacial sublayer between z_0, the roughness length for momentum flux, and the lower roughness length for heat flux, which is denoted as z_{0h}. Note that

$$\ln\left(\frac{z_{eff}}{z_{0h}}\right) = \ln\left(\frac{z_{eff}}{z_0}\right) + \ln\left(\frac{z_0}{z_{0h}}\right), \tag{2.48}$$

and therefore one finds that

$$r_H = \frac{1}{ku_*}\left[\ln\left(\frac{z_{eff}}{z_0}\right) - \psi_h\left(\frac{z_{eff}}{L}\right)\right] + \frac{1}{ku_*}\ln\left(\frac{z_0}{z_{0h}}\right) = r_a + r_b \tag{2.49}$$

represents the total resistance to sensible heat flux from the surface to height z for both neutral and non-neutral conditions, where r_b is the added resistance of the interfacial sublayer.

Typically, evidence suggests that $z_{0h} \ll z_0$ (Garratt and Hicks 1973). Unfortunately, it still is not clear how one calculates z_{0h}. Several possibilities for the determination of z_{0h} are shown by Garratt and Hicks (1973) and are supported by observations, namely

$$z_{0h} = \frac{z_0}{7} \tag{2.50}$$

or

$$z_{0h} = \frac{\alpha\kappa_m}{ku_*}, \tag{2.51}$$

where $\kappa_m = 0.18 \times 10^{-4}\,\mathrm{m^2\,s^{-1}}$ is the molecular thermal diffusivity of air and α is a fudge factor that is typically set to 1. It also is advocated that z_{0h} should depend upon the sensible heat flux, canopy height, vegetation biomass, incoming solar radiation, or perhaps the local flow conditions (Brutsaert 1982; Garratt 1992; Kubota and Sugita 1994). Zilitinkevich (1995) proposes that the ratio of the two roughness lengths should depend upon the roughness Reynolds number (Re^*), such that

$$\frac{z_0}{z_{0h}} = \exp(kC\sqrt{Re^*}), \tag{2.52}$$

where C is a constant, assumed by Chen *et al.* (1997) to be 0.1, and

$$Re^* = \frac{u_* z_0}{\nu}.$$ (2.53)

Chen *et al.* (1997) suggest that this formulation of z_{0h} is most useful, since it allows for variability in the roughness length for heat as a function of the atmospheric flow conditions. Hopwood (1995) suggests that the ratio of z_0 / z_{0h} is near 80 for inhomogeneous, vegetated surfaces, indicating that a substantial variability in z_{0h} needs to be included in models.

For model grid cells that contain a mixture of bare ground and sparse vegetation, Blyth and Dolman (1995) show using a dual-source model, which solves the energy balance over vegetation and soil separately, that the value of z_{0h} used in a model that combines the vegetation and soil components (as is true in many numerical atmospheric models) is dependent upon the humidity deficit, the available energy, the vegetation fraction, and the surface resistance of soil and vegetation. They further show that errors in the estimated values of z_{0h} from single-source models can be as large as a factor of 30! This error has a larger influence on the resistance calculations as the roughness length increases, and can produce errors in the resistance values of over 20%. This result argues for smaller grid spacing in models, in order to have each grid cell contain only bare ground or only vegetation. However, the computational cost of such a grid would be prohibitive. An alternative is the use of subsections, or tiles, within each model grid cell that represent the major vegetation types observed and upon which fluxes are calculated individually for each vegetation type before being combined to a single flux for the entire grid cell. However, it is not certain how often these large errors in the estimated values of z_{0h} occur. As emphasized by Zeng and Dickinson (1998), there is no generally accepted guidance available on how to provide a priori estimates of z_{0h} over heterogeneous surfaces for use in numerical modeling. Further research on this important question is certainly warranted.

2.5 Latent heat flux

Similarly to the calculation of sensible heat flux, the latent heat flux, or moisture flux, is defined as

$$Q_E = \frac{\rho L_v (q_{z_a} - q_{z_b})}{r_{V(a,b)}},$$ (2.54)

where L_v is the latent heat of vaporization, and z_a and z_b again refer to specific heights over which the flux is calculated. Often one would prefer to use vapor pressure instead of specific humidity. We know that

$$q \approx \frac{0.622e}{p},$$ (2.55)

where the vapor pressure term in the denominator is neglected. Using this approximation, one can derive

$$Q_E = \frac{\rho c_p (e_{z_a} - e_{z_b})}{r_{V(a,b)} \gamma},$$ (2.56)

where

$$\gamma = \frac{c_p p}{0.622 L_v} \approx 0.66 \text{ hPa K}^{-1},$$ (2.57)

and the total resistance to latent heat flux under both neutral and non-neutral conditions is

$$r_{V(0,z_{eff})} = \frac{1}{ku_*} \left[\ln \left(\frac{z_{eff}}{z_0} \right) - \psi_h \left(\frac{z_{eff}}{L} \right) \right] + \frac{1}{ku_*} \ln \left(\frac{z_0}{z_{0v}} \right).$$ (2.58)

No one has determined a correction to the moisture flux owing to non-neutral conditions, likely owing to the difficulties of observing small moisture gradients accurately, and hence it is assumed that the correction to the heat flux equation for non-neutral conditions is good enough to approximate the corrections to the moisture flux. It also is assumed that the "origin" of the moisture flux at height z_{0v} is near the height of the heat flux origin z_{0h}. But what is z_{0v} exactly? The answer is no one really knows for certain and it is an artifact of the derivation. Several options for estimating this quantity have been suggested. The first is that since

$$z_{0h} = \frac{\alpha \kappa}{ku_*},$$ (2.59)

then the roughness length for moisture might be represented by a similar formula, but with the thermal diffusivity κ replaced by the moisture diffusivity D_v, yielding

$$z_{0v} = \frac{\alpha D_v}{ku_*}.$$ (2.60)

The second option suggested is that

$$\frac{1}{ku_*} \ln \left(\frac{z_0}{z_{0v}} \right) = \frac{1}{ku_*} \ln \left(\frac{z_0}{z_{0h}} \right) \left(\frac{\kappa}{D_v} \right)^{2/3},$$ (2.61)

where κ/D_v is the ratio of the two diffusivities and equals approximately 0.93.

Selecting (2.61) for representing the additional moisture flux resistance below z_0, one obtains the complete equation for the resistance to moisture flux

$$r_{V(0,z_{eff})} = \frac{1}{ku_*}\left[\ln\left(\frac{z_{eff}}{z_0}\right) - \psi_h\left(\frac{z_{eff}}{L}\right)\right] + \frac{1}{ku_*}\ln\left(\frac{z_0}{z_{0h}}\right)\left(\frac{\kappa}{D_v}\right)^{2/3} = r_a + r_{bv},$$
(2.62)

where r_{bv} is the added resistance of the interfacial sublayer. Then to summarize, for moisture flux we have

$$Q_E = \frac{\rho L_v(q(z_{0v}) - q(z_{eff}))}{r_a + r_{bv}}.$$
(2.63)

When examining (2.63), an obvious question is what is $q(z_{0v})$? Is this value of specific humidity at some level just barely above the ground surface measured routinely? What does it indicate? The answer is that this value of specific humidity is not measured and it likely cannot be measured. One could even put forth the argument that no one really knows what it is meant to be. So to be able to calculate moisture flux, an artifact of our inability to measure $q(z_{0v})$ requires that one somehow defines a surface value for q, which leads to

$$Q_E = \frac{\rho L_v(q_S(T_g) - q(z_{eff}))}{r_a + r_{bv} + r_c} = \frac{M\rho L_v(q_S(T_g) - q(z_{eff}))}{r_a + r_{bv}},$$
(2.64)

where $q_S(T_g)$ is the saturation specific humidity at ground temperature T_g, r_c is the canopy resistance, and

$$M = \frac{r_a + r_{bv}}{r_a + r_{bv} + r_c},$$
(2.65)

which is often called the moisture availability. From the definition of M in (2.65) it is easily seen that M is constrained to vary between 0 and 1. As the canopy resistance goes to zero, M goes to 1. And as the canopy resistance becomes very large, M goes to zero. The canopy resistance represents the resistance the plant canopy must work against to obtain water from the soil.

Unfortunately, there is no real reason to suppose that $q_S(T_g)$ is proportional to $q(z_{0v})$. Perhaps the worst-case scenario to examine is a rain forecast in which the very tall tree canopy, perhaps 70 m above the ground, has a temperature and moisture flux that is quite unrelated to what happens at the ground and to the ground temperature. Typically, models that use moisture availability overestimate Q_E in humid regions (Sato *et al.* 1989).

2.5.1 Moisture availability

If one is interested in using M as a method to help calculate the latent heat flux, then are there any physical variables that can be related to M? Perhaps the soil volumetric water content (Θ) and field capacity (Θ_{fc}) could be related to M. Volumetric water content is defined as the ratio of the cubic meters of water in the soil to the cubic meters of dry soil. Soil field capacity is defined as the maximum volume of water a given unit volume of soil can hold after drainage. This can be determined experimentally by putting soil into a container with a sieve at the bottom, and pouring water into the soil until it is saturated and water is dripping out the bottom of the container. When the water stops dripping, the soil is at its field capacity. Thus, one can relate M to the volumetric water content and the field capacity using the formulas from Lee and Pielke (1992), where

$$M(z) = \frac{1}{4}\left\{1 - \cos\left[\frac{\Theta(z)\pi}{\Theta_{fc}}\right]\right\}^2 \quad \text{for} \quad \Theta < \Theta_{fc}, \qquad (2.66)$$

$$M(z) = 1 \quad \text{for} \quad \Theta \geq \Theta_{fc}. \qquad (2.67)$$

In models that use M, typically the value of M is fixed or varies only slightly during the course of a day. However, since M is defined in terms of r_c, one should be aware that r_c varies during the daytime hours. In fact, r_c is a maximum at sunrise and sunset, and reaches a minimum near solar noon if the vegetation is not stressed (Monteith and Unsworth 1990).

Another approach to determining a value for M is to use an antecedent precipitation index (API). Chang and Wetzel (1991) show an easy approach that uses daily values of API, calculated over a several month period, to specify a value for M. They define

$$API_i = k_{API}\, API_{i-1} + P_i, \qquad (2.68)$$

where k_{API} is a decay coefficient ($\sim$0.92 is typical) for a given soil layer and P_i is the precipitation that fell on day i. The precipitation total for a single day is not allowed to exceed a selected maximum value, such that if $P_i > P_m$ then $P_i = P_m$. In Chang and Wetzel (1991), P_m is set to 4 cm. If this formula is used iteratively for several months, the final value should be independent of the initial value of API selected to start the calculations. The best hope for this technique is that the volumetric water content is related to the API value, and from there a soils database can be used to get the field capacity of the soil and thus get an approximation for M.

For historical significance, the bucket approach (Manabe 1969) also was used to define M in many models dating from the 1960s. The bucket approach assumes that a bucket exists at each model grid point. Rain falling into this bucket increases the depth of the water within the bucket. Water is removed from the bucket owing to the amount of latent heat released into the atmosphere. The bucket has a maximum depth, typically assumed to be 15 cm, and any amount of water exceeding this depth runs out the top of the bucket. The value of M is then defined as the height of the water in the bucket divided by the total depth of the bucket. This approach was used in the nested grid model (NGM) and many earlier global circulation models.

2.5.2 Penman–Monteith approach

Is there any other way to determine the latent heat flux, since the original definition as shown by (2.64) includes terms that are not really physical? Assume that there is a need to calculate the latent heat flux using conventional surface data. The Penman–Monteith approach is one way to estimate the latent heat flux using surface data (Penman 1948; Monteith 1965) and is derived as follows. We know that, through simple manipulation,

$$e_s(T_g) - e_a = e_s(T_g) - e_a + e_s(T_a) - e_s(T_a)$$
$$= e_s(T_g) - e_s(T_a) + [e_s(T_a) - e_a], \qquad (2.69)$$

where e is the vapor pressure, $e_s(T)$ is the saturated vapor pressure at temperature T, T_g is the ground temperature, T_a is the air temperature at 2 m, and e_a is the vapor pressure at 2 m. We also know that we can rewrite

$$e_s(T_g) - e_s(T_a) = \frac{[e_s(T_g) - e_s(T_a)](T_g - T_a)}{T_g - T_a} = \frac{\partial \bar{e}_s(T)}{\partial T}(T_g - T_a),$$
$$= \bar{\Delta}(T_g - T_a), \qquad (2.70)$$

where $\bar{\Delta}$ is the change of saturation vapor pressure with respect to temperature and can be approximated by a constant. Combining the results from (2.70) into (2.69), one obtains

$$e_s(T_g) - e_a = \bar{\Delta}(T_g - T_a) + e_s(T_a) - e_a. \qquad (2.71)$$

Now, the latent heat flux can be written as

$$Q_E \cong M \frac{e_s(T_g) - e_a}{\gamma R_H}, \qquad (2.72)$$

where

$$R_H = \frac{r_H}{\rho c_p}, \tag{2.73}$$

assuming that $r_H = r_V$. Therefore, using (2.71) one can define the latent heat flux as

$$Q_E = \frac{[\bar{\Delta}(T_g - T_a) + e_s(T_a) - e_a]M}{\gamma R_H}. \tag{2.74}$$

This derivation can be taken a step further by noting that

$$Q_H = \frac{T_g - T_a}{R_H} \quad \Rightarrow \quad T_g - T_a = Q_H R_H, \tag{2.75}$$

which leads to

$$Q_E = M\left[\frac{\bar{\Delta}Q_H}{\gamma}\left(\frac{R_H}{R_H}\right) + \frac{e_s(T_a) - e_a}{\gamma R_H}\right]. \tag{2.76}$$

If it is further assumed that the ground temperature does not change over time, as is true under highly vegetated canopies where this equation is designed to work, then

$$Q_E \cong M\left[\frac{\bar{\Delta}}{\gamma}(R_{NET} - Q_G - Q_E) + \frac{e_s(T_a) - e_a}{\gamma R_H}\right], \tag{2.77}$$

where R_{NET} is the total net radiation from both short and longwave components of the radiation budget. Combining like terms, this equation is simplified a bit further to develop the expression

$$Q_E = \frac{M\{[e_s(T_a) - e_a]/R_H + (R_{NET} - Q_G)\bar{\Delta}\}}{\gamma + \bar{\Delta}M}, \tag{2.78}$$

which is often called the Penman–Monteith equation. The first term of (2.78) can be viewed as representing the drying power of the air, since it is proportional to the surface layer vapor pressure deficit. The second term of (2.78) can be viewed as the drying power of the sun, since it is dominated by the incoming solar radiation on most days. Potential evapotranspiration is often defined using (2.78) when $M = 1$.

The Penman–Monteith equation circumvents the problem associated with using the saturation vapor pressure at ground temperature to represent the vapor pressure conditions at the Earth's surface, which is particularly severe over desert regions with very high ground temperature values. However, the Penman–Monteith formulation does not remove the importance of the moisture availability parameter M to the calculation of latent heat flux. And since it was developed for highly vegetated surfaces, it should not be applied over the entire planet without due consideration.

2.5.3 Priestley–Taylor approach

Another approach to calculating the latent heat flux from surface observations is the Priestley–Taylor equation (Priestley and Taylor 1972). They assume that the latent heat flux is proportional to the drying power of the sun, yielding

$$Q_E = \frac{\alpha(R_{NET} - Q_G)\bar{\Delta}}{\gamma + \bar{\Delta}}. \tag{2.79}$$

This is the same as neglecting the first term in the Penman–Monteith formulation (2.78), or assuming that the atmosphere near the ground surface is saturated. Values of α less than 1 indicate that the soil is dry (0.4 being the driest), while values of α greater than 1 indicate that the soil is reasonably well watered. Davies and Allen (1973) relate the soil volumetric water content to α, such that

$$\alpha = a\left(1 - e^{-b\Theta/\Theta_{fc}}\right), \tag{2.80}$$

where $a = 1.28$ and $b = 10.563$. Thus, $\alpha = 1.28$ for saturated soil conditions as expected, but the value of α can decrease to near zero as Θ/Θ_{fc} decreases below 0.2. Williams *et al.* (1978) and Barton (1979) use the same approach, but find different values for a and b, suggesting that the relationship between α and Θ/Θ_{fc} is not very general. This approach is most appropriate for use over vegetation with shallow roots, or bare soil, although it is not the best choice in truly desert regions. Stull (1988) and Brutsaert (1982) further indicate that the Priestley–Taylor equation can yield incorrect flux values when advection occurs.

2.6 Ground heat flux

The final term in the surface energy budget is the ground heat flux. When averaged over a full 24 h cycle, this term often averages to near zero. This observation led many pioneers using general circulation models to set this term to zero. Another alternative that was used in early studies that forecast the diurnal cycle was to relate the ground heat flux to the total net radiation. Observations indicate that Q_G is typically between 5% and 15% of the net radiation during the daytime, although this fraction increases to approximately 50% at night. Others tried relating Q_G to the sensible heat flux by defining $Q_G = 0.3 \, Q_H$. Deardorff (1978) presents a good summary of these early methods of neglecting or defining Q_G. However, these approximations did not last very long and soon multi-level soil modules were in use in all numerical weather prediction models.

Multi-level soil models include an equation for the ground surface, or skin, temperature plus several specific soil levels below the ground surface. Since the heat transport process is conduction, the amount of heat transferred is proportional to the vertical temperature gradient. Therefore,

$$Q_G = -k_g \frac{\partial T}{\partial z},\tag{2.81}$$

where k_g is the thermal molecular conductivity, it is assumed that the ground heat flux $Q_G = Q_g$ at the ground surface ($z = 0$), and z increases in the downward direction. Assuming no other sources or sinks of heat, the second law of thermodynamics yields

$$\frac{\partial T}{\partial t} = -\frac{1}{c_g} \frac{\partial Q_g}{\partial z},\tag{2.82}$$

where c_g is the soil heat capacity and is equal to the soil density multiplied by the soil specific heat ($c_g = \rho_{soil} \cdot c_{soil}$). Assuming that the thermal conductivity is independent of soil depth, then these two equations can be combined to yield the classic heat conduction equation for the ground heat flux,

$$\frac{\partial T}{\partial t} = -\nu_g \frac{\partial^2 T}{\partial z^2},\tag{2.83}$$

where ν_g is the soil thermal diffusivity ($\nu_g = k_g/c_g$ and varies from 1×10^{-6} to $1 \times 10^{-7} \, \mathrm{m^2 \, s^{-1}}$). Therefore, one can set up equations for calculating the heat transfer into the ground using selected vertical levels below ground for computation, and given an initial soil temperature profile (Deardorff 1978).

A slightly different approach to predicting the ground heat flux is to simplify the soil model to just two vertical slabs (Bhumralkar 1975; Blackadar 1976). The first slab is in contact with the ground surface and typically represents a soil depth of a few cm. The second slab is considered to be a reservoir and has a constant temperature. Thus, the upper slab responds to both changes in the surface energy balance and conduction from the lower reservoir below.

If the depth of the upper slab is d_s, then a soil heat capacity per unit area can be calculated for the upper slab as $c_G = c_g \cdot d_s$. The equation for the temperature of the slab is then

$$c_G \frac{\partial T_g}{\partial t} = R_{NET} - Q_H - Q_E - \kappa(T_g - T_r),\qquad(2.84)$$

where T_r is the reservoir temperature and κ is a soil conductivity. Results indicate that the simulations using this approach are sensitive to the depth of the upper soil layer, with shallower soil layers yielding larger responses, as one would expect. Applications of this approach to handling soil temperatures are found in the land surface parameterizations of Noilhan and Planton (1989) and Mahfouf (1991).

When the thermal conductivity varies with depth, as one might expect when soil moisture is considered, then (2.82) is used (Pan and Mahrt 1987; Viterbo and Beljaars 1995) and the soil heat capacity and thermal conductivity are allowed to vary with soil moisture. The basic equation for the ground heat flux is then

$$c_g \frac{\partial T}{\partial t} = \frac{\partial}{\partial z}\left(k_g \frac{\partial T}{\partial z}\right).\qquad(2.85)$$

The heat capacity of the soil is defined as a function of soil volumetric water content, such that

$$c_g = (1 - \Theta_S)c_{soil} + \Theta c_{water},\qquad(2.86)$$

where Θ_S is the soil porosity (the maximum amount of water the soil can hold), c_{soil} is $1.26 \times 10^6\,\mathrm{W\,m^{-3}\,K^{-1}}$ and c_{water} is $4.2 \times 10^6\,\mathrm{W\,m^{-3}\,K^{-1}}$. Values of soil porosity vary between 0.3 and 0.6 depending upon the soil type. Since the heat capacity for water is over three times the value for soil, it is clear that the volumetric water content of the soil has a large role in determining the total heat capacity. Chen and Dudhia (2001) further include a third term to represent the effect of air space within the soil on the heat capacity, although this term is small.

Similarly, the thermal conductivity also is strongly dependent upon the soil volumetric water content. Many models appear to use the formulas suggested by Al Nakshabandi and Kohnke (1965), such that

$$k_g(\Theta) = \begin{cases} 420\exp(-P_f + 2.7), & P_f \le 5.1, \\ 0.1722, & P_f > 5.1, \end{cases}\qquad(2.87)$$

where

$$P_f = \log_{10}\left[\psi_s\left(\frac{\Theta}{\Theta_s}\right)^{-b}\right] \tag{2.88}$$

has units of cm and the values of ψ_s, Θ_s, and b are functions of the soil type and are discussed more fully in the next chapter and in Clapp and Hornberger (1978) and McCumber and Pielke (1981). Mahrt and Pan (1984) emphasize that the values of soil thermal conductivity can vary over orders of magnitude with depth or over time, strongly indicating the need to include the effects of soil moisture in the parameterization.

More recent results that compare the sensitivity of sensible and latent heat fluxes to the soil thermal conductivity suggest that (2.87) and (2.88) may not be appropriate for the entire range of volumetric water content (Peters-Lidard et al. 1998). An alternative method developed by Johansen (1975) is proposed and agrees better with observations. However, this method requires knowledge of the quartz content and soil particle size, which are not often known. Thankfully, the quartz content can be related to the sand content of the soils, and the soil particle size appears to have a negligible effect on the results. While the details are presented in Peters-Lidard et al. (1998), the general idea is that the soil thermal conductivity is a factor of both the dry (κ_{dry}) and saturated (κ_{sat}) thermal conductivities, such that

$$k_g = K_e(\kappa_{sat} - \kappa_{dry}) + \kappa_{dry}, \tag{2.89}$$

where K_e is the Kersten number and weights the dry and saturated conductivities. The dry thermal conductivity is calculated using

$$\kappa_{dry} = \frac{0.135(1-n)2700 + 64.7}{2700 - 0.947(1-n)2700}, \tag{2.90}$$

where n is the value of soil porosity. The expression $2700(1-n)$ represents the dry soil density in $kg\,m^{-3}$, where the soil unit weight is $2700\,kg\,m^{-3}$.

The saturated thermal conductivity of the soil is calculated using

$$\kappa_{sat} = \kappa_s^{1-n}\kappa_i^{n-x_u}\kappa_w^{x_u}, \tag{2.91}$$

where x_u is the unfrozen volume fraction (0 to 1), the ice thermal conductivity $\kappa_i = 2.2\,W\,m^{-1}\,K^{-1}$, the water thermal conductivity $\kappa_w = 0.57\,W\,m^{-1}\,K^{-1}$, and the solid's thermal conductivity is determined from

$$\kappa_s = \kappa_q^q\kappa_o^{1-q}. \tag{2.92}$$

Here q is the quartz content (0 to 1), $\kappa_q = 7.7\,W\,m^{-1}\,K^{-1}$ is the thermal conductivity of quartz, κ_o is the thermal conductivity of other minerals, and is assumed to be $2.0\,W\,m^{-1}\,K^{-1}$ for $q > 0.2$ and $3.0\,W\,m^{-1}\,K^{-1}$ otherwise.

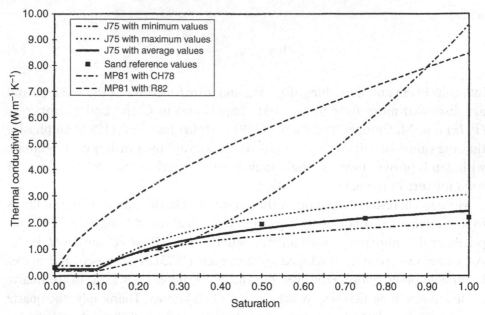

Figure 2.21. Thermal conductivity versus saturation fraction for sand. J75 is the Johansen method, MP81 is the McCumber and Pielke (1981) application of the Al Nakshabandi and Kohnke method, and reference values are indicated by solid squares. CH78 indicates use of the Clapp and Hornberger (1978) soil water retention functions, and R82 indicates use of the Rawls *et al.* (1982) soil water retention functions. From Peters-Lidard *et al.* (1998).

The Kersten number is a function only of the degree of soil saturation S_r. Peters-Lidard *et al.* (1998) provide expressions for the Kersten number for both coarse and fine soils, but consider only fine soils in their analyses and suggest that the differences in predicted values are small. For fine soils, the Kersten number is defined as

$$K_e = \log S_r + 1.0, \qquad (2.93)$$

while for frozen soils $K_e = S_r$. Table 2 in Peters-Lidard *et al.* (1998) provides the values of quartz content for the standard 12 United States Department of Agriculture (USDA) soil textures.

A comparison of the thermal conductivities from several approaches indicates that the Johansen (1975) approach appears to compare best with the reference values (Fig. 2.21), whereas the Al Nakshabandi and Kohnke (1965) values are too large when a cap value is not used. Even when the thermal conductivity is capped, the curve from Johansen (1975) appears to fit the reference values better.

Multi-level soil models, with soil heat capacity and thermal conductivity dependent upon the soil volumetric water content, represent one approach to accounting for the detail needed to accurately account for soil processes in land surface parameterizations. The main differences between parameterizations typically involve specification of the various parameter values and the number of vertical soil levels. Mahrt and Pan (1984) illustrate the importance of the thickness of the soil layers to the behavior of soil models, while Santanello and Carlson (2001) suggest that thin near-surface soil layers are needed to capture rapid soil drying. Many soil models in use within operational numerical weather prediction models have four or more vertical soil layers.

2.6.1 Frozen soil

When ground surface temperatures are below freezing, then some of the water in the soil freezes. The effects of this phase change need to be included in the heat conduction equation. It is generally assumed that the only phase change that occurs is between liquid water and ice, the soil does not deform due to the soil freezing and thawing, and there is no transport of ice within the soil. A heat balance equation can then be written (Koren et al. 1999; Smirnova et al. 2000) such that

$$c_g \frac{\partial T}{\partial t} - L_f S_{li} = \frac{\partial}{\partial z}\left(k_g \frac{\partial T}{\partial z}\right),$$ (2.94)

where L_f is the latent heat of fusion ($3.34 \times 10^5\,\mathrm{J\,kg^{-1}}$ at $0\,°\mathrm{C}$) and S_{li} is the rate of liquid mass transformation into ice. This equation is identical to (2.85) except for the $L_f S_{li}$ term that accounts for the liquid to solid phase change. The rate of liquid mass transformation into ice is defined as

$$S_{li} = -\rho_l \frac{\partial \Theta}{\partial t} = \rho_l \frac{\partial \Theta_i}{\partial t},$$ (2.95)

where Θ_i is the volumetric ice content. This equation indicates that as the soil volumetric water content decreases due to freezing, the value of S_{li} is positive and ice is created. With some modifications, Smirnova et al. (2000) show that the heat balance equation can be rewritten as

$$c_a \frac{\partial T}{\partial t} = \frac{\partial}{\partial z}\left(k_g \frac{\partial T}{\partial z}\right),$$ (2.96)

where c_a is the apparent heat capacity and is defined as

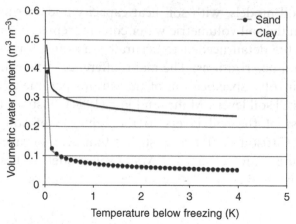

Figure 2.22. Soil-freezing characteristic curves calculated using (2.98) for soil parameters appropriate for sand (gray line) and clay (black line).

$$c_a = c_g + \rho_l L_f \frac{\partial \Theta}{\partial T}. \qquad (2.97)$$

The slope of the soil-freezing characteristic curve $\partial \Theta / \partial T$, the relationship between frozen volumetric water content and temperature, is obtained empirically (Cary and Maryland 1972; Flerchinger and Saxton 1989). Smirnova *et al.* (2000) suggest that the slope of the soil-freezing characteristic curve can be represented using

$$\Theta = \Theta_S \left[\frac{L_f (T - 273.15)}{g T \psi_S} \right]^{-1/b}, \qquad (2.98)$$

where ψ_S is the soil moisture potential for a saturated soil (units of m), has a value less than zero, and depends upon the soil type, and b and Θ_S also are constants that depend upon soil type (see Chapter 3). The general shape of the soil-freezing characteristic curve is shown in Fig. 2.22 and indicates that most of the liquid water in sandy soils becomes ice when the temperature is only 1 °C below freezing; it takes much colder temperatures for the same amount of water in clay soils to freeze and some fraction of the soil water may never freeze. Other models assume that all the water in the soil freezes when the temperature is below a given threshold value several degrees below freezing.

Once ice is present in the soil, it influences the soil heat capacity. Thus, the definition of the soil heat capacity must be modified to include the effects of ice, yielding

$$c_g = (1 - \Theta_S)c_{soil} + \Theta c_{water} + \Theta_i c_{ice}. \qquad (2.99)$$

The presence of ice also modifies the thermal conductivity k_g. Koren *et al.* (1999) use a simple linear adjustment suggested by Kutchment *et al.* (1983) to define

$$k_g(\Theta, \Theta_i) = k_g(\Theta)(1 + \Theta_i), \tag{2.100}$$

whereas Smirnova *et al.* (2000) define the thermal conductivity following Pressman (1994) as

$$k_g(\Theta, \Theta_i) = k_g(\Theta)\left(1 + \frac{\rho_i}{\rho_l}\Theta_i\right), \tag{2.101}$$

where ρ_i is the density of ice. Luo *et al.* (2003) compare the results of frozen soil parameterizations from 21 land surface schemes over Valdai, Russia, and find that the inclusion of soil-water freezing improves the simulations of soil temperature on seasonal and interannual timescales.

Parameterizations that are concerned with frozen soil also typically include some representation of snow cover and its effects on heat transport as well. Both Koren *et al.* (1999) and Smirnova *et al.* (2000) include simple snow models. These models both approximate the effects of snow on heat transport, including the effects of snow accumulation and snow melting from the snow–air interface and from the snow–soil interface. Koren *et al.* (1999) further include an approximation to snow compaction. Neither approach presently includes the refreezing of snow meltwater and the storage of liquid water within the snow. Slater *et al.* (2001) compare the simulations of snow from 21 land surface schemes as part of the Project for the Intercomparison of Land–Surface Parameterization Schemes (PILPS) and find systematic differences between the parameterizations over the 18 years of simulation.

Frozen water within the soil also has a significant influence on the ability of water to infiltrate into the soil, thereby influencing runoff. A discussion of these effects may be found in Chapter 3 where a more complete discussion is conducted on the determination of soil volumetric water content.

2.7 Surface energy budget equation

All the terms in the surface energy budget have now been defined. To conserve energy at the Earth's surface, all gains and losses of energy must balance. Thus, the surface energy budget valid at the Earth–atmosphere boundary can be defined as

$$Q_S - Q_{Lu} + Q_{Ld} = Q_H + Q_E + Q_G, \tag{2.102}$$

where Q_S already includes the effects of albedo as defined by (2.1). While this balance is maintained over time, the individual terms vary in magnitude and the values of Q_H, Q_E, and Q_G can change sign. During the daytime with large Q_S, the signs of Q_H, Q_E, and Q_G typically are positive in (2.102), indicating upward directed surface heat fluxes and a downward directed ground heat flux. Recall that the specification of these fluxes as having positive or negative values during the daytime as energy is transferred from the ground surface to the atmospheric surface layer varies in the literature, so pay attention to the direction of the energy flux to determine what sign convention is used within a given model.

2.8 Representation of terrain

The terrain field developed for a numerical weather prediction model is important for a number of reasons. The terrain height defines the depth of the atmosphere above the ground, influencing the amount of solar radiation that reaches the ground surface. As solar radiation heats the ground surface in regions of complex terrain, atmospheric circulations may develop, such as mountain–valley flow systems. These terrain-induced flows may develop clouds, thereby providing a negative feedback to the solar radiation. Terrain also can act to block or channel atmospheric flows, and can lead to the development of mountain waves and downslope winds. Finally, the terrain field is often used to separate the model grid points that are over land versus those that are over water. Thus, it is very important for numerical weather prediction models to have an accurate terrain height defined at each grid point.

Terrain data are generally obtained from digital elevation models (DEMs) created by government agencies, such as the United States Geological Survey (USGS). Global data are available at approximately 1 km spacing, although data with spacing as small as 30 m are available over some regions. However, a variety of data are used to construct global terrain data sets. In addition to point measurements of terrain height at regular horizontal intervals, the terrain heights in some regions of the globe are interpolated from terrain contour maps produced by a variety of agencies. Thus, the accuracy of the terrain data varies somewhat from location to location. In general, terrain heights at 1 km spacing have an estimated root-mean-square error as small as 9 m over well-sampled regions and as large as 300 m when derived from maps with 1000 m height isolines. It is important to understand the source of the terrain data used in the model so that the uncertainties in terrain height over the forecast area of interest are recognized.

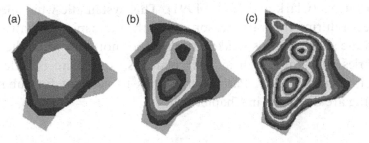

Figure 2.23. Terrain field over the island of Hawaii represented using
(a) 30 km, (b) 10 km, and (c) 5 km grid spacing. Terrain heights are shaded
every 400 m in all panels. Note the much greater terrain detail as the grid
spacing is decreased, as well as the increase in maximum terrain height as
indicated by the greater number of alterations in shading. At 30 km, the
highest terrain point is 1892 m, whereas at 10 km it is 2974 m and at 5 km it
is at 3918 m. The actual peak of Mauna Loa is at 4169 m. Also note the
northern displacement of the highest terrain in (a) compared to (b) and (c).
In addition, the higher northern peak of Mauna Kea (4205 m) is seen using
5 and 10 km grid spacing, but is not seen at all using 30 km grid spacing.
Finally, the Kohala Mountains stretching southeast to northwest along the
northern tip of Hawaii are clearly seen using 5 km grid spacing, but are missed
even at 10 km grid spacing.

Once the terrain data are available, and the grid spacing for the numerical
weather prediction model is defined, interpolations are made from the terrain
data to the model grid. Some averaging of the terrain data is often necessary to
represent the terrain field in the model appropriately. As the model grid
spacing decreases, the terrain field represented in the model becomes more
realistic (Fig. 2.23).

However, the subgrid variability of terrain is also important. Results from
medium-range forecast models and general circulation models during the
1970s and 1980s for the northern hemisphere cold season indicate that the
zonally averaged upper-tropospheric flow was too strong in the midlatitudes
(Lilly 1972; Palmer et al. 1986). This systematic error influences the predicted
values of sea-level pressure, low-level winds, geopotential heights, and the
evolution of extratropical cyclones. This error is due to the lack of sufficient
drag from rugged mountain ranges, which influences the stationary waves in
the atmosphere (Wallace et al. 1983). As described by Palmer et al. (1986), in
stratified flow a drag force can be imposed on the atmosphere via internal
gravity waves. These waves have horizontal wavelengths of ∼6 km for flow
over a hill of width 1 km. Thus, if the model grid spacing is greater than 10 km,
there can be a significant underestimation of the drag force exerted by

mountain ranges (Clark and Miller 1991). This systematic wind speed error does not exist during the warm season, when the atmospheric stratification is lower and the surface winds weaker. It also does not exist when the elevated terrain varies smoothly, such as over plateaus, and there is little subgrid variation in terrain height. Parameterizations to account for subgrid orographic drag are discussed in Chapter 10.

2.9 Discussion

One of the key pieces of information that influences the surface fluxes of sensible and latent heat is the wind profile near the surface. As seen throughout this chapter, this log wind profile is paramount in the calculation of resistances. Yet the land surfaces for which this profile has been observed are limited. Many of the observational results are from surface and boundary layer experiments over relatively flat terrain and with uniform vegetation coverage (see Garratt and Hicks 1990; Kaimal and Wyngaard 1990). It is not at all clear that the log wind profile relationship, including corrections and adjustments for non-neutral conditions, is appropriate for regions where vegetation changes are dramatic or terrain changes are rapid. Dyer (1974) clearly shows that changes in surface roughness, as could be created by changes in vegetation type, produce fluxes that do not correspond to the environmental profiles directly above. The inaccuracy that this type of situation produces in the calculated values of resistance is uncertain. While there have been several planetary boundary layer experiments in regions of complex terrain (e.g., Schneider and Lilly 1999), there remains much to be learned. This is especially true since the surface fluxes set the stage for what happens later in the day in terms of convective development.

It also is evident that there are several other uncertainties involving the calculation of surface fluxes, especially the latent heat flux. Some approaches use the ground temperature to obtain a saturation specific humidity, while others are based upon the air temperature instead. Both approaches are valid, but yield very different values for potential evaporation (when $M = 1$) and have different sensitivities, and are likely to have different error characteristics. This highlights the complexities one faces when using a numerical model, and the choices that have to be made in how to parameterize processes that the model cannot resolve explicitly. These types of choices are seen again and again in numerical prediction models.

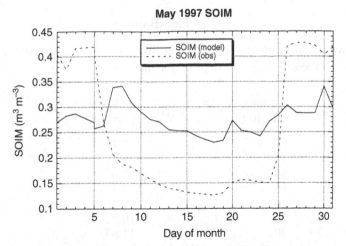

Figure 2.24. Time series comparison of model and observed 5 cm soil volumetric water content (SOIM) at Norman, Oklahoma, during May 1997. Values plotted are daily averages of hourly observations and Eta model data. From Marshall *et al.* (2003).

When thinking about how numerical weather prediction models are initialized, the meteorological community has invested a lot of time, effort, and expense towards better observations of the atmosphere. Satellite, aircraft, and radar data are all now being assimilated into models. But there is precious little information on what is occurring below the ground surface. Measurements of soil temperature and soil moisture are few and far between. Marshall *et al.* (2003) show that the Eta model initial values of soil moisture are very different from observations provided by the Oklahoma Mesonet (Fig. 2.24). Similar results are suggested when comparing the Eta model initial soil temperatures with observations (Godfrey *et al.* 2005). Since soil temperature and moisture play an important role in the calculations of sensible, latent, and ground heat flux, it is not clear why more resources are not directed towards soil measurements. The uncertainty in the soil initial states clearly represents one source of error in numerical weather prediction today.

The discussions in this chapter highlight the ever-increasing need for collaboration across the many sciences that study processes that influence the atmosphere. Whether botany, soil science, oceanography, or chemistry, the need for multidisciplinary communication has never been greater. The atmosphere can no longer be viewed and studied in isolation, and teaching new generations of meteorologists to be aware of and interact with scientists in other disciplines is of enormous importance. The motivation behind this viewpoint becomes even clearer in the following chapters.

2.10 Questions

1. The following mean wind data, sampled at various heights, occur above a vegetated canopy. Find the displacement depth d for this canopy.

Height (m)	Wind speed (m s^{-1})
10	7.11
12	8.01
14	8.63
16	9.11
20	9.81
30	10.92
40	11.63

Typically, if $\ln(z - d)$ is plotted versus wind speed, the line will be concave upward if d is too small and concave downward if d is too large.

2. Given that $T_g = 302$ K, $T_{10\,m} = 294$ K, $Q_H = 300$ W m^{-2}, and $\rho = 1.0$ kg m^{-3}, calculate the resistance to sensible heat flux in the layer from the surface to 10 m above the surface. Assuming $T_{2\,m} = 294.5$ K and $z_0 = 0.01$ m, what is the resistance from the surface to 2 m? Using this knowledge, estimate the resistance to sensible heat flux in the lowest 20 m above the surface and determine the temperature at this level. Is this temperature profile realistic? State any assumptions made and show all work.

3. Calculate the resistances for sensible heat flux assuming that $u(20\,m) = 10$ m s^{-1} and that there are three different uniform vegetation types: $h = 1.6$ m for one vegetation height, $h = 0.8$ m for another, and $h = 0.1$ m for the third. State any assumptions made. Which resistance is larger? Assuming a 10 K temperature potential, calculate the differences in sensible heat flux from these three resistances. What does this say about the importance of the roughness length in models?

4. Using the same three vegetation types as described in Question 3, calculate the resistances to latent heat flux. Assuming a 0.02 kg kg^{-1} specific humidity potential $(\Delta q = q_s(T_g) - q(z))$ and a canopy resistance of 200 s m^{-1}, what are the latent heat fluxes for the three vegetation heights? Do the changes in the relative values of latent and sensible heat fluxes over the three different vegetation types make sense? What is the value of moisture availability?

5. Draw a typical diurnal cycle for each of the terms $(R_{net}, Q_G, Q_E, \text{ and } Q_H)$ of the surface energy budget. Then draw an energy flux diagram and show the magnitudes and signs of each term at 1200 local noon and 0400 local morning (before sunrise). Why are these values realistic? Explain.

6. Write a computer program to calculate the change in surface temperature over an entire daytime heating cycle (sunrise to sunset) using the force-restore method. Assume that the calculation is for near Amarillo, Texas on 30 June

1999 (35.2° latitude, −102.0° longitude), and begin the calculations at sunrise and end the calculations at sunset. To actually do this without attaching the surface to a boundary layer model, we have to make some pretty stringent assumptions. Assume the atmospheric transmissivity is constant at 0.8, the solar constant is 1368 W m^{-2}, the thermal capacity of the slab (c_g) is 1.4×10^5 J m^{-2} K^{-1}, the skies are clear, the Bowen ratio is 0.7, the temperature at 40 hPa above the ground surface is equal to 25 °C, the initial temperature of the ground surface (skin temperature) is 23 °C, the reservoir temperature is 25 °C, the precipitable water is 2.5 cm, the emissivity is 0.95, and the albedo is 20%. Run two experiments. First, assume the sensible heat flux is 15% of the net radiation received at the surface, and second assume that the sensible heat flux is 30% of the net radiation received at the surface. Plot the radiation amounts on one graph, show the change in ground (skin) temperature on another, and plot the various surface fluxes (sensible, latent, and ground) on yet another graph. What is the phase relationship between the ground (skin) temperature and the heat flux into the ground? What are the times of sunrise and sunset? What is the maximum value of incoming solar radiation?

Remember that

$$\frac{\partial T_g}{\partial t} = \frac{1}{c_g} \left[Q_S + Q_{Ld} - Q_{Lu} - Q_H - Q_E - \kappa(T_g - T_r) \right],$$

where κ is assumed to be constant at 11 J m^{-2} K^{-1} s^{-1} and T_r is the reservoir temperature (K). It is assumed that the energy loss due to the surface albedo is already included in Q_S.

7. Seguin and Gignoux (1974) made wind speed measurements at a number of heights over two adjacent, yet very different, regions. One region consisted of cypress hedges 7 m in height and separated by short grass roughly 10 cm tall. The other region consisted of grass only with a height of roughly 40 cm (see Fig. 2.25). Both regions were large and separated by sufficient distance to allow them to be considered independent regions. The wind speeds at the top of the surface layer at 50 m

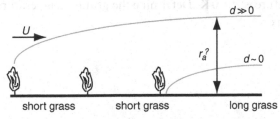

Figure 2.25. Two regions: one of regularly spaced hedges with short grass in between the hedges, and the other of longer grass. Note that the internal boundary layers formed within these two regions are different, and have log wind profiles with different displacement depths (d), and that the internal boundary layer from the hedge region is advected over the internal boundary layer of the grass-only region. The opposite would occur if the wind direction shifts 180°. After Seguin and Gignoux (1974).

above ground level (over which the flux profile laws apply) were virtually identical above both the cypress hedges and the 40 cm grass regions. However, the profiles down to the ground were different. The log wind profile in the grass-only region simply responded to a surface of grass 40 cm high. The log wind profile in the hedge region first responded to a vegetation canopy at the height of the cypress hedges (7 m), while the log wind profile below the hedge top responded to the short grass (10 cm). The height of the hedge top is where they observed this change in the log wind profile behavior in the region with hedges.

(a) Calculate and show graphically the wind speed profiles in the two regions and the difference in wind speed at hedge top level and at 2 m between the hedge and grass-only regions for a wind speed of $10 \, \mathrm{m \, s^{-1}}$ at 50 m. Assume neutral static stability. (Note that the two wind regimes in the hedge region must be meshed. The upper regime has a different roughness from the lower.) Calculate the three relevant friction velocities for the problem. Helpful hint: $z_0 \approx h/8$, where h is the average height of the grass or hedge. Comment on the differences in wind speed between the two regions.

(b) Calculate the values of resistance to sensible heat flux from the surface to 50 m for the two regions. Assuming a potential difference of 10 K, what are the values of heat flux over these two regions? If the grass-only region, as depicted at the far right of Fig. 2.25, has the cypress hedge internal boundary layer above it, how might one estimate the change in resistance to heat flux above the grass-only region? Using this estimation technique, how large a difference does this make to the value of resistance?

(This question is based upon course notes from a bioclimatology class taught by Dr. Toby Carlson at The Pennsylvania State University in the late 1980s. I thought the question was excellent at the time and still think it illustrates many important points.)

8. At a given surface observing site, the roughness length is 0.1 m, and at a particular time during the daytime the friction velocity $u_* = 0.85$. The site has two heights at which temperature is observed: 10 and 2 m. The 10 m temperature is 295.4 K and the 2 m temperature is 298.0 K. Determine the ground temperature and the surface sensible heat flux.

3

Soil–vegetation–atmosphere parameterizations

3.1 Introduction

Evolution teaches us that organisms must come to terms with their environment in order to grow and reproduce successfully. This suggests that many, if not all, successful organisms have strategies that allow them to adapt to changing environmental conditions. Unfortunately, when an organism interacts with its environment, the physical processes are rarely simple and the physiological mechanisms often are poorly understood. This is the challenge one faces when trying to incorporate vegetation into numerical models.

The atmosphere and vegetation interact in a number of different ways, and many of these need to be parameterized. The main five ways in which they interact are via the biophysical control of evapotranspiration, momentum transfer, soil moisture availability, radiation, and insulation (Sellers *et al.* 1986). The discussion of soil–vegetation–atmosphere parameterizations is organized around these five interaction types.

3.1.1 Biophysical control of evapotranspiration

Plants use photosynthetically active radiation, part of shortwave radiative energy, to combine water and carbon dioxide (CO_2) into sugars and other organic compounds. Thus, plants must allow for the transfer of CO_2 from the atmosphere to the cellular sites of photosynthesis. By doing so, plants expose their saturated tissues to the atmosphere and water loss occurs. Plants regulate the amount of CO_2 exchanged, and therefore regulate water loss, by means of valve-like structures on the leaf surface called stomates. Some plants have stomates on both sides of the leaf; others on only one side. Plants try to maximize the ratio

$$\frac{|CO_2 \text{ going in}|}{|H_2O \text{ going out}|},$$

(3.1)

in order to capture the CO_2 needed for photosynthesis without losing too much water. This ratio is called the water use efficiency.

Many plants conserve moisture when necessary by closing their stomates. In addition, vegetation intercepts precipitation on leaf surfaces. There also is evidence that some plants will become more efficient at conserving water as atmospheric CO_2 increases owing to the burning of fossil fuels (Eumas and Jarvis 1989; Bazzaz and Fajer 1992). Henderson-Sellers *et al.* (1995) indicate that such changes can lead to decreased evapotranspiration and increased temperatures, especially in boreal and tropical forests.

Plant roots extend into the soil for depths of up to several meters and can draw water from these depths for use in photosynthesis. The vertical distribution of plant roots can influence the seasonal cycle of transpiration (Desborough 1997) and is important to summer dryness (Desborough 1997; Milly 1997). Root-zone soil moisture is strongly correlated to the daily-maximum values of surface sensible and latent heat fluxes through the effects of the vegetation (Basara and Crawford 2002). These studies illustrate that vegetation strongly influences the latent heat flux, and hence the surface energy budget, at the Earth's surface.

The importance of variations in the surface sensible and latent heat fluxes to short-range, seasonal, and climate predictions is described at the beginning of Chapter 2. However, variations in vegetation type or land use also are important. Historical changes in land use are suggested to account for most of the observed surface warming of the past several hundred years (Chase *et al.* 2001). Adding the effects of estimated vegetation changes to climate change scenarios produces substantial regional differences in climate. Results from Feddema *et al.* (2005a, b) indicate that projected vegetation changes can produce additional warming over the Amazon region, influencing the Hadley circulation and monsoon circulations. Projected agricultural expansion in the midlatitudes produces cooling and decreases the diurnal temperature range.

Plants can be categorized into either C_3 or C_4 species based upon their method of carbon fixation, the conversion of carbon dioxide and water into organic compounds during photosynthesis. While over 95% of the plant species are C_3, the different methods of carbon fixation also influence evapotranspiration. C_3 plants have an advantage over C_4 plants in cooler climates, since the energy required to fix carbon is less for C_3 plants. C_3 plants also are more water-use efficient and nitrogen efficient than C_4 plants. Since nitrogen is often the limiting nutrient for plant growth, C_3 plants tend to grow faster. However, in very warm climates C_4 plants have an advantage, since they can store higher levels of oxygen without affecting the rate of carbon fixation compared with C_3 plants, and thus can close their stomates during the daytime to conserve water and still produce energy.

3.1.2 Momentum transfer

Vegetation canopies are a rough surface, with larger roughness lengths implying larger surface fluxes, and also producing larger surface drag forces. As discussed in Chapter 2, most field studies of the behavior of the atmospheric surface layer, from which the log wind profile is derived, avoid rough surfaces and make their measurements in areas of relatively flat terrain with few obstacles. However, vegetation is spatially highly variable, responding to soil conditions and seed dispersion, creating patches of obstacles of varying height that influence the low-level winds. Vegetation also changes on seasonal timescales, such that its roughness varies throughout the year. In some locations, the start of monsoon rains can change the surface vegetation from near desert-like to a tropical rain forest in just a few weeks (Douglas 1993), significantly altering the surface roughness and the surface energy budget. Prolonged periods of dryness can lead to widespread wildfires, destroying vegetation over large areas. Hail-producing thunderstorms also can destroy vegetation over their paths, altering surface conditions and hence the surface energy budget, and leading to feedbacks to the atmosphere (Segele *et al.* 2005).

3.1.3 Soil moisture availability

The depth and density of the rooting zone determines the soil moisture available for evapotranspiration. Typically evaporation from bare soil occurs only through a shallow depth of soil. Once the upper soil layers become very dry they can act as a barrier to further upward moisture transport through the soil. Vegetation can overcome this barrier, since the rooting zone typically extends over a deeper layer. This ability of vegetation to tap deeper moisture sources illustrates the importance of the amount, the type, and the variability of vegetation to the latent heat flux. For example, Milly (1997) suggests that a 14% decrease in the soil volume used for plant water uptake (as caused by a decrease in plant roots) can generate the same summer dryness in midlatitudes as the doubling of atmospheric CO_2 over present levels. Conversely, a 14% increase of this same soil volume (produced by an increase in plant roots) could offset the tendency for more frequent summer dryness in a world with double the levels of CO_2. These results further underscore the important role that vegetation plays in defining the surface energy budget and its ability to influence the atmosphere over both long and short timescales.

3.1.4 Radiation

The spectral properties of leaves and the multiple reflections that occur as sunlight passes through a vegetation canopy make leaves highly absorbent in the visible wavelength interval between 0.4 and 0.72 μm, where absorption of radiation by chlorophyll in the leaves generally occurs, and moderately reflective in the near infrared region from 0.72 to 4.0 μm. Because of this, the more complex vegetation schemes (e.g., Sellers *et al.* 1986; Dai *et al.* 2003) require the computation of up to five components of incident radiation: direct photosynthetically active radiation (PAR), diffuse PAR, direct near-infrared (NIR), diffuse NIR, and diffuse infrared. These quantities are needed because the absorption and transmission coefficients of radiation interacting with vegetation are highly dependent upon the angle of the incident radiative flux.

3.1.5 Insulation

Under dense vegetation the soil surface receives less radiation and is aerodynamically sheltered, which greatly influences the soil energy budget. As one might expect, this is particularly a concern within forests. The effects of insulation generally are accounted for by alterations to the momentum transfer and radiation, so they are not discussed explicitly in the following sections.

The recognition that plants are not passive sponge-like structures, but act in ways to maximize their long-term prospects for survival, has led to the development of parameterization schemes for the effects of vegetation. These schemes generally do not allow for the development and dispersion of vegetation, at least not yet (see Pan *et al.* (2002), Sitch *et al.* (2003), Bonan *et al.* (2003), and Krinner *et al.* (2005) for examples of large-scale dynamic vegetation models), but once the type and amount of vegetation has been specified at a given location, these schemes attempt to reproduce the influences of the vegetation on the environment. Thus, the vegetation itself is not predicted, rather the interaction between the vegetation and the atmosphere. These types of schemes are commonly called soil–vegetation–atmosphere transfer schemes (SVATS) in the literature.

3.2 Describing vegetation in models

Three important vegetation parameters that are used within numerical weather models are the green vegetation fraction (σ_f), the leaf area index (*LAI*), and the vegetation type or class. The green vegetation fraction, also referred to as the greenness fraction or fractional vegetation cover, is the model grid cell fraction where midday downward solar radiation is intercepted by a photosynthetically active green canopy (Chen *et al.* 1996) and it acts as the weighting factor between bare soil and canopy transpiration. This, in turn, affects surface temperature and moisture forecasts through the alteration of surface fluxes. The *LAI*, a measure of the vegetation biomass, is defined as the sum of the one-sided area of green leaves above a specified area of ground surface and it plays a major role in determining the amount of transpiration from the vegetation canopy. More leaves imply a greater amount of transpiration that is possible. Holding all other parameters constant, a larger *LAI* value produces greater canopy transpiration than a lower *LAI* value. The vegetation type specifies the dominant vegetation that covers the ground within a model grid cell. Together, the vegetation fraction, *LAI*, and vegetation type can be used to describe the state and the health of the vegetation covering the land surface.

The difficulty with these vegetation parameters is that they are nearly impossible (and prohibitively expensive!) to observe routinely from the ground. Thus, some method of observing the state of the vegetation remotely is absolutely essential. Thankfully, one measurement that is helpful in trying to calculate σ_f and *LAI* and other measures of biomass, and in defining vegetation type, is the normalized difference vegetation index (*NDVI*) (Tucker *et al.* 1984). This index was developed owing to the difference in the albedo characteristics of bare soil and vegetation-covered surfaces. The albedos of bare soils are fairly constant across the visible (red, 0.62–0.75 µm) and near-infrared (NIR, 0.75–1.4 µm) portions of the spectrum, while vegetation-covered surfaces show a dramatic increase in albedo in the NIR region when compared with the visible portions of the spectrum (Fig. 3.1). Thus, the *NDVI* is a measure of the difference between two specific wavelength bands as seen from either airplanes or satellites, and is defined as

$$NDVI = \frac{\text{NIR} - \text{red}}{\text{red} + \text{NIR}}. \tag{3.2}$$

When there is no vegetation, the albedo in the red and NIR are nearly equal and *NDVI* is approximately 0.1. When there is a lot of vegetation present, then *NDVI* is at its maximum value, approaching 0.9. When clouds are present, then the red reflectance is greater than the NIR reflectance and $NDVI < 0$.

Typically, *NDVI* is calculated from the National Oceanic and Atmospheric Administration (NOAA) polar orbiting satellites using the Advanced Very High Resolution Radiometer (AVHRR) or from the National Aeronautic and Space Administration (NASA) polar orbiting satellites using the Moderate

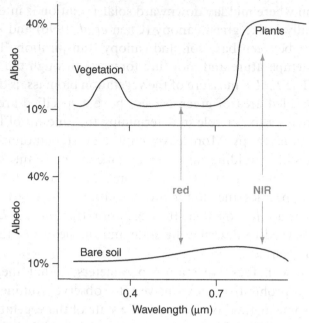

Figure 3.1. Plot of the surface reflectance (albedo) as a function of wavelength for a vegetated surface (top) and a bare soil surface (bottom). Note that the difference in the albedos between the two surfaces for the visible (red) wavelength band is small, whereas the difference in the albedos for the near-infrared (NIR) band is large.

Resolution Imaging Spectroradiometer (MODIS). While AVHRR data have a spatial resolution of ∼1 km, the newer MODIS data have a spatial resolution of ∼250 m. Thus, using either data source it is possible to provide vegetation information to present operational models at spatial resolutions smaller than the model grid spacing. These data sources have been used to develop global land cover mapping, the results of which are used to specify the values of σ_f and LAI in numerical models (DeFries and Townshend 1994; Loveland *et al.* 2000; Friedl *et al.* 2002). To remove the effects of clouds, a maximum value composite approach is often used in which the maximum $NDVI$ value observed over a multi-week period is determined for each grid cell and these maximum values are then used in any subsequent analyses. While there are many limitations to the use of $NDVI$ in deriving vegetation indices (Huete *et al.* 1996), it remains a very useful parameter in examining the state and evolution of vegetation.

When trying to define the health and the extent of the vegetation at the land surface, the most important parameter needed for land surface models is σ_f. Numerous studies also have shown that $NDVI$ can be related to σ_f. In particular, Chang and Wetzel (1991) indicate that $NDVI$ and σ_f are related by

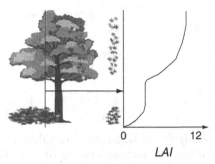

0   12
LAI

Figure 3.2. Illustration of determination of the leaf area index (*LAI*) from the surface to the top of the canopy. *LAI* is the total area of one side of all the leaves in a canopy compared to a unit surface area.

$$\sigma_f = \begin{cases} 1.5(NDVI - 0.1), & NDVI \leq 0.547, \\ 3.2(NDVI) - 1.08, & NDVI > 0.547, \end{cases} \qquad (3.3)$$

which yields a linear relationship between *NDVI* and σ_f with two slightly different regimes depending upon the value of *NDVI*. Of course, other formulas for relating *NDVI* to σ_f are available (e.g., Carlson and Ripley 1997; Gutman and Ignatov 1998) and their coefficients depend strongly upon the spatial resolution of the *NDVI* used in their development.

The other important aspect of the vegetation that needs to be known is some measure of the total vegetation biomass as determined using the *LAI* (Fig. 3.2). Typical values of *LAI* vary from 8 for forests, 7 for a mature corn crop, 4 for a mature wheat crop, and 0.5–2 for grasses. Plants with very clumped leaves can have *LAI* values above 10 and approaching 15. However, *LAI* is highly variable and can change from vegetation type to vegetation type and within a given year.

Numerous studies also have shown that the value of *NDVI* can be related, albeit imperfectly, to *LAI*. This is seen by examining the calculated values of *NDVI* versus field measurements of *LAI*, which yields a curve that slowly asymptotes to the maximum *NDVI* value as *LAI* increases (Fig. 3.3). The idea behind the calculation of *LAI* from *NDVI* is that if the maximum and minimum *NDVI* values for a given location can be determined using a multi-year time series, and if the maximum value of *LAI* attained at this same point is known based upon the vegetation type, then a linear relation between the value of *NDVI* to the expected *LAI* can be constructed. This approach leads to formulas such as that developed by Yin and Williams (1997) in which

$$LAI_i = LAI_{max} \frac{(NDVI_i - NDVI_{min})}{(NDVI_{max} - NDVI_{min})}. \qquad (3.4)$$

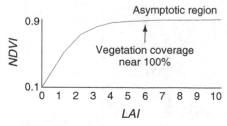

Figure 3.3. Curve typically found when plotting values of *NDVI* versus *LAI*. Saturation is reached when *LAI* increases past 3 or 4 as the value of *NDVI* is effectively constant while *LAI* continues to increase. Adapted from Sellers (1987).

Here i refers to the day of interest, the maximum and minimum values of *NDVI* are calculated over many years worth of data, and the value of LAI_{max} is determined from a vegetation database. While not a perfect relationship, this does provide useful and helpful information on the variations of the vegetation biomass for some midlatitude regions where the *LAI* is zero during the cold season. As seen in (3.4), *LAI* is zero when $NDVI_i$ equals $NDVI_{min}$. This is a reasonable relationship for midlatitude deciduous forests and grasslands, but does not apply for coniferous forests or tropical regions. For these areas *LAI* is almost never zero and a constant is probably needed on the right-hand side of (3.4) along with a corresponding modification to the value of LAI_{max}. Carlson and Ripley (1997) indicate that the relationship between *NDVI* and *LAI* is only valid up to values of *LAI* between 3 and 4. Beyond a *LAI* value of 4, *NDVI* is effectively constant.

Many models use the 5 year climatology of σ_f derived from 1 km AVHRR data by Gutman and Ignatov (1998) at 0.15° resolution. Some models use a constant value for *LAI* over the entire model grid, while others use values for *LAI* that are derived from satellite data (Buermann *et al.* 2001; Mitchell *et al.* 2004). These values of σ_f and *LAI* generally are provided once a month, and the daily values are interpolated from these monthly mean values.

However, it is clear that vegetation is greatly affected by the amount of precipitation, sunshine, and temperature and responds to changes in any of these inputs relatively rapidly. Differences between 14 day composite maximum values of σ_f compared to the multi-year mean of Gutman and Ignatov (1998) are calculated by Kurkowski *et al.* (2003) and are substantial (Fig. 3.4). Differences between the assumed model values of σ_f and the near real-time observed values exceed 30% at some locations. Thus, it is important to consider the initial data that go into the model in addition to the parameterization schemes used. Buermann *et al.* (2001) indicate that values of *LAI* also change significantly from year to year and that these changes are important to the partitioning of the sensible and latent heat fluxes.

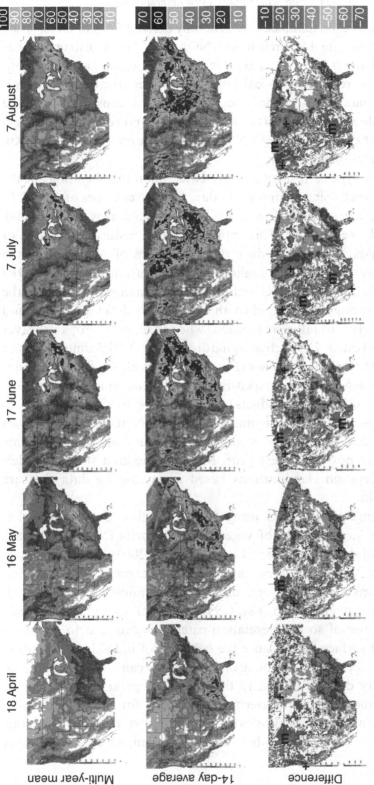

Figure 3.4. Multi-year mean values of σ_f from Gutman and Ignatov (1998) as implemented in the Eta model (top), near real-time 14 day composite maximum values of σ_f (middle), and the difference (real-time minus multi-year mean, bottom) for 5 days during the warm season of 2001. Scale for the σ_f values shown in the upper right, with the scale for the difference field in the lower right. The symbols + and m in the bottom panel denote the larger, cohesive areas with positive and negative differences, respectively. Note that the southeastern USA in April is much greener in 2001 than in the multi-year mean, as is the high plains region in July and August. The value of the real-time σ_f over the west is generally less than the multi-year mean, although the monsoon region of western Mexico shows up as greening up earlier than the multi-year mean in July and August. From Kurkowski et al. (2003).

Vegetation type or class databases are generally derived from polar-orbiting satellite data. Again, the 1 km resolution NOAA AVHRR instrument has provided most of the data used by the United States Geological Survey to create several 1 km resolution global land cover categorizations. One is a simple biosphere model vegetation categorization that consists of 16 land cover classes, while another consists of a 25 category land cover classification (Loveland *et al.* 2000). Hansen *et al.* (2000) also have developed a global 1 km land cover classification (Fig. 3.5).

These vegetation type categorizations are based upon the phenology of the vegetation as observed using composite 14 day, 1 km pixel values of *NDVI*. In these composites, the maximum value of *NDVI* observed over the 14 day period is assigned to each pixel. This approach greatly reduces the effects of clouds, since clouds generally produce negative values of *NDVI*. Once the 14 day composites are available throughout an entire year, time series of *NDVI* are constructed for each 1 km pixel and depict the annual progression of the vegetation from spring greenup, when the vegetation develops leaves and crops begin to emerge, to fall retrogression, when vegetation drops its leaves and crops are harvested. Distinctive signatures in the *NDVI* time series are found for different vegetation classes (Fig. 3.6), which allows for the vegetation class to be assigned without expensive and time-consuming ground surveys. However, the USGS also conducts ground surveys in different regions to verify the satellite-based class assignments. While satellite-based techniques require less time and effort than ground surveys, they are still very time consuming and are not done every year. Many of the land surface models presently use vegetation classifications based upon satellite data that are nearly a decade old.

Another challenge found when using vegetation class databases is that while there are a large number of vegetation categories (>200) that have been observed and documented (see Loveland *et al.* 2000), a total of fewer than 30 different vegetation types are defined for use in land surface models. This means that similar, yet different, vegetation types are aggregated into a single category. Xue *et al.* (1996) and McPherson and Stensrud (2005) show that use of so few vegetation categories can lead to problems with the resulting surface fluxes, since the behavior of individual vegetation types that are grouped together into a single category can be vastly different. However, it is very difficult to specify the vegetation parameters needed by soil–vegetation–atmosphere parameterization schemes for a vast number of vegetation types, and it is unclear whether the effort required to define these parameters will be repaid by substantially improved surface flux predictions.

(a)

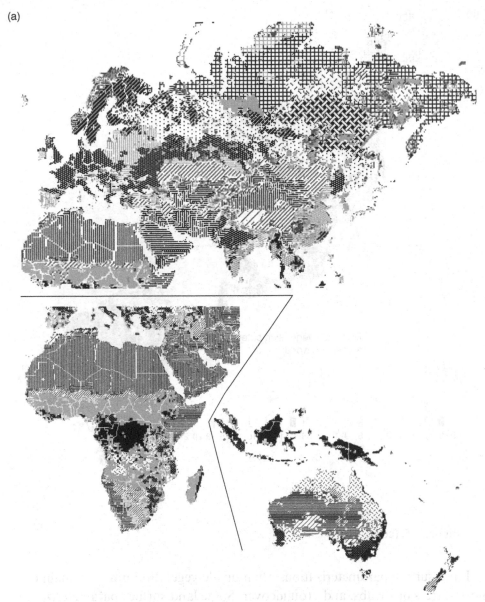

Figure 3.5. Predominant vegetation type in each 60 km grid cell over the globe, with urban = 1, dryland crop pasture = 2, irrigated crop pasture = 3, mixed dryland and irrigated crop pasture = 4, cropland and grassland mosaic = 5, cropland and woodland mosaic = 6, grassland = 7, shrubland = 8, mixed shrubland and grassland = 9, savanna = 10, deciduous broadleaf = 11, deciduous needleleaf = 12, evergreen broadleaf = 13, evergreen needleleaf = 14, mixed forest = 15, water and ice = 16, herbaceous wetland = 17, wooded wetland = 18, barren or sparse vegetation = 19, herbaceous tundra = 20, wooden tundra = 21, mixed tundra = 22, and bare ground tundra = 23. A color version of a similar global vegetation mapping can be found in Rodell *et al.* (2004).

(b)

Dominant vegetation categories
across the world

Figure 3.5. (cont.)

   Land surface parameterizations often divide vegetation into two main categories: trees or shrubs, and groundcover. Some land surface parameterizations
represent only the dominant vegetation type within the grid cell, while others
represent perhaps the first few dominant vegetation types. However, regardless
of which approach is taken, there is little doubt that each 1 km pixel observed
by the AVHRR instrument has a number of different vegetation types within
it. The vegetation classification applies most appropriately to the dominant
vegetation class within each pixel. The lack of ability to use the satellite-
derived vegetation class information fully is an area in which progress can
be made.

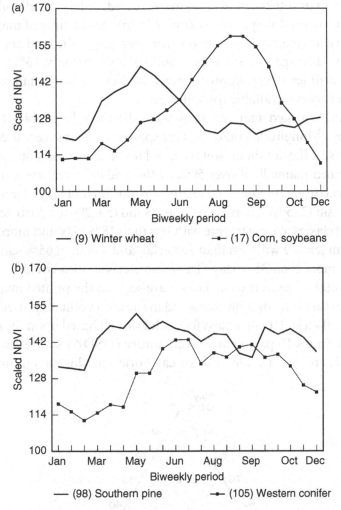

Figure 3.6. *NDVI* time series during 1990 for selected seasonal land-cover regions. Winter wheat and corn–soybeans are shown in the top panel, while southern pine and western conifer are shown in the bottom panel. From Loveland *et al.* (1995).

3.3 Describing soils in models

Worldwide information on soil type is derived from the Soil Map of the World, produced during the 1970s by the Food and Agriculture Organization (FAO) of the United Nations Educational, Scientific, and Cultural Organization (UNESCO). The original map is contained within 10 print volumes at a scale of 1:5 000 000 (1 cm on the map equals 50 km). The map is based upon data compiled over a 15 year period using input from field surveys and map collections. Roughly 11 000 individual soil maps were reviewed to produce the worldwide map and these individual maps varied widely in reliability, detail, scale, and the methodologies used (Zobler 1986). At the time of map creation, only Europe had systematic soil surveys for over half of the land area; the other continents on average had systematic soil surveys for only 16% of the total land area (Gardiner 1982). Approximately 40% of the world had only general information surveys available for soil type.

The original printed map volumes were digitized by Zobler (1986) into 1° latitude by 1° longitude grid cells. Transparent overlaps were used to define the grid cells and the dominant soil type, soil texture, and the slope for each cell were determined manually if over 50% of the grid cell area was land. The soil texture information represents the relative proportions of clay (2 μm or smaller particle size), silt (2–50 μm particle size), and sand (50–200 μm particle size) using three general classes: coarse texture with less than 18% clay and more than 65% sand, medium texture with less than 35% clay and less than 65% sand, and fine texture with more than 35% clay. These three classes also could be combined, such that a total of seven textural classes are seen on the printed map pages.

The soil texture information contained in the map volumes is used by Zobler (1986) to specify a soil texture class for each grid cell based upon the commonly used United States Department of Agriculture (USDA) soil textural triangle (Fig. 3.7). There are 12 soil texture categories in this taxonomy that are

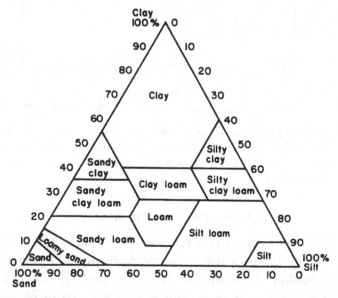

Figure 3.7. The USDA soil texture triangle, showing the percent by weight of clay, silt, and sand particles. The 12 soil categories are outlined, and indicate the range of composition of the soils within each of the classes. Other classifications of soil are available (see Marshall *et al.* 1996). From Cosby *et al.* (1984).

(a)

Figure 3.8. Global soil texture classes in each 60 km grid cell with sand = 1, loamy sand = 2, sandy loam = 3, silt loam = 4, silt = 5, loam = 6, sandy clay loam = 7, silty clay loam = 8, clay loam = 9, sandy clay = 10, silty clay = 11, clay = 12, organic materials = 13, bedrock = 14, and other = 15. See Rodell *et al.* (2004) for a color figure of global soil texture from a slightly different database.

separated based upon the relative amounts by weight of clay, silt, and sand in the soil. These data are now available in digital form and are often used to specify the soil type and texture for use in numerical weather prediction models with some soil data interpolated down to every 30 seconds. For example,

(b)

Figure 3.8. (cont.)

Reynolds *et al.* (2000) provide a 5 min, 16-category, two-layer soil texture database for the entire globe (Fig. 3.8). However, the detail and data quality of the original maps used to generate these types of databases has not changed and users should be aware of these limitations.

Information from the FAO Soil Map of the World also has been used to derive soil profiles and to estimate the water holding capacity of the soils (Webb *et al.* 1993). In addition, an international effort is underway to update the Soil Map of the World. The Soil and Terrain (SOTER) project is a joint effort of many groups to produce an updated, generalized map of the world soil resources.

Since the creation of the FAO Soil Map of the World many other soil surveys have been conducted and their data made available in digital form. In the USA, each of the individual 50 states have conducted detailed soil surveys down to a depth of 203.2 cm (80 in) or until bedrock is reached. If needed, soil samples were taken every 5.7 acres ($\sim$362 m^2) or often enough to verify a geomorphic landscape and draw out common soil areas. These data were combined into a State Soil Geographic (STATSGO) database by the USDA in which the more detailed soil surveys were transected and sampled to represent the soil information well on a regional scale. STATSGO data are available down to 30 second resolution, or grid cells of roughly 0.9 km in size and can be ingested easily into geographic information systems. Miller and White (1998) report on a 1 km, multi-layer 16-category (12 USDA soil texture classes plus organic material, water, bedrock, and other) soil texture database for 11 soil layers down to a 2 m depth that is available over the USA.

It also is important to consider how soil surveys are conducted. Some soil surveys quantify the proportion of clay, silt, and sand using a particle size analysis to obtain a very precise determination of the relative amounts of each particle type. However, the 12 soil texture classes marked off in Fig. 3.7 show the range of composition of the soils that represent each of the textural classes. It is clear that the variability in soil composition within individual soil classes often exceeds the variability in soil composition between soil classes. In addition, as discussed by Marshall *et al.* (1996), soil classification in the field is sometimes determined by working moistened soil between the fingers to sense the coarseness or fineness of the non-clay particles and to sense the strength and plasticity imparted by the clay particles. Thus, when the particle size distributions of a soil sample are plotted on the texture triangle, it is possible that the texture class into which the soil sample falls does not correspond to the texture class given in the field. Marshall *et al.* (1996) mention studies in which only half of the field texture class determinations fall within the same texture class when determined using the particle distributions. This occurs owing to the presence of clay minerals, organic matter, and cementing agents that strongly influence the field determination of texture. Field determinations of soil texture correspond to the particle size analysis in an average way, but not on a one-to-one basis. These uncertainties in the quality of the soil surveys and the range of soil compositions within a single soil textural class suggest that defining the soil conditions accurately at any specific location is difficult.

Many uncertainties arise when using the soil textures provided by global databases. In addition, many land surface schemes ignore the change in soil texture as a function of soil depth, instead assuming a constant texture. While a soil texture database is likely to provide a reasonable picture of soil texture and its variation over broad regions, it is not at all certain or guaranteed that the soil texture observed at a given location on the ground would agree with that specified from a global soil texture database. Other challenges are found when converting from the original soil database format to one that has a uniform grid cell (which is most convenient for meteorological models) as discussed by Miller and White (1998).

3.4 Biophysical control of evapotranspiration

The parameterization of evapotranspiration in most land surface schemes is divided into contributions from the bare soil, vegetation, and water on the canopy. Thus, we typically have

$$Q_E = Q_{EB} + Q_{EV} + Q_{EW}, \tag{3.5}$$

where Q_{EB} is bare soil evaporation, Q_{EV} is transpiration from the vegetation, and Q_{EW} is evaporation of liquid water from leaves on a wet canopy. The formulas for Q_{EB} and Q_{EW} are relatively straightforward and there is not as much difference in them among the various models. Thus, we examine these terms first and save Q_{EV} for last, since the parameterization of this term has the most variability from scheme to scheme.

3.4.1 Bare soil evaporation

The evaporation from bare soil is often parameterized as a fraction of the potential evaporation, where this fraction depends upon the value of the near-surface soil moisture. Thus, a number of schemes (e.g., Chen and Dudhia 2001) define

$$Q_{EB} = (1 - \sigma_f)\beta E_p, \tag{3.6}$$

where E_p is the potential evaporation and

$$\beta = \frac{\Theta_1 - \Theta_w}{\Theta_{fc} - \Theta_w}. \tag{3.7}$$

Here Θ_1 is the volumetric water content within the topmost soil layer, Θ_{fc} is the volumetric water content field capacity, and Θ_w is the wilting point (the volumetric water content at which plants can no longer bring water out of the soil). These two values depend upon the soil type and are determined from a soil database. Ek *et al.* (2003) indicate that this leads to evaporation falling off too slowly as soil moisture declines, and recommend a modified equation in which β is replaced by β^2. This modification may be related to the process of rapid soil drying as described by Santanello and Carlson (2001).

Another approach to parameterizing the bare soil evaporation is seen in Noilhan and Planton (1989), Mahfouf (1991), and Viterbo and Beljaars (1995), where they define

$$Q_{EB} = (1 - \sigma_f)\rho_a L_v \frac{(h_u q_s(T_g) - q_a)}{r_a}, \tag{3.8}$$

in which

$$h_u = \begin{cases} \frac{1}{2}\left[1 - \cos\left(\frac{\Theta\pi}{1.6\Theta_{fc}}\right)\right], & \Theta < \Theta_{fc}, \\ 1, & \Theta \geq \Theta_{fc}. \end{cases} \tag{3.9}$$

Here the dependence upon soil moisture is multiplied by $q_s(T_g)$, instead of by E_p as in (3.6). Thus, the h_u parameter is used to approximate the surface relative humidity, in contrast to β in (3.6) which is used to approximate the fraction of potential evaporation. It is unclear which approach is more realistic, and although these two methods generally behave in a similar fashion, they also produce very different answers.

It also is important to point out at this juncture that there are many ways to calculate the resistance r_a. For example, Viterbo and Beljaars (1995) use the results of Beljaars and Holtslag (1991) to define their expression for resistance, which uses a wind speed that incorporates an estimate of the vertical motion as well as the horizontal wind. They also use a transfer coefficient that depends upon the sensible heat flux. In general, the forms of the resistances closely resemble those defined in Chapter 2, but the devil is always in the detail.

If we focus only upon the potential difference (Δq) and the β and h_u parameters, ignoring the different contributions from resistances, and set $\Theta_w = 0.10$ and $\Theta_{fc} = 0.38$, then the β approach produces a linear increase in Δq whereas the h_u approach produces a more gradual, non-constant increase in Δq (see Fig. 3.9). One of the important differences between these three approaches is that Δq remains positive below $\Theta = 0.10$ from the h_u approach, whereas the β

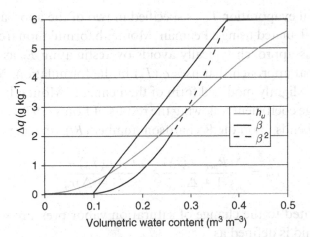

Figure 3.9. Values of Δq versus Θ for the expressions $[h_u q_s(T_g) - q_a]$ (gray line), $[\beta(q_s(T_g) - q_a)]$ (black line), and $[\beta^2(q_s(T_g) - q_a)]$ (dashed line), assuming that T_g and q_a are constant and the same for all expressions. It is assumed that $q_s(T_g) - q_a = 6\,\mathrm{g\,kg}^{-1}$, $\Theta_w = 0.10$ and $\Theta_{fc} = 0.38$ for this example.

approach allows no evaporation below the wilting point. In addition, the β approach reaches the potential evaporation at field capacity, while the h_u approach fails to reach the potential evaporation over the range of volumetric water contents evaluated. Thus, while evaporation increases in both schemes as the volumetric water content increases, the value for which evaporation first becomes non-zero and the actual values of evaporation can be quite different from the two schemes for identical values of soil moisture.

A third approach to bare soil evaporation is taken by Chen *et al.* (1996), in which they define

$$Q_{EB} = (1 - \sigma_f) \min\left(\left[-D_w \frac{\partial \Theta}{\partial z} - \kappa_w \right]_{z_1}, E_p \right), \tag{3.10}$$

where D_w and κ_w are the soil water diffusivity and thermal conductivity, respectively, and are evaluated for the topmost soil layer z_1. When the soil is moist, evaporation occurs at the potential rate E_p, but when the soil is drier evaporation only occurs at the rate by which the topmost soil layer can transfer water upward from lower soil levels. This approach certainly appears to be the most physical, in that evaporation only occurs at the rate by which moisture can be supplied by the upward water transfer in the soil. However, errors in this water transfer rate then directly influence the amount of latent heat flux, whereas in the other approaches the effect is more indirect. In addition, it is possible that the number of vertical soil levels influences the water transfer rate, so the choice of these levels is likely to influence the values of latent heat flux.

The potential evaporation E_p as specified in two of the approaches to calculating Q_{EB} is calculated using a Penman–Monteith formulation from Mahrt and Ek (1984). This approach typically avoids overestimating E_p as is seen in the more traditional approach that uses $q_s(T_g)$ in the formulation. Mahrt and Ek (1984) derive a slightly modified form of the Penman– Monteith equation, and use an exchange coefficient following the results of Louis (1979) and Louis *et al.* (1982) that depends upon the Richardson number (Ri), arriving at

$$E_p = \frac{\Delta(R_{net} - Q_G)}{1 + \Delta} + \frac{\rho L_v(q_s(T_a) - q_a)}{(1 + \Delta)r_a}, \tag{3.11}$$

where Δ is related to the change of saturation vapor pressure with respect to temperature and is defined as

$$\Delta = \frac{0.622}{p} \frac{L_v}{c_p} \frac{de_s(T)}{dT}, \tag{3.12}$$

and so is not assumed to be a constant as in the original Penman–Monteith derivation, $r_a = 1/(C_q u)$, u is the surface wind speed, and for the unstable case $(Ri < 0)$ they use

$$C_q = \left[\frac{k}{\ln\left(\frac{z+z_0}{z_0}\right)} \right]^2 \left(1 - \frac{15\,Ri}{1 + C\sqrt{-Ri}} \right), \tag{3.13}$$

with

$$C = \frac{75k^2 \left(\frac{z+z_0}{z_0}\right)^{1/2}}{\left[\ln\left(\frac{z+z_0}{z_0}\right) \right]^2}, \tag{3.14}$$

and for the stable case $(Ri > 0)$,

$$C_q = \left[\frac{k}{\ln\left(\frac{z+z_0}{z_0}\right)} \right]^2 \left[(1 + 15\,Ri)(1 + 5\,Ri)^{1/2} \right]^{-1}. \tag{3.15}$$

Here the Richardson number is defined as

$$Ri = \frac{g}{\theta} \frac{(\theta - \theta_{sfc})z}{u^2}, \tag{3.16}$$

where z is the height of the observation (θ and u), and θ_{sfc} is the potential temperature of the air at the lowest reference level. They find that this formulation of E_p is very sensitive to the diurnal variation of C_q, owing to the dependence upon Ri, and can lead to a factor of 2 difference between the formulations with and without this diurnal variation on days with moderate instability.

If we recall the definition of resistance to sensible heat flux from Chapter 2, and use our knowledge of the log wind profile under neutral conditions, then we can develop a modified definition of resistance where

$$r_H = \left(\frac{\ln\left(\frac{z+z_0}{z_0}\right)}{k} \right)^2 \frac{1}{\bar{u}}, \tag{3.17}$$

which is basically the inverse of the first term on the right-hand side of the definition of C_q in (3.13) and (3.15) multiplied by $1/\bar{u}$. Thus, the resistance used in the Mahrt and Ek (1984) formulation of potential evaporation is basically the same resistance derived previously in Chapter 2, but multiplied by an expression that depends upon the Richardson number. The use of the Richardson number is yet another approach to account for non-neutral conditions and the effects of atmospheric stability on the resistance calculations.

The previous formulations for bare soil evaporation are from land surface parameterizations that have a single combined bare soil and vegetation layer. However, there are schemes that use separate equations for the ground surface and the canopy layer, and one such scheme is the simple biosphere parameterization (Sellers *et al.* 1986). In their formulation, the evaporation from the bare soil is expressed as

$$
Q_{EB} = \frac{\rho c_p \left[e_s(T_g) - e_a \right]}{\gamma} \left[\sigma_f \left(\frac{\sigma_w}{r_d} + \frac{1 - \sigma_w}{r_d + r_g} \right) + \frac{(1 - \sigma_f)(f_h e_s(T_g) - e_a)}{(r_s + r_d)(e_s(T_g) - e_a)} \right],
$$

$$(3.18)$$

where r_d is the resistance between the ground and the canopy level, r_g is the bulk resistance of the groundcover, r_s is the bare soil surface resistance, σ_w is the wetted fraction of groundcover, and f_h is the relative humidity of the air at the soil surface. Definitions for the various resistances are found in Sellers *et al.* (1986). The main point is that, while the processes included in this expression are much greater than in the earlier bare soil evaporation equations, the number of parameters needed also has increased. Several of the resistances are expressed simply in terms of a surface-dependent constant multiplied by the near-surface wind speed. It is not clear if the added complexity provides additional accuracy to the calculations.

3.4.2 Canopy water evaporation

The parameterization for the evaporation of liquid water from on top of the leaves on a wet canopy is somewhat more consistent across the literature. Many of the formulations follow Noilhan and Planton (1989) and Jacquemin and Noilhan (1990) who define

$$
Q_{EW} = \sigma_f \left(\frac{W_c}{S} \right)^n E_p,
$$

$$(3.19)$$

where W_c is the intercepted canopy liquid water content, S is the maximum intercepted canopy water capacity (typically 0.5–2 mm), and $n = 0.5$. Mahfouf *et al.* (1995) define the value of S as proportional to *LAI*, while other approaches relate it to the vegetation type. The intercepted canopy water budget is determined by

$$\frac{\partial W_c}{\partial t} = \sigma_f P - D - \frac{Q_{EW}}{\rho_w L_v}, \tag{3.20}$$

where P is the precipitation rate, D is the rate of precipitation that drips off the canopy leaves and falls onto the ground, and ρ_w is the density of water. These schemes typically account for both precipitation and dewfall on the canopy and follow the approach outlined by Rutter *et al.* (1971, 1975). For schemes with a separate canopy temperature equation, interception and evaporation of water from groundcover also may be included (Sellers *et al.* 1986).

A slightly different expression is used by Viterbo and Beljaars (1995), where the potential evaporation is defined using the ground temperature, instead of the Penman–Monteith formulation, and the fraction of the model grid cell covered by the interception reservoir is defined as

$$C_l = \min\left(1, \frac{W_c}{S}\right), \tag{3.21}$$

where

$$S = [\sigma_f LAI + (1 - \sigma_f)] S_{max}, \tag{3.22}$$

and S_{max} is a constant that indicates the maximum amount of water on a single leaf or as a thin film over the ground (chosen as 0.0002 m in Viterbo and Beljaars (1995)). Thus, their final expression for canopy water evaporation is

$$Q_{EW} = C_l E_p. \tag{3.23}$$

They argue that, while the use of the $n = 0.5$ provides for a faster depletion of canopy water and that using $n = 1$ may mean that the canopy water never entirely disappears, they find no difficulties with using $n = 1$ in their formulation.

3.4.3 *Transpiration from vegetation*

The parameterization of the vegetation transpiration is accomplished using several approaches. Most of the canopy models described below can be called big-leaf models since they map the properties of the entire vegetation canopy onto a single leaf to calculate the transpiration. In addition, these canopy models generally only calculate the transpiration and do not worry about carbon exchange or photosynthesis rates. For example, Pan and Mahrt (1987) propose that

$$Q_{EV} = E_p k_v \sigma_f \left[\frac{z_1}{z_2} g(\Theta_1) + \frac{(z_2 - z_1)}{z_2} g(\Theta_2)\right] \left[1 - \left(\frac{W_c}{S}\right)^n\right], \tag{3.24}$$

where z_1 and z_2 are the soil layer depths, k_v is the plant resistance factor or plant coefficient (typically assumed to be 1, although it can be specified as a function of plant type), E_p is calculated as in Mahrt and Ek (1984) using a modified Penman–Monteith approach, and g is defined as

$$g(\Theta) = \begin{cases} 1, & \Theta > \Theta_{ref}, \\ \dfrac{\Theta - \Theta_w}{\Theta_{ref} - \Theta_w}, & \Theta_w \leq \Theta \leq \Theta_{ref}, \\ 0, & \Theta \leq \Theta_w, \end{cases} \tag{3.25}$$

where Θ_{ref} is assumed to be 0.25 in Pan and Mahrt (1987). Smirnova *et al.* (1997) use the same approach. This approximation draws water out of the soil layers in which the roots are assumed to exist in proportion to their depths and the volumetric water content of each layer. It also is assumed that the water on the canopy has to evaporate before maximum evaporation from the vegetation is reached. Desborough (1997) highlights how changes in surface root fraction influence transpiration, while Zeng (2001) develops a global vegetation root distribution for the most widely used land cover classifications.

Noilhan and Planton (1989) use a slightly different approach, and express the canopy evaporation as

$$Q_{EV} = \sigma_f \rho_a \frac{L_v [q_s(T_g) - q_a]}{r_a + r_c} \left[1 - \left(\frac{W_c}{S}\right)^{2/3} \right], \tag{3.26}$$

where $r_a = 1/(C_h u)$, C_h is a transfer coefficient depending upon the thermal stability of the atmosphere, and the canopy resistance is defined as

$$r_c = \frac{r_{cmin}}{LAI \cdot F_1 F_2 F_3 F_4}. \tag{3.27}$$

The F-factors account for the effects of solar radiation, vapor pressure deficit, air temperature and soil moisture in the calculation of canopy resistance (Noilhan and Planton 1989). The F_1-factor is based upon the studies of Dickinson (1984) and Sellers *et al.* (1986) and is defined as

$$F_1 = \frac{f + (r_{cmin}/r_{cmax})}{1 + f}, \tag{3.28}$$

with

$$f = 0.55 \frac{Q_s}{Q_{GL}} \frac{2}{LAI}, \tag{3.29}$$

and where $r_{c\,min}$ is the minimum stomatal resistance, $r_{c\,max}$ is the maximum stomatal resistance (typically set to $5000\,\mathrm{s\,m^{-1}}$ as in Dickinson 1983), Q_S is the incoming solar radiation that reaches the surface, and Q_{GL} is a scaling parameter that typically depends upon the vegetation type and varies from $30\,\mathrm{W\,m^{-2}}$ for trees to $100\,\mathrm{W\,m^{-2}}$ for crops (Jacquemin and Noilhan 1990). The minimum stomatal resistance also depends upon the vegetation type, highlighting the importance of our knowledge of the underlying vegetation to the calculations of transpiration. As one can see from (3.28), F_1 is less than 1 and decreases as Q_S decreases. Thus, owing to the influence of F_1 only, r_c is likely to reach its minimum value at local noon when Q_S is largest and reaches its maximum value at sunrise and sunset, and retains this value throughout the night. This diurnal evolution of r_c agrees well with observations of canopy resistance summarized in Monteith and Unsworth (1990).

The F_2-factor represents the effects of the vapor pressure deficit on the canopy resistance, and is expressed in Noilhan and Planton (1989) as

$$F_2 = 1 - g\left[e_s(T_g) - e_a\right], \tag{3.30}$$

where g is a vegetation-dependent empirical parameter. Using a different formulation, Jacquemin and Noilhan (1990), based upon the studies of Jarvis (1976) and Avissar *et al.* (1985), define F_2 as

$$F_2 = \frac{1}{1 + \alpha[q_s(T_a) - q_a]}, \tag{3.31}$$

where α is a vegetation-dependent parameter that typically varies from 36 to 60. Under typical conditions, F_2 varies from 0.5 to 1.0, with the smaller values occurring for larger vapor pressure deficits. Thus, r_c increases as the vapor pressure deficit increases, and represents the stomatal response to the environmental conditions.

The influence of the ambient temperature upon canopy resistance is taken from Dickinson (1984), such that F_3 is expressed as

$$F_3 = 1 - 1.6 \times 10^{-3}(T_{ref} - T_a)^2, \tag{3.32}$$

where T_{ref} is 298 K in temperate zones. Since the temperature difference is squared, F_3 decreases and r_c increases when the temperature departs from T_{ref}. Note that when $|T_{ref} - T_a| = 25\,\mathrm{K}$, $F_3 = 0$, indicating that the canopy resistance becomes infinitely large and all transpiration ceases. In practice, F_3 generally is required to be a very small positive number (0.0001).

The final F-factor represents the influence of soil moisture on the canopy resistance. There are several different approaches to this term, but they all attempt to determine the amount of available soil moisture for transpiration. Jacquemin and Noilhan (1990) define

$$F_4 = \frac{\Theta_2 - \Theta_w}{\Theta_{fc} - \Theta_w},\tag{3.33}$$

where Θ_2 is the volumetric water content of the bulk soil layer in which the majority of the roots are located. Other schemes compute a root-layer average soil moisture (Chen *et al.* 1996; Chen and Dudhia 2001) for which a two-layer rooting zone yields

$$F_4 = \sum_{i=1}^{2} \frac{(\Theta_i - \Theta_w)d_{z_i}}{(\Theta_{fc} - \Theta_w)(d_{z_1} + d_{z_2})},\tag{3.34}$$

where d_{z_i} is the depth of the ith soil layer. Thus, F_4 decreases and r_c increases as the soil moisture is depleted towards the wilting point when $F_4 = 0$ and transpiration ceases.

A third expression for the transpiration from vegetation is found in Viterbo and Beljaars (1995) in which

$$Q_{EV} = (1 - C_l)\sigma_f \frac{\rho L_v}{r_a + r_c} \left[q_s(T_g) - T_a \right],\tag{3.35}$$

where $r_a = 1/(C_h u)$, C_h is a transfer coefficient, u is the near-surface wind speed, C_l is the interception reservoir fraction for the wet canopy, and

$$r_c = \frac{r_{c\,min}}{LAI} f_1(PAR) f_2(\overline{\Theta}).\tag{3.36}$$

In this expression there are only two environmental factors that influence the canopy resistance, the amount of incoming photosynthetically active radiation (PAR) and the mean soil volumetric water content within the rooting zone. Following Sellers (1985), they define the f_1 adjustment factor as

$$\frac{1}{f_1(PAR)} = 1 - a_1 \log\left(\frac{a_2 + PAR}{a_3 + PAR}\right),\tag{3.37}$$

where the values of a_i are related to the vegetation type. The expression for the f_2-factor is basically the same as for g in the Pan and Mahrt (1987) scheme as shown in (3.25).

Finally, Chen *et al.* (1996) and Chen and Dudhia (2001) use the following expression for transpiration, namely

$$Q_{EV} = \sigma_f E_p B_c \left[1 - \left(\frac{W_c}{S} \right)^n \right],$$ (3.38)

where

$$B_c = \frac{1 + \Delta/r_r}{1 + r_c C_h u_a + \Delta/r_r}.$$ (3.39)

Here Δ is again related to the change of saturation vapor pressure with respect to temperature as defined by (3.12) by Mahrt and Ek (1984), r_c is the canopy resistance as defined by Noilhan and Planton (1989), including the F_1 through F_4 factors, r_r is a resistance that includes an adjustment factor that depends upon the air temperature and surface pressure and is defined as

$$r_r = \frac{4\sigma T_a^4 R_d}{p c_p C_h u_a} + 1,$$ (3.40)

where u_a is the surface layer wind speed and in which the transfer coefficient

$$C_h = \frac{k^2}{R} \frac{G_2}{\ln(z/z_0) \ln(z/z_{0h})},$$ (3.41)

where

$$G_2 = \begin{cases} e^{-aRi}, & \text{stable,} \\ 1 - \dfrac{15Ri}{1+A}, & \text{unstable,} \end{cases}$$ (3.42)

and

$$A = \frac{70.5 k^2 \sqrt{-Ri(z/z_0)}}{\ln(z/z_0) \ln(z/z_{0h})},$$ (3.43)

in which $R\ (=1)$ is the ratio of exchange coefficients, Ri is the Richardson number as defined previously, and a is a constant often assumed to be 1.0. These expressions are from Mahrt (1987) for the stable case, and Holtslag and Beljaars (1989) for the unstable case.

The canopy evapotranspiration from schemes that have separate equations for the ground surface and the vegetation canopy is often calculated directly using the following equation from Sellers *et al.* (1986):

$$Q_{EV} = \frac{[e_s(T_c) - e_a] \rho c_p}{(\bar{r}_c + \bar{r}_b)\gamma} \left(\frac{\sigma_w}{\bar{r}_b} + \frac{1 - \sigma_w}{\bar{r}_b + \bar{r}_c} \right),$$ (3.44)

where the overbars refer to resistances that are bulk values. This again is the same basic equation for latent heat flux as seen in Chapter 2, but using canopy temperature T_c and incorporating the reduction in Q_E from the interception of precipitation from leaves and using somewhat different resistance formulations.

The canopy resistance used by Sellers *et al.* (1986) also is different from the previous expressions, and is a rather complicated function that depends upon a large number of factors. The canopy resistance is defined as

$$\frac{1}{r_c} = \sigma_f N_c f_3 f_4 f_5 \int_0^{LAI} \int_0^{\pi/2} \int_0^{2\pi} \frac{O(\xi, \theta)}{r_l(F, \kappa, \xi, \theta)} \sin(\theta) \, d\xi \, d\theta \, dLAI, \qquad (3.45)$$

where σ_f is the fractional canopy coverage (0 to 1), N_c is the fraction of *LAI* that consists of live, photosynthesizing leaves, f_3 is an adjustment factor for water stress, f_4 is an adjustment factor for leaf temperature, f_5 is the adjustment factor for vapor pressure deficit, $O(\xi, \theta)$ is the leaf angle distribution function, where θ is the leaf angle above a horizontal plane and ξ is the leaf azimuth angle, and r_l is the resistance of an individual green leaf, where F is the photosynthetically active radiation incident upon the leaf and κ is the leaf conductivity.

Many other approaches to the calculation of transpiration exist in the literature. In addition to the big-leaf canopy models described above (see also Dai *et al.* 2003), multi-layer canopy models are available that integrate the fluxes from each canopy layer to give the total flux (Wang and Jarvis 1990; Leuning *et al.* 1995). Dai *et al.* (2004) develop a two-big-leaf canopy model that differentiates between sunlit and shaded leaves to overcome problems with the overestimation of transpiration seen in single big-leaf canopy models.

Unfortunately, canopy models that predict canopy or stomatal resistance directly from environmental factors, while neglecting the fundamental underlying mechanisms, have a general problem in that they tend to be specific to a particular vegetation type and must be readjusted for different vegetation types and perhaps even different climate conditions (Collatz *et al.* 1991). Thus, physiological schemes for estimating stomatal resistance have also been developed (Ball *et al.* 1987; Kim and Verma 1991; Collatz *et al.* 1991). Although empirically based, these schemes mimic the physiological response of the vegetation to determine the appropriate regulation between the conflicting roles of permitting CO_2 to diffuse to the leaf to support photosynthesis and restricting the diffusion of water vapor out of the leaf. Dai *et al.* (2004) and Daly *et al.* (2004) develop other examples of canopy models that include consideration of photosynthesis in their formulation. Comparisons between the stomatal resistance calculation of Noilhan and Planton (1989) and those from physiological approaches by Niyogi and Raman (1997) indicate that the physiological schemes tend to produce a better trend in stomatal resistance, but also have a large amount of variability. Similar results affirming the claim that photosynthesis-based schemes reproduce the measured stomatal resistances more closely are seen in Collatz *et al.* (1991), Baldocchi and Rao (1995), and Su *et al.* (1996). Unfortunately, these more complex physiological models are computationally more expensive than the simple big-leaf canopy models discussed above and require information on photosynthesis rates and CO_2 concentrations at the leaf surface that is not readily available (Niyogi *et al.* 1998). Thus, these more sophisticated and more realistic schemes are not presently used in operational numerical weather prediction models.

3.4.4 Subgrid variability

Some of the land surface models allow for heterogeneous land surfaces within each grid cell of the model domain. This subgrid variability in the surface forcing is represented in the model by dividing the grid cell into a number of homogeneous subregions. The number of subregions depends upon either the number of different vegetation types or soil types, or may be specified in advance. As first discussed by Avissar and Pielke (1989), it is assumed that the horizontal fluxes between the homogeneous land surface patches are small compared to the vertical fluxes and that the atmospheric conditions above the grid cell are the same for each of the subgrid patches. Thus, within each grid cell, the horizontal areas of each type of surface class are summed even if they are not adjacent to each other. This aggregation procedure greatly reduces the total number of subgrid classes needed for the grid cell, since it depends upon the surface class only and not the patch location within the grid cell. The total flux of energy from the surface to the atmosphere within each grid cell is defined as

$$Q_x = \frac{\sum_{i=1}^{n} A_i Q_x}{\sum_{i=1}^{n} A_i},$$

(3.46)

where Q_x is the flux (latent, sensible, ground, longwave radiation, etc.), A_i is the area of the *i*th surface class, and *n* is the number of surface classes, or patches, within the grid cell. Avissar and Pielke (1989) suggest that while mesoscale atmospheric patterns can be simulated without including the subgrid variability in land surface conditions, the inclusion of subgrid variability allows for the prediction of the micrometeorological conditions within each land class or patch that may be important to end users of forecast information.

3.5 Momentum transfer

In simple terms, the roughness length z_0 is increased when a ground surface is vegetated. This leads to a rougher surface, such that the vertical mixing is enhanced. Another way of looking at it is to note that as the roughness length increases, the resistance decreases, and thus the surface fluxes of sensible and latent heat are larger.

In land surface schemes that include a single temperature for the combined bare soil and vegetation surface, the roughness lengths typically are defined through information on the dominant vegetation type within the grid cell. These roughness lengths thus already incorporate the gross effects of vegetation on the vertical mixing. However, in land surface schemes that have separate equations for ground and vegetation canopy temperatures, there are several additional roughness lengths that are needed to determine the fluxes. In these schemes, there are typically two different types of vegetation: ground cover and canopy. This requires the addition of one extra resistance and some reconfiguring of the layers over which the resistances are calculated (Sellers *et al.* 1986). With a canopy, the resistances that need to be specified are:

1. resistance from the top of the ground cover and soil surface to the canopy height (r_d);
2. resistance within the canopy layer ($\bar{r}_b$);
3. resistance from the canopy height to within the atmospheric surface layer (r_a) so that the flux calculations can be completed.

The first resistance r_d is defined in Sellers *et al.* (1986) as

$$r_d = \int_0^h \frac{1}{K_m}\, dz = \frac{c_2}{u_c}, \tag{3.47}$$

where K_m is the eddy viscosity ($\mathrm{m^2\,s^{-1}}$), h is the height of the canopy, c_2 is a surface-dependent constant, and u_c is the wind speed at the canopy level. For the second resistance $\bar{r}_b$, note that if r_{bi} is assumed to be the resistance of an individual leaf, then $\bar{r}_b$ is the resistance of all the leaves in the canopy. Using various scaling arguments, one can arrive at the resistance of an individual leaf

$$r_{bi} = \frac{c_s}{L_i \sqrt{u_i}}, \tag{3.48}$$

where c_s is another transfer coefficient, L_i is the leaf area ($\mathrm{m^2}$), and u_i is the wind speed on the individual leaf i. This can then be summed over all leaves to obtain

$$\frac{1}{\bar{r}_b} = \sum_{i=1}^{n} \frac{L_i\sqrt{u_i}}{c_s P_s}, \tag{3.49}$$

where P_s is a shelter factor defined as the ratio of the observed transfer coefficient within the canopy to the transfer coefficient of an individual leaf in isolation (Monteith and Unsworth 1990). If one further assumes that the leaf density is constant within the canopy, then

$$\frac{1}{\bar{r}_b} = \int_{z_{c\,bottom}}^{z_{c\,top}} \frac{LAI\sqrt{u}}{c_s P_s}\, dz = \frac{c_1}{\sqrt{u_c}}. \tag{3.50}$$

Finally, the last resistance that needs to be calculated is the resistance from the canopy height to within the atmospheric surface layer. This is our old friend

$$r_a = \frac{1}{ku_*}\left[\ln\left(\frac{z_{eff}}{h}\right) - \psi_h\left(\frac{z_{eff}}{L}\right)\right]. \tag{3.51}$$

Note that the resistances r_d and $\bar{r}_b$ are both defined using coefficients (c_1 and c_2) that are dependent upon the vegetation type. Thus, these schemes have yet more parameters that are dependent upon the vegetation type as specified in the database used by the numerical model, which typically only incorporates 15–30 vegetation types compared to the hundreds that are estimated from satellite data.

3.6 Soil moisture availability

It is obvious that when over a certain time period the evapotranspiration is greater than the rainfall over the same time period, then the difference represents the amount of water that must be supplied from the soil or some other source. The complicating factors in trying to calculate this water transport are that plants have developed strategies for growth and reproduction that maximize the odds of survival of their species, and that water–soil interactions may not be as simple as originally hoped. Canny (1998) suggests that managing the stream of water that plants need to survive "dominates the life process of plants." Accordingly, a lot of new terminology (at least to meteorologists) must be introduced before one can understand how models of soil and vegetation operate.

We begin with a few new terms from soil science and botany. Potential (short for potential energy) is one of the most important new terms. The basic idea is that since work is performed by water as it moves from one system to

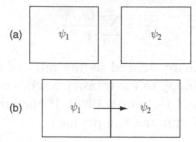

Figure 3.10. Two equilibrium soil water systems that (a) are not interacting and (b) that are interacting after coming in contact with each other. Water flows from high potential to low potential, with $\psi_1 > \psi_2$.

another, then water in an equilibrium system has the potential to do work, and hence the energy is called potential energy. This terminology replaces terms such as suction, tension, stress, head, and pressure that were used previously in soil sciences. Potential as used in meteorology generally has units of $J\,kg^{-1}$ or meters and is noted by the Greek letter ψ. Potential can be thought of as the partial specific Gibbs free energy of the water in the system. To convert potential from units of $J\,kg^{-1}$ to m, note that multiplying or dividing by gravitational acceleration g does the trick. To convert potential from units of $J\,kg^{-1}$ to bars, divide by 100.

As an example, examine two soil water systems ψ_1 and ψ_2 that are presently in equilibrium (Fig. 3.10), have different potentials with $\psi_1 > \psi_2$, and are initially separate from each other. When these two systems are brought into contact with each other, then water flows from system 1 into system 2 as water is defined to flow from higher potential to lower potential, since work must be done to move the water. Thus, the water flux density (F_w) with units of $kg\,m^{-2}\,s^{-1}$ is related to the gradients of potential, such that

$$F_w = -k_w \frac{\partial \psi}{\partial x}, \tag{3.52}$$

where k_w is the hydraulic conductivity. This basic relationship is called Darcy's law (see Houser 2003). Typically, for soils and plants it is assumed that the water flux density, F_w, is only influenced by vertical gradients in potential, although horizontal gradients are possible owing to horizontal variations in rainfall or being close to streams, lakes, or rivers.

Potential is produced by a number of different physical mechanisms, depending in part upon whether one is studying the potential for plants or for soils. One of these mechanisms is directly related to the pressure potential (ψ_p) that results from an overall pressure that is different from the reference

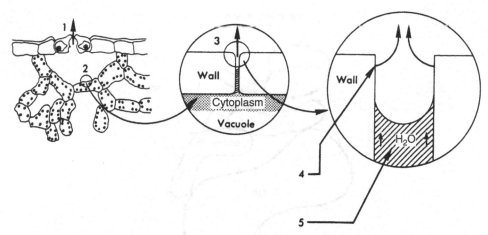

Figure 3.11. Illustration of the rise of sap within a tree as a result of transpiration. (1) Transpiration through the stomates (indicated by heavy upwardly directed arrows), (2) evaporation from microcapillaries into the intercellular space of the stomates, (3) partial drying of the microcapillary cell walls, (4) water from film attracted by surface tension, and (5) rising of whole water column because of cohesive force of water. From Levitt (1974), used with permission from Elsevier.

pressure (i.e., atmospheric pressure). If $p > p_{ref}$, then $\psi_p > 0$. For plants, pressure potential is due to the effects of surface tension (capillarity), since the water system in most plants is unbroken from roots to leaves. As described in Levitt (1974), the microcapillaries that supply water to stomates within the leaves are very thin, of the order of 0.1 μm. As the stomates open to obtain carbon dioxide from the atmosphere, transpiration occurs and to replace the water vapor in the intercellular space within the stomates, evaporation of water occurs within the microcapillaries (Fig. 3.11). As the microcapillary water evaporates, the walls of the microcapillaries partially dry and the force of adhesion between the walls of the microcapillaries and water pulls the water towards the stomates. The column of water between the stomates and the roots then moves en masse owing to the force of cohesion between the water molecules. This process is called the cohesion theory for water movement within plants. Calculations indicate that this process could supply water to trees that reach 300 m in height, much higher than any trees observed today. While there are some who contest the cohesion theory for water movement in plants, it remains the most widely accepted theory in explaining plant water movement (Taiz and Zieger 2002). Canny (1998) provides a summary of the difficulties and challenges in trying to understand water movement in plants and offers an alternative to cohesion theory. For soils, $\psi_p = 0$ unless the soil is

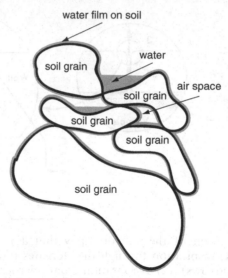

Figure 3.12. Soil grains surrounded by a thin layer of water (gray shading), due to adsorptive effects of the soil, and also supporting small areas in which water can collect.

supersaturated and for this case $\psi_p > 0$. The pressure potential is never less than zero for soils.

A different mechanism for soils is directly related to the matric potential (ψ_m). Matric potential is the portion of potential that results from the colloidal matrix, or the intercellular substance of a tissue or mass, of the soil system. Thus, matric potential includes the effects of adsorption (the gathering of water in a condensed layer on a surface), capillarity, and curved air–water interfaces. It is found that soil particulates adsorb water such that a very thin layer of water surrounds each soil grain (Fig. 3.12). This water layer may not be very thick, but it is very important to the overall moisture level in the soil. Adsorptive forces are very strong and it is difficult to remove water from the soil when the water layer is reduced to four or five molecules in thickness. If soils are examined very closely, then both air spaces in between the soil grains and wedges in which water can collect are observed. Thus, soil is highly fractal in structure.

The third type of potential is osmotic potential (ψ_o), produced by the combined effects of all solute species present in the soil or plant water system. In plants, the osmotic potential is the potential difference between water in solution and pure free water. It is defined as

$$\psi_o \cong -cRT, \tag{3.53}$$

where c is the moles of solute per unit weight of the mixture (mol kg^{-1}), R is the universal gas constant of 8.31 J K^{-1} mol^{-1}, and T is the temperature in kelvin. This expression is called the van't Hoff relation (Taiz and Zieger 2002). It is found that ψ_o is either zero or negative. In soils, osmotic potential happens mainly when salts are present in the soil; otherwise the osmotic potential is zero. For plants, the sum of the pressure potential and the osmotic potential is called the mesophyll potential ($\psi_m = \psi_o + \psi_p$) and represents the total attraction of the intercellular tissues for water.

The final potential that needs to be defined is the gravitational potential (ψ_g), which is that portion of the potential attributable to the gravitational force field and is dependent upon the vertical elevation of the water. Thus,

$$\psi_g = \rho_w g(z - z_{ref}), \tag{3.54}$$

where ρ_w is the density of water, g is the acceleration due to gravity, z is the height of the water system, and z_{ref} is the reference height to which it is being compared (typically the height of the ground surface). The total potential for plants and soil can then be defined as

$$\psi_{plant} = \psi_{m_{plant}} + \psi_g = \psi_{p_{plant}} + \psi_{o_{plant}} + \psi_g, \tag{3.55}$$

$$\psi_{soil} = \psi_{m_{soil}} + \psi_g. \tag{3.56}$$

Now that potential has been defined, it is important to understand how potential is observed and what it means physically. Both perhaps can be explained by examining one method for calculating potential. While potential can be observed using tensiometers, gas pressure devices can also be used (Marshall *et al.* 1996). The most important component of a gas pressure device is a horizontal porous ceramic plate that allows water to pass through the plate while also impeding the flow of air. Below the ceramic plate in these devices is a water reservoir at atmospheric pressure that is in direct contact with the plate. Placed on top of the ceramic plate are soil samples. And surrounding everything is a chamber into which gas can be pumped and measured (Fig. 3.13). If the air in the chamber is at atmospheric pressure, then water flows into the soil if the soil is initially dry owing to the soil matric potential – the adsorptive effects of the soil. At point A in Fig. 3.13, $\psi_p = 0$ since the soil is at the reference pressure and the soil is not saturated. However, $\psi_m < 0$ at this same point owing to the adsorptive effects of the soil matrix and the soil not being saturated, and thus water moves upward into the soil. The value of ψ_m for the soil can be determined by increasing the air pressure in the container until no water flows, yielding

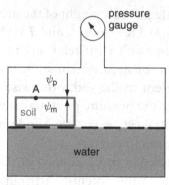

Figure 3.13. Schematic of a gas pressure device used to measure pressure and thereby matric potential. Water fills the bottom portion of the container, with a ceramic plate porous to water but not to air separating the water from the soil (dashed line). Initially, water flows into the unsaturated soil owing to the adsorptive effects of the dry soil or its matric potential. By increasing the pressure in the chamber, this upward water movement can be stopped. At this time, the pressure and matric potentials are equal.

$$\psi_{m_{soil}} = \psi_p = \frac{(p_{ref} - p)}{\rho_w} < 0. \tag{3.57}$$

As indicated by (3.57), the values of both soil and plant potential typically are negative.

Observations indicate that transpiration removes a considerable amount of water from the plant tissues as the stomates expose the water in the plant to evaporation. The rate of transpiration thus determines the rate at which the soil must supply water to the plant. For some woody plants, the plant potential decreases rapidly as transpiration increases even when the water supply to the roots is adequate (Camacho-B *et al.* 1974). In contrast, for herbaceous plants studied under identical conditions, the plant potential remains nearly constant across a fairly large range of transpiration rates. Thus, it appears that plants can be grouped into three broad categories based upon their response to environmental conditions: (1) plants with inefficient water transport systems and strong stomatal control to regulate water loss, (2) plants with both efficient water transport systems and strong stomatal control, and (3) plants with very efficient water transport systems and little stomatal control (Camacho-B *et al.* 1974). The different responses of these three broad plant categories can be generalized as being due to differences in resistance of the soil–plant system at the leaf–air interface (stomatal control) and at the root–soil interface (water transport). Therefore, what happens at these soil–plant–air interfaces can be important in defining the rate of transpiration.

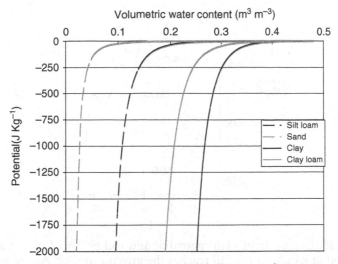

Figure 3.14. Characteristic curves of soil potential (J kg^{-1}) versus volumetric water content for four different types of soils: sand, silt loam, clay loam, and clay. Data based upon coefficients from Cosby *et al.* (1984).

The root–soil interface resistance is greatly influenced by the soil potential, which is a function of the soil water content. Observations indicate that as the volumetric water content of the soil decreases from saturated conditions (values of ~ 0.4–0.5), the soil potential at first decreases very little. However, eventually a point is reached at which the soil potential quickly becomes more negative as the volumetric water content decreases slightly owing to the adsorptive effects of the soils for water (Fig. 3.14). Enough similarity is seen within soils that characteristic curves have been identified for various soil types that relate the soil potential to the water content of the soil. Expressions that define these curves are used within the soil–vegetation–atmosphere transfer models.

The evolution of the plant and soil potentials in relation to each other over a sequence of days indicates that each morning after the sun rises, the plant potential is lowered in order for water to flow into the plant through the root system (Fig. 3.15). The plant potential reaches a minimum sometime typically in the afternoon, and then increases toward nighttime as the demand for water decreases. During this time period, enough water has been removed from the initially near saturated soil that a noticeable difference is seen, but it is not very large. During the night, the plant and soil potentials equilibrate, as the plant no longer has any need to move water. As day two dawns, the plant again reduces the potential and the uptake of water from the soil begins again. This process continues over several days, and perhaps even weeks, depending upon the water needs of the plant and the atmospheric conditions, until at some time (assuming no new rain has fallen), the plant potential has to decrease to very

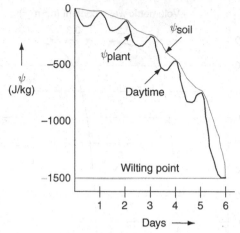

Figure 3.15. Evolution of plant potential and soil potential over a multiple day period in which no rainfall reaches the ground surface. As the sun rises, the plant potential decreases in order to move water from the soil to the plant system. The potential difference is related to the amount of water flowing into the plant. At night, the plant and soil potentials are equal as no water is flowing in the plant–soil system. After several days, the soil potential drops so low (owing to water loss due to the plant) that the plant can no longer draw more water from the soil. This is the wilting point, and indicates that transpiration has ceased.

negative values in order to continue to move water from the soil to the plant. Eventually, the plant reaches its wilting point, when capillary water becomes unavailable to the plant cells and the plant wilts. Thus, once the soil potential is equal to or less than the plant potential at the wilting point, water flow to the plant stops and transpiration ceases. The potential value at the wilting point for most plants is close to $-1500\,\mathrm{J\,kg^{-1}}$, although potentials as low as $-4000\,\mathrm{J\,kg^{-1}}$ have been observed for plants in semi-arid regions (Peláez and Bóo 1987).

To model the behavior of plants and soils and their interactions, the same resistance approach developed earlier for the calculations of the fluxes of sensible and latent heat is used. However, the potential difference for plant and soil interactions is the difference in potential of the plant and soil systems. So one needs to be able to determine the plant and soil potentials, and the required resistances, in order to calculate the water flux into the plant and evapotranspiration. Many soil–vegetation–atmosphere parameterizations avoid calculating the plant potential, assuming that the plant water needs are met as long as the soil potential is above the wilting point.

The characteristic curves of soil potential versus water content of the soil provide the empirical results necessary to model this type of behavior

(Fig. 3.14). As seen previously, the soil potential remains fairly steady as the volumetric water content is first decreased from its value at saturation, but then a point is reached where the soil potential undergoes a rapid decrease as the volumetric water content is decreased slowly. As might be imagined, the point at which the soil potential begins to decrease rapidly is important to the behavior of the soil–plant system. The characteristic curves can be fitted for the various soil texture types, yielding an expression that relates the soil potential to the volumetric water content

$$\psi_{soil} = \psi_s \left(\frac{\Theta}{\Theta_s}\right)^{-b}, \tag{3.58}$$

where ψ_s is the saturation matric potential ($\psi_s < 0$), b is an empirically deter-mined constant, and Θ_s is the soil porosity. Porosity represents the maximum amount of water that the soil can hold, assuming all the pores that contribute to a change in water storage are filled with water. It also is referred to as the volumetric water content at saturation. The parameters ψ_s, Θ_s, and b depend upon the soil texture class. For example, sand has $b \sim 4$, $\psi_s \sim -0.12 \, \text{J} \, \text{kg}^{-1}$, and $\Theta_s \sim 0.39$. In contrast, clay has $b \sim 11$, $\psi_s \sim -0.41 \, \text{J} \, \text{kg}^{-1}$, and $\Theta_s \sim 0.48$.

The shapes of the characteristic curves are very sensitive to the values of b (Fig. 3.16). According to Cosby et al. (1984), who examined 1448 different soil samples, the standard deviations in the values of b for the 11 soil types vary

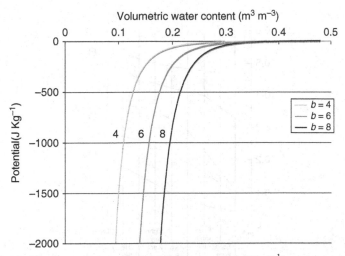

Figure 3.16. Characteristic curves of soil potential ($\text{J} \, \text{kg}^{-1}$) versus volumetric water content ($\text{m}^3 \, \text{m}^{-3}$) for a soil using three different values for the b parameter and with all other parameters constant. Here it is assumed that $\psi_s = -1 \, \text{J} \, \text{kg}^{-1}$ and $\Theta_s = 0.50$.

from a minimum of 1.38 for sand to a maximum of 4.33 for silty clay loam. This uncertainty in the values of b results in a fairly large uncertainty in the values of soil potential. The values in Cosby *et al.* (1984) show that the variations of b within the soil classes are as large as the differences in b between soil classes. For values of volumetric water content near saturation, this uncertainty in the value of b makes little difference in the curves (Fig. 3.16). However, as the volumetric water content decreases, this uncertainty can lead to differences in potential of more than $500\,\mathrm{J\,kg}^{-1}$ for moderate values of volumetric water content $(0.2\text{–}0.3\,\mathrm{m}^3\,\mathrm{m}^{-3})$ and differences of more than $1000\,\mathrm{J\,kg}^{-1}$ for smaller values of volumetric water content $(\Theta < 0.2\,\mathrm{m}^3\,\mathrm{m}^{-3})$. This can have a significant difference on determining when the wilting point is reached and transpiration ceases, potentially resulting in large errors in evapotranspiration when the value of soil potential approaches the wilting point.

The classification of the soil texture is important as it is used to define the various parameters that relate to soil–water interactions. First, the water holding properties differ between the various soil textures, leading to significantly different values of the volumetric water content at the porosity, field capacity, and wilting point (Fig. 3.17). Second, the values of b and ψ_s also depend upon soil texture and vary across the soil texture triangle. While the values of ψ_s vary across a relatively small range, the values of b vary from 2.79 for sand to 11.55 for light clay (Table 3.1). Since b is the exponent in (3.58) in which the soil potential is related to the volumetric water content, it plays a large role in determining the soil potential. The standard deviations of b found by

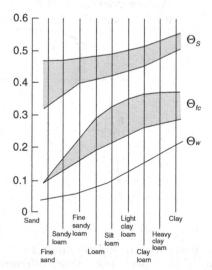

Figure 3.17. Water holding properties of various soils. Shading highlights the variation in the parameters for a given soil type. After Houser (2003).

Table 3.1. *Mean and standard deviations of the b parameter for each of 11 soil classes and the number of samples tested. From Cosby* et al. *(1984).*

Soil class	No. samples	Mean value of b	Standard deviation
Sandy loam	124	4.74	1.40
Sand	14	2.79	1.38
Loamy sand	30	4.26	1.95
Loam	103	5.25	1.66
Silty loam	394	5.33	1.72
Sandy clay loam	104	6.77	3.39
Clay loam	147	8.17	3.74
Silty clay loam	325	8.72	4.33
Sandy clay	16	10.73	1.54
Silty clay	43	10.39	4.27
Light clay	148	11.55	3.93

examining a large number of soil samples (Clapp and Hornberger 1978; Cosby *et al.* 1984) indicate a fair amount of uncertainty in the exact soil potential at a given location for any predicted or measured volumetric water content.

Recall that soil texture typically is defined from worldwide soil surveys (see Fig. 3.8) that vary in quality and degree of detail from country to country. Once the soil texture is known at a given grid point, the results of Cosby *et al.* (1984) can be used to specify soil porosity, saturated matric potential, saturated hydraulic conductivity, and the exponential parameter b. However, the volumetric water content field capacity and wilting point are not provided by Cosby *et al.* (1984) and so must be computed. Generally, the field capacity is determined either as a fraction of the porosity (75% of porosity as in Noilhan and Planton 1989) or by determining the soil moisture value that yields a hydraulic conductivity equal to a specific drainage flux. For example, Wetzel and Chang (1987) define field capacity as the volumetric soil water content that produces a value of hydraulic conductivity of $0.1 \, \mathrm{mm \, day^{-1}}$, whereas Chen and Dudhia (2001) use $0.5 \, \mathrm{mm \, day^{-1}}$. The wilting point typically is determined using a constant value of soil potential below which plants have extreme difficulty in extracting water from the soil, and then calculating the volumetric soil water content for that value of potential. As mentioned previously, a potential value of between -1500 and $-2000 \, \mathrm{J \, kg^{-1}}$ is often used for the wilting point.

Once the soil potential is calculated, the next step is to determine the water transport. The soil hydraulic conductivity κ_{soil}, which is inversely related to the soil resistance, can be specified as (Campbell 1974)

$$\kappa_{soil} = \kappa_s \left(\frac{\Theta}{\Theta_s}\right)^{2b+3} = \frac{\alpha_f}{r_{soil}} = \kappa_s \left(\frac{\psi_s}{\psi_r}\right)^{(2B+3)/B}, \tag{3.59}$$

where κ_s is the value at saturation, b and B are specified constants that depend upon soil type (the b parameter is the same as used in determining soil potential), ψ_r is the soil potential in the root zone, and α_f is a conversion factor. Values of κ_s vary from 10^{-4} to $10^{-5}\,\mathrm{m\,s^{-1}}$ for sand to 10^{-6} to $10^{-9}\,\mathrm{m\,s^{-1}}$ for clay. Larger values of b produce smaller values of conductivity and larger values of soil resistance as Θ decreases.

When thinking about soil moisture availability, it is important to define all the sources and sinks of moisture. The uptake of water by plants is a sink for soil moisture, whereas precipitation provides a source for soil moisture. Finally, given a little thought one could probably realize that soil moisture can flow between different soil layers and perhaps even drain away. This drainage term could be either a source or a sink, depending upon the direction of movement of the water in the soil.

One way to determine changes to the soil volumetric water content Θ is to use Darcy's law, which relates the change in potential to the flow rate, such that

$$\frac{\partial \Theta}{\partial t} = \frac{\partial}{\partial z}\left[\left(\frac{\kappa_{soil}}{\rho_w g}\right)\frac{\partial \psi}{\partial z}\right] = \frac{\partial}{\partial z}\left(\frac{\kappa_{soil}}{\rho_w g}\frac{\partial \psi}{\partial \Theta}\frac{\partial \Theta}{\partial z}\right), \tag{3.60}$$

where κ_{soil} is the hydraulic soil conductivity ($\mathrm{m\,s^{-1}}$) as defined earlier. Recalling the discussion of soil potential, there exist empirical relationships for the characteristic curves of soil matric potential and also a relationship describing the hydraulic soil conductivity. Thus, substituting (3.58) and (3.59) into an equation describing Darcy's law yields

$$\frac{\partial \Theta}{\partial t_{matric}} = \frac{\partial}{\partial z}\left[\frac{\kappa_s}{\rho_w g}\left(\frac{\Theta}{\Theta_s}\right)^{2b+3}\left(\frac{-b\psi_s}{\Theta_s}\right)\left(\frac{\Theta}{\Theta_s}\right)^{-(b+1)}\frac{\partial \Theta}{\partial z}\right]$$
$$= \frac{\partial}{\partial z}\left[\frac{-\kappa_s b\psi_s}{\rho_w g\Theta_s}\left(\frac{\Theta}{\Theta_s}\right)^{b+2}\frac{\partial \Theta}{\partial z}\right]. \tag{3.61}$$

The result is that water diffuses towards lower values of water content, which is just what one would expect.

We also need to develop a similar equation for the gravitational potential using Darcy's law. Recall that

$$\psi_g = \rho_w g(z - z_{ref}). \tag{3.62}$$

If this expression is plugged into the potential in Darcy's law, then

$$\frac{\partial \Theta}{\partial t_{grav}} = \frac{\partial}{\partial z}\left(\frac{\kappa_{soil}}{\rho_w g}\frac{\partial \psi_g}{\partial z}\right) = \frac{\partial}{\partial z}\left(\frac{\kappa_{soil}}{\rho_w g}\rho_w g\frac{\partial z}{\partial z}\right) = \frac{\partial}{\partial z}(\kappa_{soil}). \tag{3.63}$$

Again, using knowledge of the equation for the hydraulic soil conductivity, one finds that

$$\frac{\partial \Theta}{\partial t_{grav}} = \frac{\partial}{\partial z}\left[\kappa_s\left(\frac{\Theta}{\Theta_s}\right)^{2b+3}\right] = \frac{\partial}{\partial \Theta}\frac{\partial \Theta}{\partial z}\left[\kappa_s\left(\frac{\Theta}{\Theta_s}\right)^{2b+3}\right]$$

$$= \frac{\kappa_s}{\Theta}\left(\frac{\Theta}{\Theta_s}\right)^{2b+2}(2b+3)\frac{\partial \Theta}{\partial z}. \tag{3.64}$$

These two equations, (3.61) and (3.64), determine the flow between layers and the drainage, while uptake by plants is calculated using the plant transpiration rate (i.e., the latent heat flux) and the precipitation rate is calculated using the atmospheric model. Plant uptake typically occurs over specified soil layers. For example, most vegetation schemes assume that the lowest soil layer is below the plant rooting depth (Fig. 3.18). Water is removed from the soil layers in proportion to their depth and the available water in each layer.

The expressions for the time rate of change in the volumetric water content for a three-layer soil model are fairly consistent in the literature and are based upon Richards' equation that extends Darcy's law to non-saturated soil conditions. Chen *et al.* (1996) and Chen and Dudhia (2001) predict the volumetric water content for four soil layers using

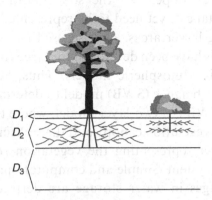

Figure 3.18. Illustration of depth over which roots are assumed to exist and can remove water from the soil. Vertical soil layers are labelled D_1, D_2, and D_3. Trees (or shrubs) and ground cover may have different rooting depths. After Sellers *et al.* (1986).

$$dz_1 \frac{\partial \Theta_1}{\partial t} = -D \left(\frac{\partial \Theta}{\partial z} \right)_{z_1} - \kappa_{z_1} + P_d - R - \left[\frac{(Q_{EB} + Q_{EV_1})}{L_v \rho_w} \right], \tag{3.65}$$

$$dz_2 \frac{\partial \Theta_2}{\partial t} = D \left(\frac{\partial \Theta}{\partial z} \right)_{z_1} - D \left(\frac{\partial \Theta}{\partial z} \right)_{z_2} + \kappa_{z_1} - \kappa_{z_2} - \frac{Q_{EV_2}}{L_v \rho_w}, \tag{3.66}$$

$$dz_3 \frac{\partial \Theta_3}{\partial t} = D \left(\frac{\partial \Theta}{\partial z} \right)_{z_2} - D \left(\frac{\partial \Theta}{\partial z} \right)_{z_3} + \kappa_{z_2} - \kappa_{z_3} - \frac{Q_{EV_3}}{L_v \rho_w}, \tag{3.67}$$

$$dz_4 \frac{\partial \Theta_4}{\partial t} = D \left(\frac{\partial \Theta}{\partial z} \right)_{z_3} + \kappa_{z_3} - \kappa_{z_4}, \tag{3.68}$$

where dz is the soil layer depth, D is the soil water diffusivity, κ is the soil hydraulic conductivity, P_d is the precipitation not intercepted by the canopy, R is the surface runoff, and the diffusivity is defined as

$$D = \frac{\kappa(\Theta)}{g} \frac{\partial \psi}{\partial \Theta}, \tag{3.69}$$

where potential has units of $J\,kg^{-1}$. As discussed by Chen and Dudhia (2001), this diffusive form of Richards' equation (Richards 1931) is derived under the assumption of a rigid, isotropic, homogeneous, and one-dimensional vertical flow domain (Hanks and Ashcroft 1986). Other forms of the equations governing soil moisture are possible (e.g., Sellers *et al.* 1986).

The calculation of runoff (R) is challenging as it depends upon the vegetation cover, the surface slope, and the soil conditions, which can vary greatly over small distances, yet need to be represented in models that have grid cells representing larger areas. As discussed by Schaake *et al.* (1996), complex runoff models have been developed for large river basins but are not appropriate for use in atmospheric models. Thus, Schaake *et al.* (1996) develop a simple water balance (SWB) model to determine runoff for atmospheric models. The SWB is a two-layer model, where the top layer is a thin layer representing the vegetation canopy and the soil surface and the bottom layer is a thicker layer representing the vegetation root zone in the soil plus the groundwater system. Simple and computationally affordable equations to govern changes in water storage are derived. Surface runoff is defined as

$$R = P_d - I_{max}, \tag{3.70}$$

where I_{max} is the maximum infiltration into the soil. Chen and Dudhia (2001) show how the infiltration is calculated in connection to a four-layer soil model, defining

$$I_{max} = \frac{P_d D_x \left[1 - \exp\left(-\kappa_{dt}\delta_i\left(\kappa_s/\kappa_{ref}\right)\right)\right]}{P_d + D_x\left[1 - \exp\left(-\kappa_{dt}\delta_i\left(\kappa_s/\kappa_{ref}\right)\right)\right]}, \tag{3.71}$$

where D_x is the soil moisture storage deficit specified as

$$D_x = \sum_{i=1}^{4} \Delta z_i(\Theta_S - \Theta_i), \tag{3.72}$$

and where δ_i is the model time in days, $\kappa_{dt} = 3.0$, κ_s is the saturated hydraulic conductivity, which depends upon soil texture, and κ_{ref} is the reference value of hydraulic conductivity ($2 \times 10^{-6}\,\mathrm{m\,s}^{-1}$ in Chen and Dudhia 2001). As discussed in Chen and Dudhia (2001), further effort is needed to calibrate these parameters for various basins across the world. In addition, if runoff occurs at one grid cell then it is possible that the runoff flows into a neighboring grid cell where it infiltrates into the soil, but this effect is not yet taken into account in these simple schemes.

When the soil becomes frozen, then the soil can become even less permeable than unfrozen soil, resulting in significant amounts of runoff. This is because ice in the soil can create nearly impermeable soil layers. Observations suggest that the area of these impermeable layers increases as the ice content of the soil increases (Koren *et al.* 1999). To account for this effect, Koren *et al.* (1999) assume that all precipitation runs off when it falls on these impermeable soil areas and develop an equation to determine the fraction of the model grid cell that contains these impermeable ice layers.

The methodology of most of the vegetation schemes generally is to first calculate the potentials and resistances, then to calculate the evapotranspiration rates and update the values of volumetric water content. But note that the soil potential depends upon the volumetric water content, which is changed by the evapotranspiration, and the evapotranspiration depends upon the ground or canopy temperature, which both depend upon the sensible heat flux. It is clear that there are feedbacks everywhere!

3.7 Radiation

The amount of vegetation above the ground surface influences the amount of radiation received at the ground and available for warming the soil, in addition to the reflective properties of the vegetation that influence the surface albedo.

While radiation is covered in greater depth in Chapters 8 and 9, a few brief points are discussed here. First, many schemes use a single temperature for the soil/vegetation surface. This approach appears to yield good results (Chen *et al.* 1997), even though the difference in temperature between the soil and plants can exceed 5 °C (Mahrt and Ek 1984). Since soil and vegetation temperatures are represented by a single temperature value on a single surface, there are no major changes needed in the model radiation scheme to account for vegetation. However, other schemes include separate temperature equations for the ground, defined as a mixture of bare soil and groundcover vegetation, and the vegetation canopy. These most complex vegetation parameterizations require some modifications to the way in which radiation is handled at the surface–vegetation–atmosphere interface.

One approach computes the effects of vegetation on the five radiation components separately (Sellers *et al.* 1986). Thus, there are modifications to the radiation calculations to produce values for direct photosynthetically active radiation (PAR), diffuse PAR, direct near-infrared (NIR), diffuse NIR, and diffuse infrared radiation. These calculations incorporate the multiple reflections of light by the leaves in the canopy and the ground surface (whether vegetated or bare). The albedo is allowed to vary during the daytime, requiring information on leaf angle, and leaf transmission and reflection of radiation. Further information may be found in Dickinson (1983) and Sellers (1985).

One important parameter that is needed to account for the effects of vegetation canopies within the radiation calculations is *LAI*. Assuming that the radiative flux at the top of the canopy is

$$Q_S = S(1 - a)\cos(\zeta)\tau_s = Q_{S_o}, \qquad (3.73)$$

then the total energy swept out by the vegetation in the vertical interval dz is

$$dQ_S = Q_{S_o}\sigma_p \, dLAI, \qquad (3.74)$$

where σ_p is an attenuation coefficient, or optical depth, that is related to the shape factor of the plant leaves, and $dLAI$ is the vertical increment of LAI. Therefore, one can calculate the radiation that reaches the surface through the canopy, yielding

$$Q_{S_{sfc}} = Q_{S_o}e^{-\sigma_p LAI_{max}}. \qquad (3.75)$$

This relationship applies to both direct solar radiation flux and diffuse solar radiation flux. For example, if $\sigma_p = 0.4$ and $LAI_{max} = 5$, then we find $Q_{S_{sfc}} = 0.14\,Q_{S_o}$ at the ground surface, indicating that the canopy has greatly reduced the incoming radiation.

3.8 Specifying soil temperature and soil moisture

Now that the soil texture, vegetation class, and vegetation health (σ_f and *LAI*) have been defined at each grid point, and assuming that sufficient information to define the atmospheric state above the ground surface and the ocean state is available, the remaining parameters needed are soil temperature and soil moisture. Unfortunately, routine observations of soil temperature and soil moisture are not commonly available. Satellites may provide observations of skin temperature, but not for soil temperatures as a function of soil depth. This creates a significant challenge for the initialization of land surface models. The most common approach to overcoming this challenge is the development and use of land data assimilation systems (LDAS).

The general idea behind an LDAS is to force a land surface model with observations in order to constrain the behavior of the land surface model so that the soil temperature and moisture values predicted by the model are reasonable. Mitchell *et al.* (2004) discuss the North American LDAS and Rodell *et al.* (2004) discuss a global LDAS (GLDAS). These projects are very important in providing much improved data for initializing land surface models, and they typically involve a number of institutions owing to the wide variety of expertise needed. Typical forcing fields that are provided to the land surface model are routine hourly surface observations (2 m temperature and specific humidity, 10 m winds, and surface pressure), precipitation (gauge, radar-estimated, satellite-estimated, or model produced), model or satellite-estimated incoming solar radiation and incoming longwave radiation. In general, the observation-based fields are used wherever possible, with the model fields used when the observations are not available. Some LDAS also use model estimates of the relative contributions of convective versus total precipitation. These systems are run both retrospectively and in real time to provide the best analyses to use for selected observational periods of the past and for operational numerical weather prediction.

3.9 Discussion

We have explored how the incorporation of vegetation changes the land surface parameterizations and the additional data sets that are needed to initialize the many soil and vegetation parameters. After all this additional complexity has been added to create a soil–vegetation–atmosphere transfer scheme (SVATS), an important question is how well these schemes reproduce the observed fluxes from the land surface to the atmosphere. Unfortunately, direct measurements of surface fluxes are few, so we instead rely on comparisons to other variables that are measured. A number of important studies have addressed this question in various ways, and several of these are examined to briefly survey the state of the science.

Comparisons between total water storage, defined as the mean soil moisture multiplied by the soil layer depth, from four different land surface parameterizations and observations are made by Schaake *et al.* (2004). The land surface parameterizations evaluated are the Noah (Chen *et al.* 1996), Mosaic (Koster and Suarez 1996), variable infiltration capacity (VIC; Liang *et al.* 1994), and Sacramento (SAC; Burnash *et al.* 1973) schemes. The schemes are forced using the North American LDAS data over a 3 year period, with the first year of forcing data being used to spin-up the schemes. The values of total water storage are averaged over 17 locations in Illinois where soil moisture observations are made bimonthly. Results from the final 2 year period (1997–1999) indicate a strong correlation between the simulated and observed values of total water storage (Fig. 3.19). In general, the simulated total water storage values from the Noah and SAC schemes agree well with the observations, although there is a fair amount of scatter in the data points. The Mosaic and VIC schemes do not compare well with the observations, both underestimating the total water storage. These results suggest that land surface schemes may be able to simulate mean soil moisture values over relatively large geographic areas, but it is not clear how well these schemes can do at providing accurate point measurements

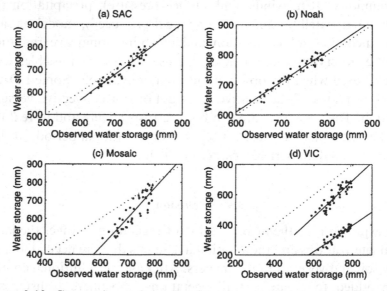

Figure 3.19. Comparison of simulated total water storage (mm) from four land surface schemes with bimonthly observations averaged over 17 locations in Illinois. The schemes are (a) SAC, (b) Noah, (c) Mosaic, and (d) VIC. From Schaake *et al.* (2004).

of soil moisture that are needed for accurate numerical weather prediction. These model results also suggest that some users may find the model predictions valuable if they are most concerned with large river basins.

To examine how well land surface schemes can reproduce local measurements of soil moisture, Liang *et al.* (1996) compare one year of weekly soil moisture measurements from a single location against results from VIC and find promising results. Crawford *et al.* (2000) use the Soil Hydrology Model (SHM; Capehart and Carlson 1994, 1997) to compare simulated soil moisture values against daily average observations of soil moisture at four depths at 38 Oklahoma Mesonet sites during July 1997. Results suggest that the SHM is able to simulate changes in soil moisture at the 5 cm level, whereas at deeper levels the model is more sluggish in responding to changes in soil moisture. They hypothesize that cracks in the soil, which form through repeated wetting and drying episodes, may be responsible for rapid increases in soil moisture at deep levels (below 50 cm) and this type of infiltration process is not represented in the SHM.

Direct comparisons between modeled soil moisture valid for a numerical model grid cell (say of 10 km × 10 km) and point soil moisture observations are a little ambiguous (Chen and Mitchell 1999; Robock *et al.* 2003), and so spatial averaging is often used to extract a more robust signal. Robock *et al.* (2003) compute daily average soil moisture observations from within the top 40 cm of soil from the Oklahoma Mesonet and compare against four land surface schemes that are again forced with the North American LDAS data. Results indicate that the land surface schemes follow the same trend as the observations (Fig. 3.20), but all the schemes clearly have biases, some of which appear to be systematic while others are related to reduced variability compared to the observations. Needless to say, obtaining accurate values of spatially averaged soil moisture is challenging, and obtaining accurate point values is even more challenging.

Part of the challenge in accurately reproducing observed soil moisture values is the heterogeneity of soil texture. Soil texture can change dramatically across very small land areas. Basara (2000) reports standard deviations approaching 10% in the silt content of soil within a small 20 m × 20 m area in central Oklahoma, similar to the differences within a single soil texture of the USDA soil texture triangle (Fig. 3.7). Robock *et al.* (2003) also compare the observed soil texture at the Oklahoma Mesonet sites to those assigned by the land surface schemes at 1/8° resolution. They find fair agreement at the 5 cm soil depth, but only rough agreement when the soil textures are integrated to get a bulk soil texture for the entire soil column (Fig. 3.21). These differences, and in particular the specification of the soil hydraulic parameters for each soil texture, play a large role in how well the land surface schemes can

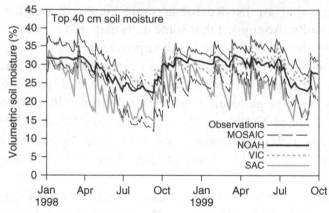

Figure 3.20. Time series from January 1998 through September 1999 of spatially and vertically averaged soil moisture in Oklahoma. Values are from observations and four land surface schemes forced by the North American LDAS. Soil moisture is for the top 40 cm layer. The schemes used are Mosaic, Noah, variable infiltration capacity (VIC), and Sacramento (SAC). Adapted from Robock *et al.* (2003).

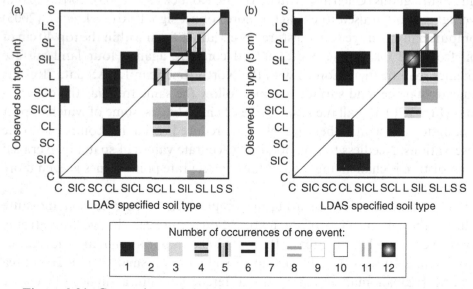

Figure 3.21. Comparison of the observed soil types at Oklahoma Mesonet stations with the land surface scheme assigned soil type at 1/8° grid spacing. Shaded squares indicate the number of occurrences of each event. A perfect correspondence between the Mesonet and assigned soil types would be indicated by a diagonal line of shaded squares. The vertical percentage values of sand, silt, and clay at the four depths at which soil moisture is observed are used to obtain a bulk soil texture for the entire column as shown in (a). The top 5 cm layer soil types are shown in (b). Adapted from Robock *et al.* (2003).

(a)

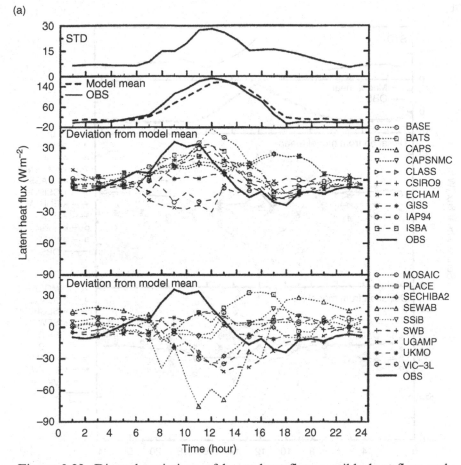

Figure 3.22. Diurnal variations of latent heat flux, sensible heat flux, and ground heat flux averaged over 10 days during September 1987 from both observations and 23 land surface schemes forced using the same input data. STD is the standard deviation of all the model values. From Chen *et al.* (1997).

reproduce the observations. In addition, Basara (2000) compares soil moisture values calculated within a $20\,m \times 20\,m$ area with each other and finds coefficients of variation (standard deviation of soil moisture divided by the mean value) of over 33% in the 5 cm layer on 19 July 1999. While the values of the coefficients of variation decrease with soil depth, the variability of soil moisture within this small $400\,m^2$ area is large and underscores the difficulty in both trying to predict and to observe soil moisture.

One important scientific endeavor intimately related to land surface schemes is the Project for Intercomparison of Land-Surface Parameterization Schemes (PILPS; Henderson-Sellers *et al.* 1993). This project provides opportunities to compare more than 20 land surface schemes against each other and

(b)

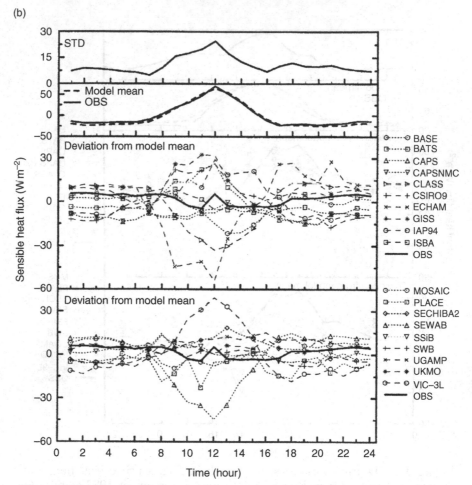

Figure 3.22. (cont.)

observations at several different locations across the globe. Results from the Cabauw, Netherlands, comparisons appear to be characteristic and indicate that there is a fairly large difference in the predicted surface fluxes from the different schemes (Fig. 3.22). The average value of all the land surface schemes differs by more than $30\,W\,m^{-2}$ from observations of latent heat flux near the time of local noon, which represents a 20% error in mean flux magnitude. The average value of sensible heat flux is in much closer agreement with the observations, yet the variability from scheme to scheme remains large and represents over half of the observed flux magnitude for the days evaluated. The situation for the ground heat flux is no different, with the average value from all the schemes departing from observations by nearly $40\,W\,m^{-2}$. Similar

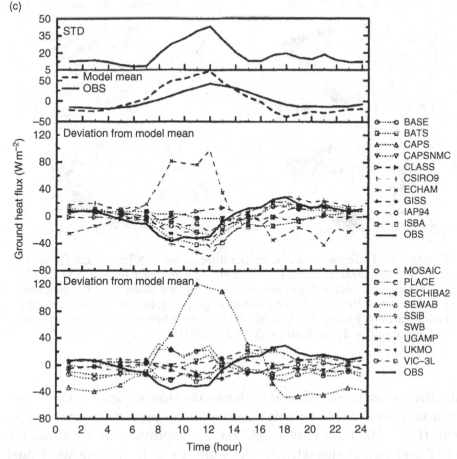

Figure 3.22. (cont.)

results are seen in Robock *et al.* (2003) for four land surface schemes that are forced using the North American LDAS and in Marshall *et al.* (2003) for the operational Eta model. These results clearly indicate that simulating or forecasting accurate surface fluxes is a very difficult challenge.

These errors in the specification of soil moisture, soil texture, vegetation fraction, and even the scheme physical process representations all lead to errors in variables that are more familiar to us and to the public. Kurkowski *et al.* (2003) compare Eta model forecasts using climatological values for the green vegetation fraction (Eta) with forecasts using near real-time satellite-derived green vegetation fraction (VegEta). Results show that the use of near real-time satellite-derived vegetation fraction leads to differences approaching 1 °C for midday 2 m temperatures and exceeding 2 °C for midday dewpoint temperatures (Fig. 3.23) and leads to reduced errors in these predicted values.

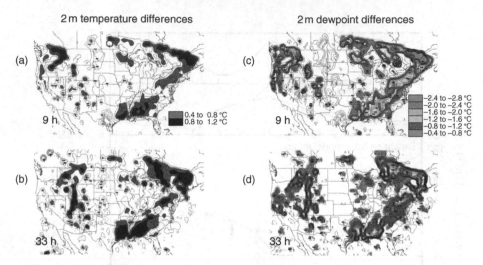

Figure 3.23. Differences in 2 m temperatures from two Eta model forecasts from 13 May 2001 at the (a) 9 h and (b) 33 h forecast times (2100 UTC). Differences are defined as Eta model using satellite-derived vegetation fraction minus Eta model using climatological vegetation fraction, and the isolines are every 0.2 K. Keys highlight the color schemes used for the various difference values. From Kurkowski *et al.* (2003).

This effect is seen more clearly early during the growing season, when vegetation departures from climatology appear to be larger and more likely to persist (Fig. 3.24). These results highlight that improvements in our specification of land surface characteristics lead directly to improvements in model forecasts.

These results all strongly suggest that the accurate specification of the land surface, in terms of soil and vegetation properties, is an important and timely issue for the meteorological community. It is clear that the addition of improved data leads to improved forecasts, even with land surface schemes that are far from perfect. This only makes sense, as we have seen how additional data have improved our atmospheric models for decades. For land surface schemes, initial data to truly specify the state of the land and vegetation surface at the model initial time are few and far between. Most are based upon using climatological data or calculations from land data assimilation systems. While these systems are significant improvements over what was available in the past, they also are not equivalent to having actual observations. Efforts to improve this situation are certain to yield good results and help the meteorological community in general to provide better services to the public.

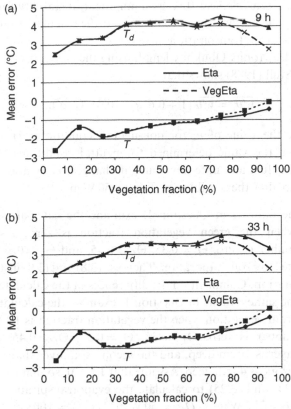

Figure 3.24. Mean temperature (T) and dewpoint temperature (T_d) errors vs. satellite-derived vegetation fraction for 12 cases of Eta model forecasts for (a) 9 h and (b) 33 h forecast times (both 2100 UTC). Solid lines are for the Eta forecasts and dashed lines are for the Eta forecasts with the satellite-derived vegetation fraction (VegEta). Data are averaged over 10% vegetation fraction bins. From Kurkowski *et al.* (2003).

3.10 Questions

1. The following data were taken over a tall grass field at Norman, Oklahoma on 17 May 2004 at local noon.

IR surface temperature	37 °C
Air temperature	28 °C at 2 m
Specific humidity of air	14 g kg^{-1} at 2 m
Wind speed	7.0 m s^{-1} at 10 m
Vegetation height	1.0 m
Leaf area index	2.0
Net radiation	600 W m^{-2}
Ground heat flux	100 W m^{-2}

Rainfall totals of slightly more than 0.25 in occurred the day before.

(a) Calculate the sensible heat flux, assuming neutral stability for the log profile laws. Using the surface energy balance and the calculated sensible heat flux, solve for the evapotranspiration flux.

(b) Calculate the Monin–Obukhov length using the value of u_* calculated above. Note that Stull (1988) shows that

$$\overline{w'\theta'_v} = \overline{w'\theta'}(1 + 0.61\bar{q}) + 0.61\bar{\theta}(\overline{w'q'}).$$

(c) Recalculate the value of u_* for non-neutral conditions. How much does it change from the value determined for neutral conditions? Recalculate the sensible heat flux, and note how much it changes. How much would u_* have to change to alter the sensible heat flux by $50\,\mathrm{W\,m^{-2}}$?

2. Using the same data as in Question 1, examine the calculations for bare soil evaporation when the green vegetation fraction is zero. Assume that the Richardson number is zero, $\Theta_1 = 0.35$, $\Theta_{fc} = 0.45$, and $\Theta_w = 0.18$. Use (3.6) and (3.7) to calculate bare soil evaporation. Then use (3.8) and (3.9) instead to calculate bare soil evaporation. Comment on the differences in the calculated fluxes.

3. Again, using the same data as in Question 1, examine the calculations for evapotranspiration from vegetation when the vegetation fraction is 1. Assume that the Richardson number is zero, $\Theta_1 = 0.35$, $\Theta_2 = 0.25$, $\Theta_{fc} = 0.45$, and $\Theta_w = 0.18$. The first soil layer is $10\,\mathrm{cm}$ deep, and the second soil layer that incorporates the rest of the root zone is $40\,\mathrm{cm}$ deep. Assume that $W_c = 0$, so that the canopy surface is dry. Use (3.24) and (3.25) to calculate the evapotranspiration. If one further assumes that $Q_S = 700\,\mathrm{W\,m^{-2}}$, $Q_{GL} = 50\,\mathrm{W\,m^{-2}}$, $r_{c\,min} = 100\,\mathrm{s\,m^{-1}}$, and $\alpha = 40$ in the definition of F_2, then use (3.38)–(3.43) to calculate evapotranspiration assuming $r_r = 15\,\mathrm{s\,m^{-1}}$. Comment on the differences in the calculated fluxes.

4. sing the same data as in Question 1 and the results of Questions 2 and 3, calculate the total latent heat flux as a function of the vegetation fraction as the vegetation fraction varies from 0 to 1 in increments of 0.1. If one assumes that errors of 10% in vegetation fraction are reasonable, how large is the uncertainty in the latent heat flux?

5. For a 50% vegetation fraction at a particular grid cell, assume errors of 10% in the vegetation fraction and errors of $0.05\,\mathrm{m^3\,m^{-3}}$ in volumetric soil moisture measurements. Using (3.6) and (3.7) for the bare soil evaporation and (3.24) and (3.25) for evapotranspiration from the vegetation, determine the range in uncertainty for the total latent heat flux owing to these two factors. Comment on the importance of these two measurements.

6. Calculate the values of soil potential for sand, loam, and sandy clay from $\Theta = 0.05$–0.40 with $\Theta_s = 0.40$ assumed for all three soil textures. Assume that $\kappa_s = 1 \times 10^{-5}\,\mathrm{m\,s^{-1}}$ and that $\psi_s = -0.5\,\mathrm{J\,kg^{-1}}$. Using the results in Table 3.1, incorporate one standard deviation in the values of b and recalculate the values of soil potential. Using (3.69), calculate the values of diffusivity D for these nine values of soil potential. Comment.

7. Using the same basic data as in Question 1, examine the benefits of mosaic tiling. The latent heat flux for 1 m tall grasses has already been calculated. Now assume that 70% of the model grid cell is represented by these grasses, but that 30% of the model grid cell is represented by oak trees that are 8 m tall and have a *LAI* of 10. Determine the latent heat flux using the tiling approach of (3.46). Comment on the differences in the values of latent heat flux for a homogeneous versus hetero-geneous land surface as described.

4
Water–atmosphere parameterizations

4.1 Introduction

Pictures of the Earth from space often show a small blue and white ball on a black background. While we often think of the Earth as being green in color, in reality most of our planet's surface has a bluish color owing to the prevalence of the oceans. Oceans cover 70% of the Earth's surface, with an average depth of nearly 4 km, and as a major component in the global hydrological cycle liquid water is an essential resource for the survival of all animals and plants. It is perhaps not surprising to learn that to predict the Earth's climate, or to provide an accurate weather forecast, one must understand how the atmosphere and the ocean interact, and be able to represent these processes in numerical weather prediction models.

Air–sea interactions occur across a wide range of spatial and temporal scales. The ocean influences the atmosphere predominantly through variations in sea surface temperature (SST) and its effects on surface fluxes into the atmosphere, while the atmosphere influences the ocean predominantly through variations in the stress exerted upon the ocean surface by low-level winds. However, the interactions occurring on any given day are often much more complicated.

Surface fluxes of sensible and latent heat from the ocean surface to the atmosphere can play a significant role in the development and evolution of large-scale atmospheric disturbances such as extratropical cyclones, tropical cyclones, and polar lows. Bosart (1999) summarizes our present understanding of the importance of air–sea interaction to extratropical cyclones by stating that fluxes from the ocean surface can act to favorably precondition the atmosphere for cyclogenesis by enhancing the low-level baroclinicity. This is particularly true near warm ocean currents such as the Gulf Stream in the Atlantic Ocean and the Kuroshio Current in the Pacific Ocean, where climatologies indicate that explosively deepening cyclones are most common (Sanders and Gyakum 1980). Bosart (1999) further suggests that the role of ocean surface fluxes in cyclogenesis is time, region, and life cycle dependent, with the fluxes most important early in the life cycle of a cyclone when environmental preconditioning occurs.

With regard to tropical cyclones, Anthes (1982, p. 49) remarks that one of the "remarkable relationships of tropical cyclone climatology is the existence of a threshold sea-surface temperature below which tropical cyclones do not form." This threshold temperature is 26.5 °C (Anthes 1982) and clearly suggests the importance of the oceans to these very intense cyclones. Indeed, Emanuel (1986) proposes an air–sea interaction instability mechanism in which tropical cyclones develop and are maintained by self-induced anomalous surface fluxes from the ocean, requiring only an initial finite-amplitude disturbance to begin the process. Hong *et al.* (2000) examine the interactions between Hurricane Opal (1995) and a circular region of warm water (called a warm core ring and associated with the so-called loop current) within the Gulf of Mexico. They show that the warm core ring assisted in the intensification of Opal, while the oceanic response to Opal resulted in a negative feedback that limited the intensification of the hurricane.

Sensible and latent heat fluxes from the ocean also are important to polar lows (Rasmussen 1985; Bresch *et al.* 1997) and small, quasi-tropical cyclones that form over the western Mediterranean in the fall (Homar *et al.* 2003). Polar lows are a special type of marine cyclogenesis event, where the low is not associated with a frontal boundary, is of relatively small horizontal scale (1000 km diameter or less) and shallow vertical extent, and forms near strong sea and air temperature gradients. They are most frequently observed over the North Atlantic and North Pacific Oceans. Thus, in areas where cyclones are observed over water, the fluxes of sensible and latent heat from the water surface can play a large role in the development and evolution of these systems.

However, even in regions and seasons where cyclones are infrequent, surface fluxes from the ocean play a large role in the structure of the overlying marine atmospheric boundary layer and in mesoscale structures within the boundary layer. Strong gradients in SST can help to produce atmospheric frontal zones (Holt and Raman 1992; Sublette and Young 1996), low-level jets (Doyle and Warner 1990), clouds (Carson 1950), and rain bands (Hobbs 1987; Holt and Raman 1992). Waters cooler than the overlying atmospheric boundary layer tend to stabilize the atmosphere, while waters warmer than the overlying atmospheric boundary layer tend to destabilize the boundary layer (Sweet *et al.* 1981). In coastal zones, the land–sea surface temperature contrast can lead to the development of sea and land breezes (Atkinson 1981).

On much longer timescales, the persistent and extensive warming of the eastern tropical Pacific Ocean called El Niño produces a global atmospheric response (Philander 1989). The El Niño Southern Oscillation (ENSO) is perhaps the best known example of air–sea interaction. Every 3 to 7 years, the typically cool waters of the eastern tropical Pacific warm by several degrees Celsius as the easterly trade winds weaken. This eastward extension of the warm ocean waters leads to an eastward shift in the region of tropical atmospheric deep convection and its associated upper-level divergence, shifting the source of Rossby wave train generation that can influence global circulation patterns (Tribbia 1991), especially during the winter months. The SST patterns in the tropical Atlantic Ocean also are known to influence seasonal precipitation amounts in sub-Saharan Africa (Lamb 1978a, b; Lamb and Peppler 1992).

The oceans also play the key role in defining the amount of water vapor in the atmosphere. Cold polar air masses are significantly modified as they move over the warmer waters of the Gulf of Mexico (Henry and Thompson 1976; Liu *et al.* 1992). The warming and moistening of the air mass over the gulf waters leads to a doubling of the boundary layer depth as well as a dramatic moistening of the boundary layer in less than a day. These studies illustrate how quickly oceans can modify air masses that originate over land. On seasonal timescales, Hastenrath (1966) and Rasmussen (1967) both show that there is a strong moisture flux from the Caribbean Sea over the Gulf of Mexico and into North America during the summertime. These results should not be a surprise, since water vapor is released mainly in the subtropical oceans and is transported both equatorward and poleward to supply the observed precipitation regions in the Tropics and the midlatitudes (Peixoto and Oort 1992). In the midlatitudes, most of the poleward transport of water vapor occurs in atmospheric rivers (Zhu and Newell 1998) or moisture conveyor belts (Bao *et al.* 2006), filaments of air with high water vapor content associated with extratropical cyclones. Further underscoring the importance of the oceans to the hydrological cycle, Peixoto and Kettani (1973) indicate that evaporation from the oceans is six times larger than evaporation from the land masses.

While these studies indicate the importance of the ocean to the atmosphere, the reverse is also true. As mentioned earlier, ENSO events begin with the weakening of the easterly trade winds in the eastern Pacific Ocean (Tribbia 1991). This weakening allows the warmer waters in the western Pacific to flow eastward. Also on these longer timescales, water vapor that is transported poleward precipitates either directly into the oceans or over land, where a portion of the precipitation flows into rivers and is eventually discharged back into the oceans, influencing the temperature and salinity of the ocean waters.

On shorter timescales, the atmosphere influences the ocean due to both the forcing of local surface winds and heat transfer. The presence of cold continental air and strong winds over the Gulf Stream, in association with the passage of an extratropical cyclone, lead to enhanced mixing of the ocean mixed layer. In addition, the heat transferred by the water to the cold low-level air leads to a decrease in Gulf Stream water temperature (Bane and Osgood 1989; Xue *et al.* 1995) as the heat transfer to the atmosphere is balanced by the cooling of the upper water column of the ocean (Xue *et al.* 1995, 2000; Xue and Bane 1997). Over shallow shelf waters, extended cold air outbreaks can lead to decreases in water temperature of 5 °C or more over a 10 day period (Nowlin and Parker 1974). This decrease in water temperature becomes smaller as the shelf water depth increases. The stronger atmospheric surface winds and the cooling of the ocean waters also can lead to enhanced mixing in the ocean mixed layer and a deepening of this layer (Bane and Osgood 1989; Adamec and Elsberry 1985), which produces a small negative feedback on the surface fluxes to the atmosphere.

Tropical cyclones also influence ocean temperatures, with lower temperatures observed following the passage of tropical cyclones over a region (Fisher 1958; Jordan 1964; Leipper 1967). Sea surface temperature decreases of 5 °C are not unusual, and may be produced by upwelling of cooler ocean water, wind-induced mixing of ocean water, heat lost to the atmosphere, and horizontal advection of cooler water.

In coastal zones, strong surface winds also can produce upwelling of ocean water. Under certain atmospheric conditions, gaps in the local coastal terrain can produce very strong offshore winds (gusts in excess of $60\,\text{m s}^{-1}$), leading to enhanced vertical mixing of the ocean waters and upwelling along the coast (Schultz *et al.* 1997). Sea surface temperature decreases of 8 °C have been observed in a few hours in association with these gap winds in the Gulf of Tehuantepec off the southern coast of Mexico (Stumpf 1975; Schultz *et al.* 1997).

In addition to the mixing and cooling of ocean waters, the atmospheric surface winds also can influence ocean currents. In a cold air outbreak over the Gulf Stream, the upper ocean responds strongly to the local wind field producing a $20\,\text{cm s}^{-1}$ flow in a coupled atmosphere–ocean model (Xue *et al.* 2000). In response to the even stronger winds from a tropical cyclone over the Gulf of Mexico, Brooks (1983) shows that a southward current surge reaching $91\,\text{cm s}^{-1}$ is observed at 200 m depth coincident with a 4 °C temperature increase and a 3% salinity increase. A wake oscillation forms after the passage of the cyclone eye having a near-inertial period and currents approaching $50\,\text{cm s}^{-1}$. This wake decays with a timescale of ~5 days.

Owing to the importance of the oceans to the Earth's climate, the latest generation of global climate models are coupled ocean–atmosphere models. Models are developed separately for both the oceans and for the atmosphere and linked through the interactions that occur between the ocean and the atmosphere. For operational numerical weather prediction models, the ocean often has a prescribed SST, based upon observations, that is held constant throughout the forecast period or relaxed towards climatology. Regardless of the approach used, the parameterizations of the air–sea interactions are very important to the resulting forecasts or simulations.

4.2 Observing sea surface temperature

The most important variable needed in meteorological models over the oceans is the SST. Thankfully, satellite data can be used to observe SSTs routinely, since *in situ* observations over the oceans are sparse. The approaches used most commonly are similar to that of Reynolds and Smith (1994), who develop an optimum interpolation approach to produce weekly and daily 1.0° analyses of SST from both satellite and *in situ* data. Presently, analyses of SST with horizontal grid spacings of 0.5° are common (Thiebaux *et al.* 2003) and grid spacings as small as 14 km are available over some regions. Seas and large lakes, such as the Mediterranean, Red, Black, and Caspian Seas and the Great Lakes, also are included in these analyses. Smaller lakes are not included.

The satellite observations are from the National Oceanic and Atmospheric Administration's Advanced Very High Resolution Radiometer (AVHRR) on board polar-orbiting satellites. Combined, these satellites pass over any given location on the Earth several times a day. Two infrared channels within the atmospheric window are used to produce a multichannel SST retrieval with a horizontal resolution of approximately 8 km (McClain *et al.* 1985; Walton 1988). Different algorithms are used for daytime and nighttime satellite observations. While these satellite data have very good horizontal resolution and global coverage (Fig. 4.1), they are limited to sampling in cloud-free areas, generally provide sparse data near ice edges in the poles, and can be biased by the presence of atmospheric aerosols.

The *in situ* data are from ships and buoys. The distribution of these observations depends upon shipping traffic and is most dense in the northern hemisphere and least dense in the southern hemisphere (Fig. 4.2). A number of the buoy locations have been selected to improve data coverage in regions where shipping traffic is minimal (Fig. 4.3). These *in situ* data also

(a)

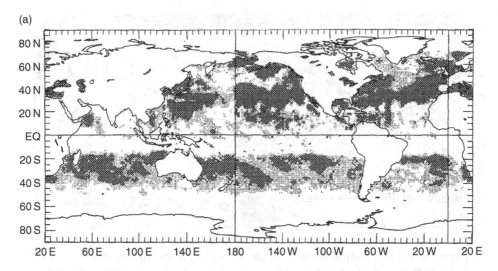

(b)

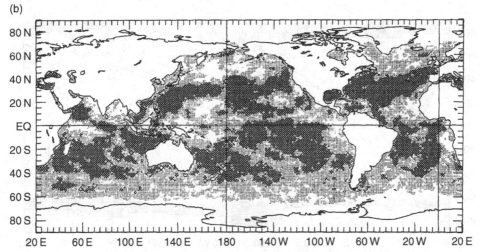

Figure 4.1. The distribution of AVHRR daytime (top) and nighttime (bottom) retrievals within 1° latitude–longitude grid cells for the week of 4–10 August 1991. A dot indicates that one to nine observations are present within the grid cell, while an X indicates ten or more observations. From Reynolds and Smith (1994).

provide sparse coverage near the ice edge, requiring additional information on sea ice based upon satellite observations of brightness temperature (e.g., Cavalieri *et al.* 1991) from the special sensor microwave imager (SSM/I) on board the defense meteorological satellite program (DMSP) polar-orbiting satellites.

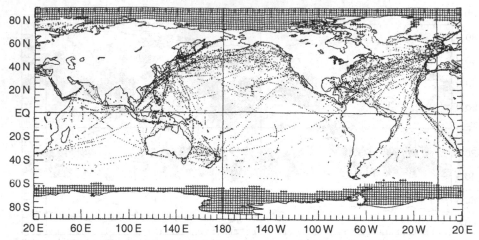

Figure 4.2. Distribution of ship observations available during the week of 4–10 August 1991. Gridded sea ice observations are denoted by a plus sign. From Reynolds and Smith (1994).

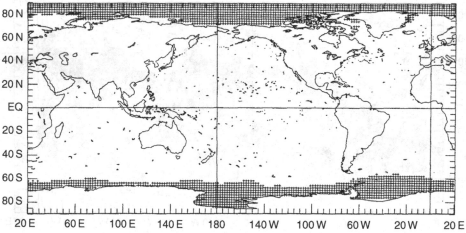

Figure 4.3. The distribution of buoy observations during August 1991. Gridded sea ice observations indicated by a plus sign. From Reynolds and Smith (1994).

Once the satellite and *in situ* data are processed and quality controlled, they generally are combined into a single analysis of SST once every day following the optimum interpolation method of Reynolds and Smith (1994). This approach includes the previous SST analysis as a first guess, incorporates seven days worth of data, and uses a Poisson technique to correct for any biases in the satellite observations as determined from comparisons with the *in situ* SST observations from ships and buoys. The importance of the bias

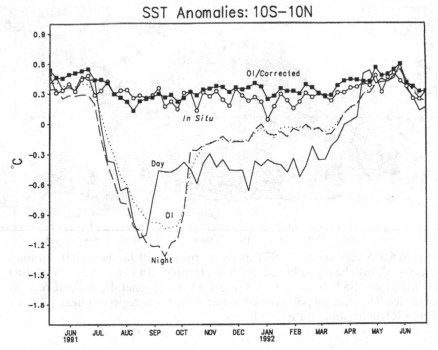

Figure 4.4. Zonally averaged anomalies of SST for the latitude band from 10° S to 10° N from May 1991 through July 1992. The curves show the *in situ* (open circles), daytime (solid) and nighttime (dashed) satellite retrievals, the optimum interpolation without satellite data bias correction (dotted), and the optimum interpolation with satellite data bias correction (solid squares) anomaly values. Mt. Pinatubo in the Philippines (latitude 15° N) erupted on 15 June 1991 and affected the satellite SST retrievals through April 1992. From Reynolds and Smith (1994).

correction to the satellite data is clearly seen when aerosols are injected into the atmosphere from volcanic eruptions (Fig. 4.4). The end result of the optimum interpolation is a global SST field that contains a fair amount of detail (Fig. 4.5). Other recent approaches use the AVHRR satellite data, but also must blend the satellite data with *in situ* observations (Thiebaux *et al.* 2003), or microwave data (Chelton and Wentz 2005). Now that the ocean SST is defined, we can turn our attention to how the surface fluxes of sensible and latent heat are calculated.

4.3 Sensible heat flux

The main differences between flow over land as discussed in Chapter 2 and flow over the ocean are the undulating, wavy nature of the ocean surface and the large heat capacity of the ocean. This means that the surface roughness

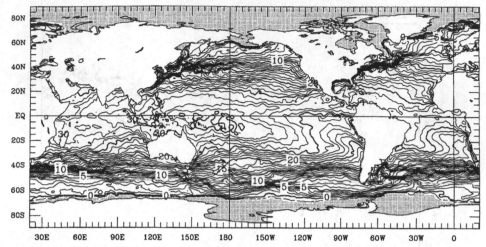

Figure 4.5. Mean weekly 1° SST analysis from 13–19 January 1991. Isolines every 1 °C, with heavy isolines every 5 °C starting with 1 °C. The −1 °C isoline is dashed, and SSTs below −1.78 °C are shaded and generally indicate regions of sea ice. The freezing point for sea water in the polar regions is near −1.8 °C. From Reynolds and Smith (1994).

length depends upon the state of the ocean surface, and that the values of SST can be held constant for relatively short numerical weather forecasts. In general, SSTs are held constant for short-range (0–4 day) and medium-range (5–14 day) weather forecasts. For seasonal and climate predictions, a coupled ocean–atmosphere modeling system is needed.

The equation for sensible heat flux is exactly the same basic form as found in Chapter 2, namely

$$Q_H = \frac{\rho c_p [T_{SST} - T(z)]}{r_H},$$
(4.1)

where T_{SST} is the sea surface temperature (K), r_H is the resistance (s m^{-1}), $T(z)$ is the temperature (K) at height z, ρ is the atmospheric density (kg m^{-3}), and c_p is the specific heat at constant pressure (J K^{-1} kg^{-1}). The resistance is again defined as

$$r_H = \frac{1}{ku_*}\left[\ln\left(\frac{z}{z_0}\right) - \psi_h\left(\frac{z}{L}\right)\right] + \frac{1}{ku_*}\ln\left(\frac{z_0}{z_{0h}}\right),$$
(4.2)

where z_0 is the roughness length (m), z_{0h} is the roughness length for sensible heat flux, u_* is the friction velocity, k is von Karman's constant ($\sim$0.4), z is the depth over which the resistance is calculated (m), ψ_h is the stability function for non-neutral conditions, and L is the Monin–Obukhov length (m).

As shown in Fairall *et al.* (2003), the main differences from the calculation of (4.1) over the ocean as compared to over land are that the roughness lengths are different and depend upon the ocean state and that the calculation of u_* incorporates another factor for conditions of no wind. If one defines the mean wind speed $S(z)$ so that it incorporates a gustiness factor u_g, in which

$$u_g = \beta \left[\frac{g z_i}{T} \left(\overline{w'\theta'_v} \right) \right]^{1/3} = \beta w^*,\tag{4.3}$$

where g is the acceleration due to gravity ($m\,s^{-2}$), T is temperature (K), z_i is the atmospheric boundary layer depth, $\overline{w'\theta'_v}$ is the surface buoyancy flux, β is a scaling parameter ($= 1.25$), and w^* is the free convection scaling velocity (Stull 1988), then

$$S(z) = \left(u^2 + v^2 + u_g^2 \right)^{1/2},\tag{4.4}$$

where u and v are the horizontal wind components at height z. Using these definitions, one can finally define a slightly modified u_* as

$$u_* = \frac{S(z)k}{[\ln(z/z_0) - \psi_m(z/L)]},\tag{4.5}$$

where $S(z)$ is replaced by $u(z)$ in the original definition when $u_g = 0$.

The only other needed parameters are the roughness lengths z_0 and z_{0h}. The definitions of roughness length over the oceans often refer back to the study of Charnock (1955), who defined a relation for rough flow over the sea. This relation is given by

$$z_0 = \alpha_c u_*^2/g,\tag{4.6}$$

where α_c is referred to as Charnock's constant. Typically, α_c has a value near 0.011 with values ranging from 0.005 to 0.025 (Fairall *et al.* 2003). Smith (1988) adds a smooth flow limit, yielding

$$z_0 = \alpha_c u_*^2/g + 0.11\nu/u_*,\tag{4.7}$$

where ν is the kinematic viscosity ($m^2\,s^{-1}$). In some parameterizations, α_c is a constant (Zeng *et al.* 1998), while in others it increases as a function of wind speed (Fairall *et al.* 2003).

Many of the parameterizations have different formulations for the stability function profiles ψ_m and ψ_h as is also seen in the parameterizations over land surfaces. It is not certain if the Monin–Obukhov similarity theory used in developing the stability function profiles over land is valid over an undulating water surface. This yields a fair amount of uncertainty in the use and definition of the stability functions ψ_m and ψ_h. In particular, Fairall *et al.* (2003) recommend using a power-law relationship for the stability function calculation when $z/L < -1$ during convective conditions.

Bourassa *et al.* (1999, 2001), Taylor and Yelland (2001), and Oost *et al.* (2002) relate the values of z_0 to aspects of the sea state, such as swell, gravity waves, and capillary waves. Capillary waves are the shortest waves (generally less than a few centimeters in wavelength) and are commonly called ripples. The primary restoring force for capillary waves is surface tension. While capillary waves can be neglected for high wind speeds, Bourassa *et al.* (2001) argue that the sum of the effects of all capillary waves to surface roughness cannot be ignored for light wind speeds because there are so many capillary waves compared to other waves in the ocean. Gravity waves are the dominant wave features we see visually on the ocean surface. Swell refers to gravity waves that are not growing or sustained by the surface winds, one example of which is waves when the local winds are calm.

The basic idea behind these parameterizations is that the roughness length is determined principally by the steepest waves. Thus, Taylor and Yelland (2001) suggest

$$z_0 = 1200 h_s \left(\frac{h_s}{L_p} \right)^{4.5}, \tag{4.8}$$

where h_s is the significant wave height (m) and L_p is the wavelength (m) associated with the peak of the wave frequency–size spectrum with its associated dominant wave period T_p. The wavelength L_p can be determined from T_p using the standard deep-water gravity wave relationship

$$L_p = \frac{g T_p^2}{2\pi}, \tag{4.9}$$

while the value of h_s is determined empirically from formulas for a fully developed sea in which

$$h_s = 0.0248 [u_n (10\,\text{m})]^2, \tag{4.10}$$

where $u_n(10\,\text{m})$ is the wind speed at 10 m under neutral conditions. Finally, the value of T_p is also determined empirically using

$$T_p = 0.729 u_n (10\,\text{m}). \tag{4.11}$$

In a similar approach, Oost *et al.* (2002) relate the phase speed C_p of the dominant wave to the friction velocity u_* in order to obtain a measure of wave age, which is then related to the roughness length. In their study using data

Table 4.1. *Values for the coefficients a_h and b_h for specifying z_{0h} and a_v and b_v for specifying z_{0v} as a function of the roughness Reynolds number R_r. Values from Liu et al. (1979).*

R_r	a_h	b_h	a_v	b_v
0–0.11	0.177	0	0.292	0
0.11–0.825	1.376	0.929	1.808	0.826
0.925–3.0	1.026	−0.599	1.393	−0.528
3.0–10.0	1.625	−1.018	1.956	−0.870
10.0–30.0	4.661	−1.475	4.994	−1.297
30.0–100.0	34.904	−2.067	30.790	−1.845

from several field campaigns in the North Sea, they find that Charnock's parameter is related to wave age by

$$\alpha_c = 50\left(\frac{C_p}{u_*}\right)^{-2.5}. \tag{4.12}$$

Finally, the determination of the roughness length for heat, z_{0h}, also requires a different relationship than over land. Liu *et al.* (1979) suggest that

$$z_{0h} = \frac{\nu}{u_*}f_h(R_r) = \frac{\nu}{u_*}a_h R_r^{b_h}, \tag{4.13}$$

where R_r is the roughness Reynolds number ($R_r = u_* z_0 / \nu$), and the values of a_h and b_h depend upon the value of R_r and range from 0.17 to 35 for a_h and −2.1 to 0.93 for b_h. Table 4.1 is taken from Liu *et al.* (1979) and specifies the details on the values for a_h and b_h as a function of R_r.

A slightly different relationship is suggested by Brutsaert (1982), in which

$$z_{0h} = \begin{cases} ae^{-bu_*^{1/4}}, & \text{for rough surfaces,} \\ \dfrac{d\nu}{u_*}, & \text{for smooth surfaces,} \end{cases} \tag{4.14}$$

where $a = 0.169$, $b = 1.53$, and $d = 0.624$. Rough surfaces occur when $R_r > 0.13$. This approach is used by Zeng *et al.* (1998). Others relate z_{0h} to z_0 directly using the values of u_* (Makin and Mastenbroek 1996) or R_r (Garratt 1992; Zilitinkevich *et al.* 2001).

Both sensible and latent heat fluxes are modified when sea spray is present. Sea spray typically is observed for wind speeds in excess of $15\,\mathrm{m\,s^{-1}}$, may be present for wind speeds as low as $7\,\mathrm{m\,s^{-1}}$, and is produced by wind gusts, bursting bubbles, and breaking waves (Kraus and Businger 1994; Andreas *et al.* 1995) as illustrated in Fig. 4.6. The presence of sea spray produces an

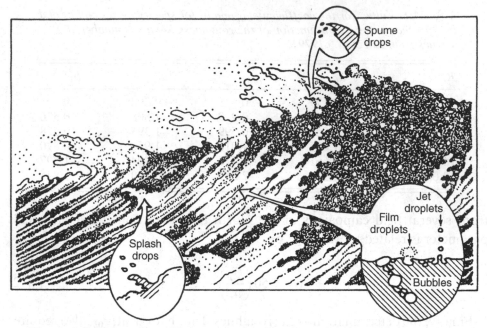

Figure 4.6. Illustration of how sea spray droplets originate. Splash droplets occur when wave crests spill, film and jet droplets occur where bubbles rise and burst, and spume droplets occur at the crests of steep waves and are torn directly from the crests by the wind. The general form of the illustration is based on one by Hokusai (*c.* 1833). From Andreas *et al.* (1995).

atmospheric layer in which the droplets are evaporating and their temperature comes into equilibrium with the environment, thereby altering the surface sensible and latent heat fluxes. Sea spray droplets initially accelerate horizontally as they are injected into the atmosphere, quickly reaching the low-level wind speed and producing a net drag on the airflow.

Relatively simple scaling relationships to parameterize the effects of the sea spray are documented in Andreas (1992), Andreas *et al.* (1995), Fairall *et al.* (1994), and Andreas and DeCosmo (1999). More recently, Andreas and Emanuel (2001) explore the effects of re-entrant spray in which the larger spray droplets fall back into the ocean prior to their complete evaporation. They show that the timescale for the spray droplet to reach its equilibrium temperature is very short, of the order of 1 s, and requires very little evaporation (Fig. 4.7). In contrast, the timescale for the spray droplet to reach its equilibrium radius is much longer, of the order of 100 s, suggesting that the droplet can fall back into the ocean prior to reaching its equilibrium radius. The net effect of this re-entrant spray is that the droplets give up heat to the atmosphere as they cool, yet many droplets fall back into the sea before extracting from the air the heat needed to evaporate. This yields a net sea-to-air

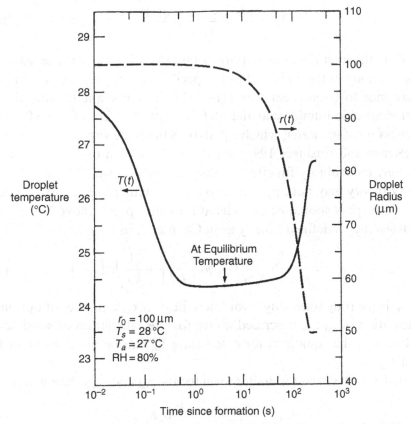

Figure 4.7. The temperature (*T*) and radius (*r*) of a sea spray droplet (100 μm initial radius and initial salinity of 34 practical salinity units) versus time (s) that is ejected from a 28 °C ocean surface into air of 27 °C with 80% relative humidity and a barometric pressure of 1000 hPa. Note that the droplet equilibrium temperature is much less than air temperature and is reached within 1 s of droplet ejection, whereas it takes over 100 s for the droplet radius to reach equilibrium. From Andreas and Emanuel (2001).

enthalpy flux that may be very important to tropical cyclone intensification as wind speeds increase to near hurricane force (Andreas and Emanuel 2001). New measuring technologies are needed to measure spray in high wind conditions to validate these spray parameterizations over the entire range of atmospheric environments (Andreas *et al.* 1995).

4.4 Latent heat flux

As with the sensible heat flux, the expression for the latent heat flux from a water surface takes on exactly the same formulation as for over land, but with a few modifications in the details. The latent heat flux can be defined as

$$Q_E = \frac{\rho L_v(q_{S\_SST} - q(z))}{r_{V(0,z)}}, \tag{4.15}$$

where L_v is the latent heat of vaporization ($J\,kg^{-1}$), q_{S_SST} is the saturation specific humidity at the SST, $q(z)$ is the specific humidity at height z, and r_V is the resistance to latent heat flux ($s\,m^{-1}$). It is important to note that the saturation specific humidity at the SST is computed from the SST and the vapor pressure of sea water, which equals 0.98 times the vapor pressure of pure water (Kraus and Businger 1994). As in the discussion of sensible heat flux over water, the stability function profiles described using Monin–Obukhov similarity theory may not apply well to an undulating ocean surface and the effects of sea spray should be considered for wind speeds above $15\,m\,s^{-1}$.

The resistance is defined exactly as in Chapter 2, using

$$r_{V(0,z)} = \frac{1}{ku_*}\left[\ln\left(\frac{z}{z_0}\right) - \psi_h\left(\frac{z}{L}\right)\right] + \frac{1}{ku_*}\ln\left(\frac{z_0}{z_{0v}}\right), \tag{4.16}$$

where z_{0v} is the roughness length for latent heat flux. A number of options for the calculation of z_0 are described above for the calculation of sensible heat flux. However, the equations for calculating z_{0v} can be quite different from those of z_{0h}.

Liu *et al.* (1979) suggest a similar form for the z_{0h} and z_{0v} relationships, with

$$z_{0v} = \frac{\nu}{u_*}f_v(R_r) = \frac{\nu}{u_*}a_v R_r^{b_v}, \tag{4.17}$$

where the parameters a_v and b_v depend upon the value of R_r. The parameter values in Table 4.1 are from Liu *et al.* (1979) and show that a_v varies from 0.29 to 30.79 while b_v varies from -1.85 to 0.83.

Fairall *et al.* (2003) suggest the following empirical fit for z_{0v} that describes well the observed fluxes calculated during the Coupled Ocean–Atmosphere Experiment (COARE) and Humidity Exchange Over the Sea (HEXOS), such that

$$z_{0v} = \min(1.1 \times 10^{-4}, 5.5 \times 10^{-5} R_r^{-0.6}). \tag{4.18}$$

Garratt (1992) relates z_{0v} to both z_0 and R_r, yielding

$$z_{0v} = z_0 \exp(2 - 2.28 R_r^{1/4}). \tag{4.19}$$

Other approaches again are found in Makin and Mastenbroek (1996) and Zilitinkevich *et al.* (2001). The effects of sea spray may also influence the latent heat flux and so must be taken into account for higher wind speeds (Andreas 1992, 1998; Fairall *et al.* 1994; Andreas *et al.* 1995; Andreas and DeCosmo 1999).

4.5 Coupled ocean–atmosphere models

For seasonal and climate simulations or predictions, and for studies of hurricanes, coupled ocean–atmosphere models often are used. In these coupled modeling systems there are actually up to three models in use: an atmospheric model, an ocean model, and a wave model (see Bao *et al.* 2000; Powers and Stoelinga 2000). However, owing to computational costs, timescales of interest, and the intended application, many systems today couple two of these models, either an ocean–atmosphere coupled modeling system (e.g., Frey *et al.* 1997; Hodur 1997; Ineson and Davey 1997; Schneider *et al.* 1997; Stockdale *et al.* 1998; Gregory *et al.* 2000; Kanamitsu *et al.* 2002; Kiehl and Gent 2004) or a wave–atmosphere coupled modeling system (e.g., Weber *et al.* 1993; Doyle 1995, 2002; Gregory *et al.* 2000). The coupling takes place at the air–sea interface. For example, when all three models are used, the atmospheric model provides the surface stress to the wave model, which uses this information to derive the two-dimensional wave energy spectrum. The wave model provides the wave-induced roughness length to the atmospheric model for use in calculating the surface fluxes, which also requires the SST provided by the ocean model. The wave-induced stress from the wave model along with the surface fluxes and radiation from the atmospheric model are used by the ocean model to derive the SST.

While the synoptic observational data generally are sufficient to start an atmospheric model, observations over the oceans are sparse below the surface. This situation leads to the ocean model being run for months or years prior to the start time of any forecast or simulation in order to develop a representative three-dimensional ocean state. During this assimilation period, the ocean model is forced by surface wind stresses provided by global analyses or an atmospheric model, observed sea surface height anomalies derived from satellite data, and the observed SSTs (e.g., Bao *et al.* 2000). The wave models generally do not need to be initialized prior to the start of the coupled model simulation unless wave characteristics during the first day or so are important.

4.6 Discussion

Observing the oceans is very difficult. Radiation does not penetrate very far into the oceans, perhaps a few tens of meters, and so observations of any quantities not near the surface must be made from direct *in situ* measurements. This requires the use of ships or buoys, the cost of which can quickly become very expensive. While research observations have been sufficient to describe the basic large-scale ocean circulation and to provide an understanding of the mechanisms for the major ocean currents, we have much less understanding of mixing and friction (Sarachik 2003).

In addition, observing sensible and latent heat fluxes over the ocean also is expensive and difficult, especially in high wind conditions associated with significant atmospheric phenomena, such as hurricanes and rapidly developing cyclones. This makes it very difficult to determine the impact of sea spray on the air–sea momentum and the sensible and latent heat fluxes. Thus, many parameterizations are developed using data within a limited range of conditions and are then extrapolated to the entire range visited by the atmosphere. For example, Bao *et al.* (2000) compare several roughness length schemes for surface sensible and latent heat flux over the oceans. All of these schemes apply to observed wind speeds of $20\,\mathrm{m\,s^{-1}}$ or less, but are extrapolated in models to much higher wind speeds. When evaluated in an idealized simulation of Hurricane Opal, the different roughness length schemes produce a large sensitivity in the intensity of the simulated hurricane (Fig. 4.8). Curiously, the use of one particular roughness length scheme results in no intensification at all, while other schemes result in differences in the minimum sea-level pressure of $17\,\mathrm{hPa}$. This study illustrates both the large disparity in the formulas for surface roughness lengths and their very different behaviors when used within a numerical weather prediction model.

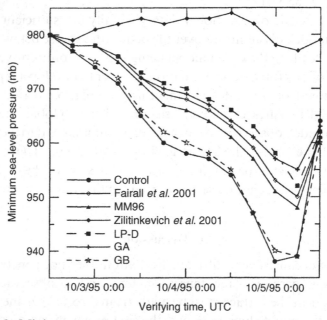

Figure 4.8. Minimum sea-level pressure of simulated hurricanes that use different roughness length schemes as compared with a control run. From Bao *et al.* (2000).

It also may be important to include the effects of sea ice in numerical weather prediction models. Sea ice is not uniform and incorporates many leads, or fractures between ice floes (ice sheets) sufficiently wide to be navigated by a ship and several kilometers in length. Leads are not very common during the winter season, but can be very prevalent during summer. While some leads are covered with a thin layer of ice, others are not. This can produce direct contact between the atmosphere and the ocean surface within widespread regions of sea ice. Several parameterizations of sea ice have been developed (Maykut and Untersteiner 1971; Ebert and Curry 1993; Hunke and Dukowicz 1997; Bitz and Lipscomb 1999) and are used in numerical models, especially for climate simulations. Finally, it is important to account for the air–sea gas transfer of CO_2 in climate simulations (Csanady 2001) as the oceans contain 56 times as much carbon as the atmosphere and dissolve roughly one-third of the CO_2 released by fossil fuels each year (Schlesinger 1997).

4.7 Questions

1. Go onto the world-wide web and search for a site that provides daily or weekly global SST analyses or regional SST analyses appropriate for your location and available over at least a 1 year period. Explore these data for a 1 year period and describe the SST evolution nearest to where you live. What is the typical range of SST values? Are there any features or structures that you can describe (e.g., currents, anomalies such as warm rings)? How do you think the oceans influence the weather where you live?
2. Using the sensible heat flux equation (4.1), examine the sensitivity to roughness lengths z_0 and z_{0h}. First, assume $\rho = 1.0 \, \text{kg m}^{-3}$, $T_{SST} = 303 \, \text{K}$, $T\,(10 \, \text{m}) = 300 \, \text{K}$, $L = -50 \, \text{m}$, $w* = 1.0 \, \text{m s}^{-1}$, the 10 m wind speed is $10 \, \text{m s}^{-1}$, $u_* = 0.8 \, \text{m s}^{-1}$, and viscosity $\nu = 1.46 \times 10^{-5} \, \text{m}^2 \, \text{s}^{-1}$. Determine z_0 using (4.7) and then z_{0h} from (4.13) and (4.14). Calculate Q_H using both values of z_{0h}. How sensitive are the results to the formulas used?
3. Using the results from Question 2, now calculate z_0 using (4.8) and assume that the neutral wind speed is equal to the assumed 10 m wind speed given. Recalculate the values of z_{0h} using (4.13) and (4.14) and again determine the values for Q_H. How sensitive is the value of Q_H to the values of z_{0h} and to the values of z_0?
4. We know that SST varies slowly in comparison to changes in the atmospheric surface layer, even over the tropical oceans far away from land. What processes could produce variability in the atmospheric surface layer over the oceans?

5

Planetary boundary layer and turbulence parameterizations

5.1 Introduction

We have seen how the various properties of the land surface can influence the fluxes of energy that occur in the atmospheric surface layer. Now that these boundary conditions have been described, the next step is to determine how these energy fluxes influence the evolution of the atmosphere. This influence first occurs within the planetary boundary layer (PBL).

The planetary boundary layer is that part of the troposphere which is influenced directly by the presence of the Earth's surface, and responds to surface forcing with a timescale of an hour or less (Stull 1988). Thus, the land surface and the boundary layer are intimately tied together. The PBL can be as shallow as a few tens of meters, and as deep as several kilometers (Fig. 5.1). We previously learned in Chapters 2 and 3 that the values of temperature and specific humidity within the surface layer have a profound effect on the surface sensible and latent heat fluxes, since they are used in determining the potential difference between the soil–vegetation–land surface and the atmosphere. Now it is seen that the boundary layer responds to these fluxes over very short time periods. It also is important to recognize that turbulence, the irregular fluctuations that occur in fluid motions, is the dominant mechanism by which surface forcing is transmitted throughout the boundary layer.

One could arbitrarily say that the evolution of the boundary layer begins with sunrise. Just before sunrise the boundary layer is stable, as indicated by the potential temperature increasing with height (line "3" in Fig. 5.2). At sunrise, the radiant energy reaches the Earth's surface and begins to warm the ground (line "9"). Heat and moisture fluxes from the ground to the atmosphere become larger. The boundary layer responds to these fluxes by slowly deepening as thermals (bubbles of buoyant air that originate at the surface and rise into the boundary layer) reach the top of the boundary layer and

138

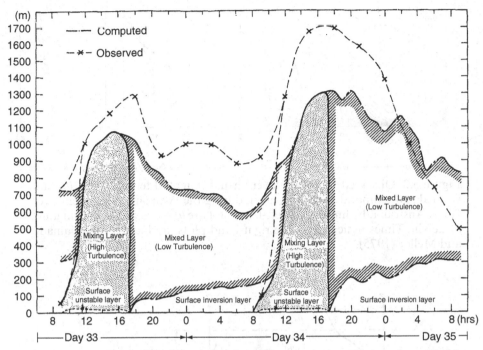

Figure 5.1. Schematic of the boundary layer evolution during days 33–35 from the Wangara experiment. During the daytime the mixing layer grows to depths above 1 km and is characterized by high levels of turbulence and a shallow and unstable surface layer. As the sun sets, the level of turbulence drops precipitously, leaving a residual layer behind. The surface inversion layer grows throughout the nighttime hours, influenced by episodic turbulence and radiational cooling. From Yamada and Mellor (1975).

overshoot their level of neutral buoyancy. Vertical velocities within thermals can reach $5\,\mathrm{m\,s^{-1}}$, although values of 1–$2\,\mathrm{m\,s^{-1}}$ are more common (Stull 1988). While thermals come in all shapes and sizes, the largest thermals typically have a horizontal length of ~ 1.5 times the boundary layer depth and are present over slightly less than half the boundary layer at any given time (Young 1988). Thermals are smaller early in the day, when the boundary layer is shallow, and become larger as the boundary layer deepens.

By the middle of the day, the boundary layer typically is near its maximum depth and often approaches a well-mixed structure (line "12" in Figs. 5.2 and 5.3). This occurs when turbulence is vigorous and it tends to uniformly mix the boundary layer, especially the potential temperature. Below the mixed layer is the surface layer, in which the potential temperature increases towards the warmer ground surface.

Above the mixed layer is the interfacial layer or inversion layer, in which the potential temperature increases with height. This layer separates the turbulent

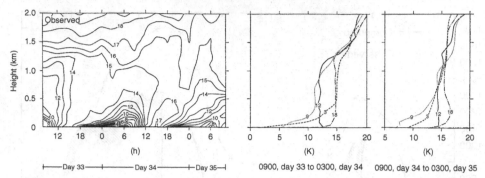

Figure 5.2. Observed boundary layer virtual potential temperature (θ_v) and vertical and temporal variations taken during the Wangara experiment over Hays, Australia. Isolines on the left-hand plot are $\theta_v - 273$ and shown in units of kelvin. Times indicated on the right-hand plots are local. From Yamada and Mellor (1975).

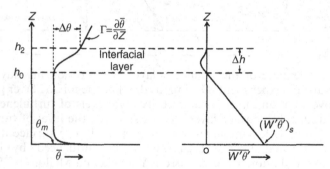

Figure 5.3. Illustration of a well-mixed boundary layer. The potential temperature is nearly constant with height within the mixed layer, and increases both below and above it. The heat flux profile is linear, with the largest value just above the ground and decreasing to zero at the top of the interfacial or inversion layer. From Deardorff (1979).

boundary layer from the less turbulent free atmosphere. The inversion layer is where entrainment occurs, the process by which the rising thermals overshoot their level of neutral buoyancy, owing to their upward momentum, and then sink back into the boundary layer bringing wisps or curtains of air from above the boundary layer into the boundary layer. Because of the vigorous turbulence in the boundary layer, the buoyant air from above is quickly mixed and becomes part of the boundary layer.

Since potential temperature increases with height, entrainment and surface sensible heat flux both act to increase the potential temperature of the boundary layer and help support the well-mixed structure of the boundary layer. However, moisture and momentum are not necessarily well mixed within the boundary

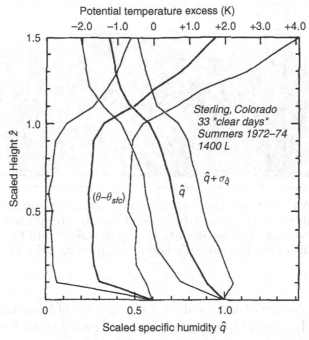

Figure 5.4. Vertical profiles of potential temperature (θ) and specific humidity (q), scaled by the surface value, at 1400 local time from Sterling, Colorado, averaged over 33 cloud-free days. Thin lines are the standard deviations, while θ_{sfc} is the surface value of potential temperature. Note that while the potential temperature is nearly constant with height below the boundary layer top near 1 km, the specific humidity clearly decreases with height in this same layer. From Mahrt (1976).

layer (Fig. 5.4). Since the free atmosphere above the boundary layer typically is drier than the boundary layer, moisture in particular often tends to decrease with height in the boundary layer. Indeed, Mahrt (1991) characterizes boundary layers into moistening and entrainment drying regimes, with moistening regimes occurring under large surface evaporation and large mean wind shear. Entrainment drying regimes are most often associated with small surface evaporation and unstable conditions. The diurnal evolution of the boundary layer often begins with a moistening regime in the morning, transitioning to an entrainment-drying regime later in the day. The boundary layer momentum field is even more complicated. Momentum may be well mixed within the boundary layer, particularly under quiescent synoptic-scale conditions, but momentum is strongly influenced by the larger-scale pressure gradients and often does not appear well mixed.

The daytime convective boundary layer is a complex phenomenon (Fig. 5.5). It is governed by turbulent processes, but also can be filled with more organized secondary features that interact with the turbulence (Brown

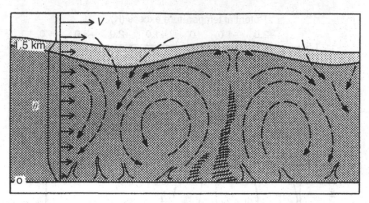

Figure 5.5. An artistic rendering of the daytime convective boundary layer, in which turbulence and organized thermals interact to vertically mix the layer. Note the entrainment of air from the free atmosphere into the boundary layer, and the horizontal changes in boundary layer depth. From Wyngaard (1985).

1980). These secondary features often take the form of boundary layer rolls, which act to help transfer the heat flux more efficiently (Stull 1988). Rolls are most often observed via satellite as cold air moves over a warm surface and creates lines of clouds (cloud streets), although clear air radar data indicate that roll-like circulations are a regular part of the boundary layer in the summertime when surface heating is strong. And, of course, rolls and other secondary circulations can occur at night just as well as the day, and occur throughout the year. During winter, boundary layer rolls are even thought to produce snowfall from time to time (Schultz *et al.* 2004).

So far this discussion makes it sound as though boundary layer evolution is governed only by the sensible heat flux from the ground surface. However, while buoyancy often is a main driver of daytime convective boundary layers, wind shear also generates turbulence and definitely plays a large role in boundary layer development. Schneider and Lilly (1999) investigate the daytime PBL just to the east of the Rocky Mountains and find that a combination of buoyancy and shear-produced turbulence controls the PBL evolution. The turbulent transfers in the upper portion of this PBL are tied to a complex field of interwoven vortical structures (Fig. 5.6) that are episodic and shift in response to local shear and buoyancy profiles. They conclude that a number of the common simplifying assumptions and scalings used for the boundary layer are inappropriate when describing this complex behavior. It is clear from their results that the boundary layer is not a simple system.

When clouds do not occur in the PBL, the boundary layer typically is fully turbulent throughout its depth during most of the daytime hours. When clouds

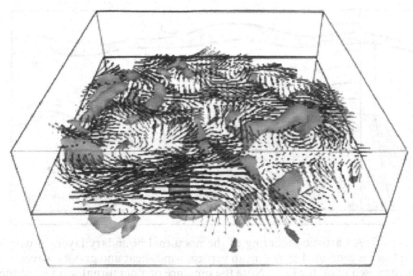

Figure 5.6. Visualization of wind vectors and isosurfaces of vorticity from 22 June 1984 during the Phoenix II experiment as derived from radar data. The reader is looking down on the analysis area and towards the north. The vorticity isosurfaces of $15 \times 10^{-3} \, \text{s}^{-1}$ show many vortical tubes within the turbulent flow, and these grow and decay over time. From Schneider and Lilly (1999).

are present the turbulence can be modified by the cloud circulations and sometimes the turbulence in the cloud becomes decoupled from the turbulence in the boundary layer below the cloud, especially when clouds completely cover the sky. Observations indicate that turbulence is an intrinsic part of the PBL and the largest eddies give a turbulent flow its distinguishing characteristics (Wyngaard 1985), so the effects of turbulence must be included in any attempt to understand and/or model the PBL. The lack of a complete understanding of a physical process, such as turbulence, does not mean that it cannot be modeled with some degree of confidence. One just needs to recognize the limitations of the models that are produced and selected for use.

As perhaps suggested by the results from Schneider and Lilly (1999), the nocturnal boundary layer, lacking the controlling influence of a strong surface buoyancy flux, is quite complex (Fig. 5.7). As the sun sets, the heat fluxes from the surface are reduced and the air near the ground surface begins to cool as radiational cooling dominates the surface energy budget. Potential temperature again increases with height, and wind shear is the main source for generating turbulence. Thus, turbulence in the nocturnal PBL is often intermittent. The air above the surface inversion layer is the residual from the daytime convective boundary layer. In addition, drainage flows occur very near the surface owing to variations in elevation, and gravity waves may

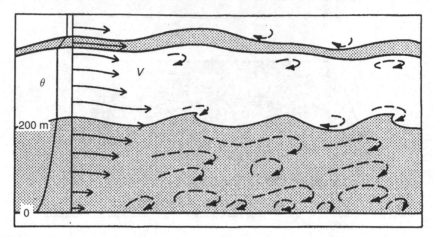

Figure 5.7. An artistic rendering of the nocturnal boundary layer, in which turbulence is generated by the mean vertical wind shear and gravity waves are superimposed upon the flow. Note the presence of a nocturnal wind jet at the top of the thin turbulent layer. From Wyngaard (1985).

develop within the stable boundary layer and further complicate the flow. Wyngaard (1985) nicely outlines the delicate and precarious balance between turbulence generated by the mean wind shear and its dissipation from viscous effects and buoyancy.

The nocturnal boundary layer also is well-known as a breeding ground for low-level jets. A low-level jet is a wind speed maximum that occurs in the lowest few kilometers of the atmosphere. These jets are observed worldwide, are important to the horizontal and vertical fluxes of temperature and moisture, and often are associated with the development and evolution of deep convection (Stensrud 1996). While low-level jets are observed during the daytime and can be caused by a number of different mechanisms (see Stensrud 1996), they are strongest and most commonly observed at night. One mechanism for producing low-level jets is a diurnal variation of eddy viscosity as discussed by Blackadar (1957). During the daytime, the boundary layer is strongly coupled to the surface layer and frictional effects cause the boundary layer winds to be subgeostrophic. When turbulent mixing ceases and a shallow nocturnal boundary layer begins to develop, the winds above the nocturnal inversion are decoupled from the surface layer and are no longer in balance. This imbalance between the Coriolis and pressure gradient forces induces an inertial oscillation of the wind that produces a wind speed maximum approximately 6–8 h after turbulent mixing ceases (see Hoxit 1975). Low-level jets are not only important to weather and climate, but also are used by birds and insects to assist in their migrations (Drake 1985).

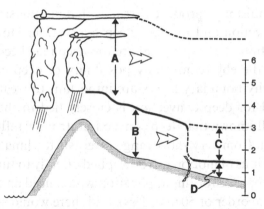

Figure 5.8. Depiction of two vastly different elevated residual layers being created, with the environmental wind blowing from left to right. Layer B is dry adiabatic, well mixed, and has low relative humidity. This layer is advected over layer D, forming an elevated residual layer (C). Layer D in this case is a moist boundary layer. Layer A, assuming that the convection is widespread and long-lived, is moist adiabatic and has high relative humidity. Thus, the airmass over layer D has a three-layer structure that is defined by boundary layer processes (active or past) over much of its depth. From Stensrud (1993).

The contrast between daytime and nighttime PBLs is dramatic, yet both must be represented within boundary layer schemes. Owing to the complexity of the boundary layer, the parameterization schemes used in operational forecast models and even most research models are focused upon representing the evolution of the mean boundary layer state through the diurnal cycle and include the effects of turbulence indirectly.

The evolution of boundary layers is important not only because these layers largely determine the characteristics of the conditions we live in each day (how hot or dry, and how windy it will be), but also because boundary layers over elevated terrain can be advected over regions with lower terrain, and thereby influence the development of other boundary layers (Fig. 5.8). This process is a frequent occurrence over central USA (Carlson *et al.* 1983; Lanicci and Warner 1991), and likely anywhere else downwind of mountain ranges. These boundary layers that are advected off elevated terrain are called elevated mixed layers or elevated residual layers, since their potential temperature profiles are often well mixed. These layers have a large effect on the convective available potential energy and often are seen within the environments of severe weather. This only further emphasizes the importance of parameterizing boundary layers correctly.

The evolution of the boundary layer often sets limits on the types of atmospheric phenomena that can be produced on a given day. Boundary layers that are shallow and moist may produce thermals that are too small to reach their level of free convection and produce deep convection. However, under the same conditions, boundary layers that are too deep may become too dry from entrainment and thereby remove any possibility of deep convection. Thus, another reason why boundary layers are important to weather prediction is that the potential for deep convection is closely tied to the boundary layer structure. It is well-known that deep convection strongly influences numerical weather prediction from very short-range forecasts to climate predictions.

While the equations of motion can be applied directly to turbulent flows, the models most appropriate for this application would need an exceedingly small grid spacing (of the order of 50 m or less), and there would still be eddies that would not be represented on the model grid. The effects of these subgrid eddies still need to be accounted for in some way, which usually is based upon a statistical approach to the eddy effects. A useful technique called Reynolds averaging is now outlined that sets the mathematical stage for looking at how turbulent flows influence boundary layer development.

5.2 Reynolds averaging

We begin the study of turbulent effects with the basic equations of motion and statistically average over the smaller eddy sizes. Each dependent variable in the equations of motion is assumed to be represented by a mean and a perturbation from the mean as defined by $A = \bar{A} + a'$. For example, assume there are two variables A and B, both defined as a mean quantity plus a perturbation. When they are multiplied together, what happens when they are averaged over some time period? The result is

$$\overline{A \cdot B} = \overline{(\bar{A} + a')(\bar{B} + b')} = \overline{\bar{A}\bar{B} + \bar{A}b' + \bar{B}a' + a'b'} = \overline{\bar{A}\bar{B}} + \overline{a'b'}, \qquad (5.1)$$

since the perturbation terms a' and b', even when multiplied by a constant such as $\bar{A}$, average to zero over the given time period by definition. However, when the perturbation terms are multiplied together, the end result is not necessarily zero. The last term on the right represents the influence of the eddy motions. Thus, even though the evolution of the mean quantities is of primary interest, the eddies clearly play a significant role in determining how these mean quantities change. The process of separating a given variable into a mean and a perturbation component, and then averaging over time, is called Reynolds averaging.

When dealing with the full equations of motion, instead of this simple two-variable example of Reynolds averaging, using summation notation to write the equations is very helpful in simplifying the derivations. Summation notation is a compact way to represent several different equations in one. Define

x_i as a generic distance, with $x_1 = x, x_2 = y, x_3 = z$

u_i as a vector, with $u_1 = u, u_2 = v, u_3 = w$

δ_i as a unit vector, with $\delta_1 = \vec{i}, \delta_2 = \vec{j}, \delta_3 = \vec{k}$

and

δ_{mn} as the Kronecker delta which equals 1 when $m = n$ and zero otherwise.

The alternating unit tensor is defined as

$$\varepsilon_{ijk} = \begin{cases} +1, & \text{if } i, j, k \text{ are in ascending order} \\ -1, & \text{if } i, j, k \text{ are in descending order} \\ 0, & \text{otherwise,} \end{cases}$$

where an ascending order means that the values of ijk are in a sequence of $1, 2, 3$, or $2, 3, 1$, or $3, 1, 2$. Descending order is the opposite, such as $3, 2, 1$, or $2, 1, 3$, or $1, 3, 2$. The alternating unit tensor is zero if any of the values of ijk are the same.

The summation notation also requires the application of two basic rules. The first rule is that one sums on repeated indices when they appear in two quantities that are multiplied together. The second rule is that all indices take on the values of 1, 2, and 3. For example, a single term in summation notation such as

$$u_i \frac{\partial \theta}{\partial x_i} = u_1 \frac{\partial \theta}{\partial x_1} + u_2 \frac{\partial \theta}{\partial x_2} + u_3 \frac{\partial \theta}{\partial x_3} = u \frac{\partial \theta}{\partial x} + v \frac{\partial \theta}{\partial y} + w \frac{\partial \theta}{\partial z}, \tag{5.2}$$

which allows us to combine a number of separate terms and even equations into a very compact format.

5.3 Turbulence closure

To explore the role of turbulence in the evolution of the planetary boundary layer, the Boussinesq equations of motion are used (Dutton and Fichtl, 1969). The momentum equations are written in summation notation as

$$\frac{\partial u_i}{\partial t} + u_j \frac{\partial u_i}{\partial x_j} = -\delta_{i3} \left(g - \frac{\theta_v'}{\theta_v} g \right) + f \varepsilon_{ij3} u_j - \frac{1}{\rho} \frac{\partial p}{\partial x_i} + \nu \frac{\partial^2 u_i}{\partial x_j^2}. \tag{5.3}$$

The next step is to expand each variable into a mean and a perturbation, where the perturbations are assumed to represent the effects of eddies, or turbulence, yielding

$$\frac{\partial(\bar{u}_i + u'_i)}{\partial t} + \frac{(\bar{u}_j + u'_j)\partial(\bar{u}_i + u'_i)}{\partial x_j}$$

$$= -\delta_{i3}\left(g - \frac{\theta'_v}{\bar{\theta}_v}g\right) + f\varepsilon_{ij3}(\bar{u}_j + u'_j) - \frac{1}{\bar{\rho}}\frac{\partial(\bar{p} + p')}{\partial x_i} + \nu\frac{\partial^2(\bar{u}_i + u'_i)}{\partial x_j^2}. \quad (5.4)$$

Now all the terms are multiplied out and separated to obtain

$$\frac{\partial \bar{u}_i}{\partial t} + \frac{\partial u'_i}{\partial t} + \bar{u}_j\frac{\partial \bar{u}_i}{\partial x_j} + \bar{u}_j\frac{\partial u'_i}{\partial x_j} + u'_j\frac{\partial \bar{u}_i}{\partial x_j} + u'_j\frac{\partial u'_i}{\partial x_j}$$

$$= -\delta_{i3}\left(g - \frac{\theta'_v}{\bar{\theta}_v}g\right) + f\varepsilon_{ij3}\bar{u}_j + f\varepsilon_{ij3}u'_j - \frac{1}{\bar{\rho}}\frac{\partial \bar{p}}{\partial x_i} - \frac{1}{\bar{\rho}}\frac{\partial p'}{\partial x_i} + \nu\frac{\partial^2 \bar{u}_i}{\partial x_j^2} + \nu\frac{\partial^2 u'_i}{\partial x_j^2}.$$

$$(5.5)$$

This equation represents the evolution of both the mean and the turbulent portions of the momentum equation. But only the evolution of the mean portion is of interest, which is what the models predict, so we average over the entire equation. This yields

$$\frac{\partial \bar{u}_i}{\partial t} + \bar{u}_j\frac{\partial \bar{u}_i}{\partial x_j} + \overline{u'_j\frac{\partial u'_i}{\partial x_j}} = -\delta_{i3}g + f\varepsilon_{ij3}\bar{u}_j - \frac{1}{\bar{\rho}}\frac{\partial \bar{p}}{\partial x_i} + \nu\frac{\partial^2 \bar{u}_i}{\partial x_j^2}. \quad (5.6)$$

How does this equation differ from the original Boussinesq equation for momentum? There are two changes apparent. One is that the term with the perturbation virtual potential temperature, which represented the effects of buoyancy, has disappeared. The second change is that a new term, $\overline{u'_j(\partial u'_i/\partial x_j)}$, has been added. This new term represents the advection of turbulence by turbulence.

It would be helpful to write the turbulence advection term in a more useful form. If the atmosphere is assumed to be incompressible in the boundary layer, i.e. assume that the shallow Boussinesq approximation is valid (which for the boundary layer is a pretty good assumption), then

$$\frac{\partial u_j}{\partial x_j} = 0 \quad \text{or} \quad \frac{\partial \bar{u}_j}{\partial x_j} + \frac{\partial u'_j}{\partial x_j} = 0. \quad (5.7)$$

If we again average over time, this results in

$$\frac{\partial \bar{u}_j}{\partial x_j} = 0, \quad (5.8)$$

and since

$$\frac{\partial \bar{u}_j}{\partial x_j} + \frac{\partial u'_j}{\partial x_j} = 0, \tag{5.9}$$

then (5.8) and (5.9) require that

$$\frac{\partial u'_j}{\partial x_j} = 0. \tag{5.10}$$

Thus, both the mean wind and the perturbation wind components are incompressible. If the divergence of the perturbation wind is multiplied by any quantity, even a perturbation, then it is still zero since it must always be zero. From this knowledge, it is apparent that

$$\frac{\partial \overline{(u'_i u'_j)}}{\partial x_j} = \frac{\overline{u'_i \partial u'_j}}{\partial x_j} + \frac{\overline{u'_j \partial u'_i}}{\partial x_j} = \frac{\overline{u'_j \partial u'_i}}{\partial x_j}. \tag{5.11}$$

Using this result to rearrange the terms in the equation for the time rate of change of the mean momentum, after Reynolds averaging, gives

$$\frac{\partial \bar{u}_i}{\partial t} + \bar{u}_j \frac{\partial \bar{u}_i}{\partial x_j} = -\delta_{i3}g + f\varepsilon_{ij3}\bar{u}_j - \frac{1}{\rho}\frac{\partial \bar{p}}{\partial x_i} + \nu \frac{\partial^2 \bar{u}_i}{\partial x_j^2} - \frac{\partial \overline{(u'_i u'_j)}}{\partial x_j}. \tag{5.12}$$

The last term on the right-hand side of this equation is called the Reynolds stress (covariance) term. Now, numerical models predict the mean variables, but what about this Reynolds stress term? Since it is not predicted explicitly, there are two options. First, an empirical relationship can be derived between this term and known model variables (parameterize). Second, additional equations can be derived to predict the Reynolds stress term explicitly.

Recall that we began with the Boussinesq equation for momentum, and then divided each variable into a mean and a perturbation component. If we now begin with the equation that represents the evolution of both the mean and turbulent portions of the momentum equation, then we can subtract off (5.6) for the mean and the result is an equation for the time rate of change of a turbulent gust. This process yields

$$\frac{\partial u'_i}{\partial t} + \bar{u}_j \frac{\partial u'_i}{\partial x_j} + u'_j \frac{\partial \bar{u}_i}{\partial x_j} + u'_j \frac{\partial u'_i}{\partial x_j} = \delta_{i3}\left(\frac{\theta'_v}{\bar{\theta}_v}g\right) + f\varepsilon_{ij3}u'_j - \frac{1}{\rho}\frac{\partial p'}{\partial x_i} + \nu \frac{\partial^2 u'_i}{\partial x_j^2} + \frac{\partial \overline{(u'_i u'_j)}}{\partial x_j}. \tag{5.13}$$

To develop additional equations to predict the Reynolds stress term explicitly, an equation for the time rate of change of $\overline{u'_i u'_j}$ is needed. Therefore, using the equation for the turbulent gust, and noting that one can change j to k in the equation since when using summation notation all indices take on the values 1, 2, and 3, it is found that

$$\frac{\partial \overline{(u_i' u_k')}}{\partial t} = \overline{u_i' \frac{\partial u_k'}{\partial t}} + \overline{u_k' \frac{\partial u_i'}{\partial t}}. \tag{5.14}$$

So to obtain the evolution of the Reynolds stress term, simply multiply the equation for the turbulent gust by u_k' to produce the second term on the right-hand side of the above equation. The first term on the right-hand side can be determined by taking the equation for the turbulent gust, changing all the i-indices to k-indices and then multiplying by u_i'. The end result is

$$\frac{\partial \overline{(u_i' u_k')}}{\partial t} + \bar{u}_j \frac{\partial \overline{(u_i' u_k')}}{\partial x_j} = - \overline{(u_i' u_j')} \frac{\partial \bar{u}_k}{\partial x_j} - \overline{(u_k' u_j')} \frac{\partial \bar{u}_i}{\partial x_j} - \frac{\partial \overline{(u_i' u_j' u_k')}}{\partial x_j}$$

$$+ \frac{g}{\bar{\theta}_v} \left(\delta_{k3} \overline{u_i' \theta_v'} + \delta_{i3} \overline{u_k' \theta_v'} \right) + f \left(\varepsilon_{kj3} \overline{u_i' u_j'} + \varepsilon_{ij3} \overline{u_k' u_j'} \right)$$

$$- \frac{1}{\bar{\rho}} \left[\frac{\partial \overline{(p' u_k')}}{\partial x_i} + \frac{\partial \overline{(p' u_i')}}{\partial x_k} - \overline{p' \left(\frac{\partial u_i'}{\partial x_k} + \frac{\partial u_k'}{\partial x_i} \right)} \right]$$

$$+ \nu \frac{\partial^2 \overline{(u_i' u_k')}}{\partial x_j^2} - 2\nu \overline{\frac{\partial u_i' \partial u_k'}{\partial x_j^2}}. \tag{5.15}$$

This represents nine different equations, including equations for $\overline{u'u'}, \overline{u'v'}, \overline{u'w'}, \overline{v'v'}, \overline{v'w'}$, and $\overline{w'w'}$. However, the total number of equations is reduced to six owing to symmetry (e.g., $\overline{u'v'} = \overline{v'u'}$).

Equations for the mean momentum variables and for the Reynolds stress terms are now available, so they can be included in a numerical model and integrated forward in time. However, closer inspection reveals that yet another unknown term is present in the equation for the Reynolds stress, namely the $\partial \overline{(u_i' u_j' u_k')}/\partial x_j$ term. This unknown is now a triple correlation term! As this pattern suggests, if equations for the triple correlation term are developed, then a quadruple correlation term is created. This cascading of unknowns is called the *turbulence closure* problem. There are always more unknowns than equations, so at some point the remaining unknown terms need to be parameterized by relating them to some combination of the known variables. Stull (1988) shows the turbulence closure problem that occurs in the remaining equations of motion.

There are two related terms that describe where in this cascade of unknowns assumptions are made, and define the most complex correlation terms that are related to known variables. These terms are "order" and "level". Both terms are found in the meteorology literature, but the term order is used here. First-order closure means that there are equations for the state variables (u, v, w, T, q), or the first moments, and the covariance terms (e.g., $u'v'$) are parameterized. Second-order closure means that there are equations for both the state variables and the covariance terms, and the triple correlation terms are parameterized. There also are examples of non-integer closures, such as half-order and one-and-a-half-order closure. Part of the motivation for higher-order closures is that, for example, "if a crude assumption for second moments predicts first moments adequately, perhaps a crude assumption for third moments will predict second moments adequately" (Lumley and Khajeh-Nouri[1] 1974, p. 171).

When the unknown terms in the equations are related to the known variables, there are two very different approaches that can be used for boundary layer parameterization. One approach is called *local closure*, and the other approach is called *non-local closure*. Local closure relates the unknown variables to known variables at nearby vertical grid points. Thus, when the model is solving for the Reynolds stress term at $z = 500\,\text{m}$, only the model variables within a small distance around $z = 500\,\text{m}$ are used for the unknown terms. In contrast, non-local closure relates the unknown variables to known variables at any number of other vertical grid points. So for the same situation, a non-local closure scheme may use all the vertical grid points within the boundary layer to determine the unknown terms and to solve for the Reynolds stress term at $z = 500\,\text{m}$. As can be seen, these approaches are conceptually very different and can result in dramatically different evolutions of the boundary layer. Some common closure approaches are now explored.

5.4 Non-local closure schemes

The boundary layer is influenced directly by what is happening at the Earth's surface and responds to changes in surface forcing very quickly. Most of the turbulent energy is found in the largest eddies, which typically are of the depth of the boundary layer. The potential benefits of a non-local interpretation are illustrated in Stull (1991), where he clearly shows that a non-local viewpoint explains the turbulent characteristics of a boundary layer above a forest canopy, whereas a local viewpoint produces expectations that do not match observations of the heat flux (Fig. 5.9). It is clear that non-local approaches have a number of advantages over local approaches, and so we begin with the simplest non-local approach – a mixed layer scheme. We then move toward more sophisticated schemes that are representative of the non-local schemes most commonly used in atmospheric models.

---

[1] Reprinted with permission from Elsevier.

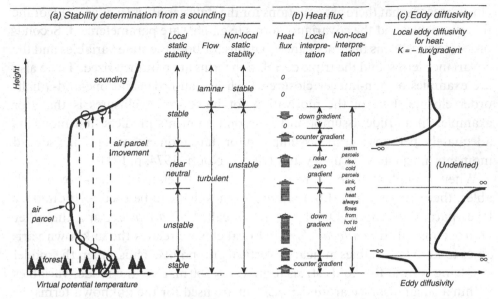

Figure 5.9. Local versus non-local interpretations of boundary layer stability and associated turbulence. A typical sounding from above a forest canopy is shown in (a), illustrating the local static stability and turbulent conditions. The heat flux is shown in (b), with the local and non-local interpretations of this flux. The calculated eddy diffusivity is shown in (c), further illustrating the challenges to local interpretations of this boundary layer that lead to incorrect expectations for turbulence. From Stull (1991).

5.4.1 Mixed layer schemes

Mixed layer models were first developed in the 1960s and were used to explore stratocumulus development by Lilly (1968) and later for general boundary layer development (Carson 1973; Betts 1973). Mixed layer models are still used in many ocean models, and are quite efficient computationally.

The basic assumptions in mixed layer models are that potential temperature is constant with height in the mixed layer, and that the mixed layer is horizontally homogeneous. Neglecting the effects of horizontal advection, one finds that

$$\frac{\partial \bar{\theta}}{\partial t} = -\frac{\partial}{\partial z}\left(\overline{w'\theta'}\right), \tag{5.16}$$

for a mixed layer model. Since the potential temperature is constant with height, $\overline{w'\theta'}$ must vary linearly with height (as often observed) and, therefore, the only fluxes that are important to the change of potential temperature with time occur at the top and bottom boundaries. This leads to

$$\frac{\partial\bar\theta}{\partial t} = \left[(\overline{w'\theta'})_S - (\overline{w'\theta'})_H\right]\left(\frac{1}{H}\right), \tag{5.17}$$

where H is the top height of the mixed layer and the subscript S represents fluxes from the ground surface.

The surface fluxes are obtained from calculations of Q_H as discussed in Chapters 2 and 3, so all that is needed is an equation for the fluxes at the top of the boundary layer. The flux at the top of the mixed layer is equal to the turbulent entrainment of potential temperature into the mixed layer as the mixed layer grows and ingests parcels of air from above. Assuming no change in the mixed layer depth due to environmental subsidence or lifting, this can be described mathematically as

$$(\overline{w'\theta'})_H = -\frac{dH}{dt}(\theta_+ - \bar\theta), \tag{5.18}$$

where θ_+ is the potential temperature of the air just above the mixed layer top as shown in Fig. 5.10.

This result can be derived using Liebniz' rule for differentiation of integrals (Dutton 1976). Recall that this rule states that

$$\frac{d}{dt}\int_a^b F(x,t)\,dx = \int_a^b \frac{\partial F(x,t)}{\partial t}\,dx + F(b,t)\frac{db}{dt} - F(a,t)\frac{da}{dt}. \tag{5.19}$$

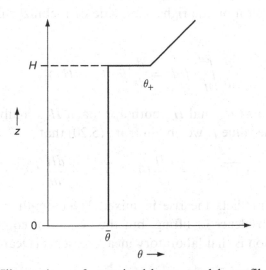

Figure 5.10. Illustration of a mixed-layer model profile of potential temperature (θ) versus height (z) in which the potential temperature is constant within the mixed layer and there is a discontinuity of potential temperature at the top of the layer (H). The value θ_+ defines the potential temperature at the bottom of the atmospheric layer that lies on top of the mixed layer.

By assuming that $F(x, t) = \theta(z, t)$ and integrating across the inversion layer it is found that

$$\int_{H_-}^{H_+} \frac{\partial \theta}{\partial t} \, dz = \frac{d}{dt} \int_{H_-}^{H_+} \theta \, dz - \theta_+ \frac{dH_+}{dt} + \bar{\theta} \frac{dH_-}{dt}, \tag{5.20}$$

where H_+ is a height level just above the inversion height H, and H_- is a height level just below the inversion height H. Also note that the terms are reordered slightly to move the partial derivative to the left-hand side of the equal sign. Recalling the definition of the time rate of change of the mixed layer potential temperature, namely

$$\frac{\partial \theta}{\partial t} = -\frac{\partial}{\partial z} \left(\overline{w'\theta'} \right), \tag{5.21}$$

and substituting this expression into the left-hand side term of Liebniz' rule in (5.20) yields

$$\int_{H_-}^{H_+} \frac{\partial \theta}{\partial t} \, dz = - \int_{H_-}^{H_+} \frac{\partial (\overline{w'\theta'})}{\partial z} \, dz = (\overline{w'\theta'})_H, \tag{5.22}$$

since by definition the flux is zero above the mixed layer at H_+. By the mean value theorem, there exists a mean value of θ, say $\tilde{\theta}$, for which $\bar{\theta} < \tilde{\theta} < \theta_+$, allowing the first term on the right-hand side of Liebniz' rule in (5.20) to be rewritten as

$$\frac{d}{dt} \int_{H_-}^{H_+} \theta \, dz = \frac{d}{dt} \left[\tilde{\theta}(H_+ - H_-) \right], \tag{5.23}$$

which goes to zero as H_+ and H_- both approach H. Finally, as H_+ and H_- approach the same value H we obtain from (5.20) that

$$\left(\overline{w'\theta'} \right)_H = -\frac{dH}{dt} \left(\theta_+ - \bar{\theta} \right) = -\frac{dH}{dt} \Delta\theta. \tag{5.24}$$

This equation predicts the rise in mixed layer depth in the absence of environmental subsidence or lifting, but the system of equations is still not closed. One solution is that laboratory measurements (Deardorff *et al.* 1969) suggest

$$\left(\overline{w'\theta'} \right)_H = -0.2 \left(\overline{w'\theta'} \right)_S = -k_e \left(\overline{w'\theta'} \right)_S, \tag{5.25}$$

where k_e is an entrainment coefficient. This relationship then closes the system of equations. As seen later, there is some uncertainty in the value of k_e. Regardless, if the following expression is used for the surface sensible heat flux,

$$\left(\overline{w'\theta'}\right)_S = \frac{(\theta_S - \overline{\theta})}{r_H},$$
(5.26)

then a set of two equations is obtained that describes the evolution of the mixed layer

$$\frac{d\overline{\theta}}{dt} = \frac{(1+k_e)(\theta_S - \overline{\theta})}{r_H H} = (1+k_e)\frac{Q_H}{\rho c_p H},$$
(5.27)

$$\frac{dH}{dt} = \frac{k_e(\theta_S - \overline{\theta})}{r_H \Delta\theta} = \frac{k_e Q_H}{\rho c_p \Delta\theta}.$$
(5.28)

When the surface skin temperature is held constant, as happens over short timescales when air moves over the oceans, the mixed layer temperature approaches the surface skin temperature (Fig. 5.11). However, the depth of the mixed layer is very much dependent upon the amount of entrainment.

One can also include mixed layer equations for moisture and momentum, such that

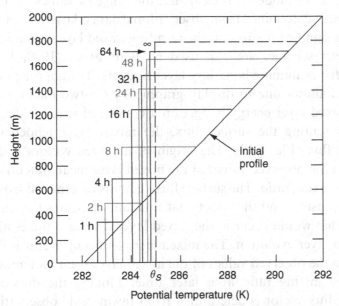

Figure 5.11. Evolution of the potential temperature versus height within a mixed layer model as a function of time. Note that as time becomes infinite, the mixed layer temperature and the surface skin temperature are equal. From Lilly (1968).

$$\frac{d\bar{q}}{dt} = \frac{1}{H}\left(\frac{Q_E}{\rho L_v} + \frac{k_e Q_H \Delta q}{\rho c_p \Delta \theta}\right),$$ (5.29)

$$\frac{d\vec{V}_H}{dt} = -f\vec{k} \times (\vec{V}_H - \vec{V}_g) + C_D \vec{V}_H \cdot \left(-\vec{V}_H + \frac{k_e Q_H \Delta \vec{V}_H}{\rho c_p \Delta \theta}\right)\frac{1}{H},$$ (5.30)

where the H subscript on the wind vector denotes the horizontal wind, C_D is a dimensionless drag coefficient, Δ refers to differences across the top of the mixed layer for either specific humidity or momentum, f is the Coriolis force, and the g subscript denotes the geostrophic wind.

Artaz and Andre (1980) compare different mixed layer schemes and find that the ones that directly relate the inversion-level heat flux to the surface sensible heat flux perform as well as any of the other approaches in predicting mixed layer depth. However, the greatest uncertainty in these schemes is the value of the entrainment coefficient. For free convective conditions, values of 0.1–0.4 appear to be reasonable (Stull 1976; Heidt 1977). However, Dubosclard (1980) suggests an entrainment coefficient near 1.0 for situations when the surface sensible heat flux is small. Betts *et al.* (1992) argue that 0.4 is a better value than the traditional value of 0.2, while Margulis and Entekhabi (2004) calculate a range of values from 0.22 to 0.54 by coupling observations with a mixed layer model. It is clear that the range of values for the entrainment coefficient determined from observational data is large and that the value chosen has a significant effect on the resulting mixed layer evolution.

Mixed layer schemes are also used by Betts *et al.* (1990, 1992) as the foundation for examining boundary layer budgets. The approach outlined in these studies allows one to display graphically (in two-dimensional vector form) the mixed layer energy budget on a conserved variable diagram, with vectors representing the surface flux, the mixed layer tendency, and the entrainment flux (Fig. 5.12). The origin of all three vectors is determined directly from the observed values of the mixed layer mean potential temperature and the mixing ratio. The surface flux vector is determined from observations of the sensible and the latent heat flux, and indicates the warming and moistening that would occur in the mixed layer if surface fluxes alone determined mixed layer evolution. The mixed layer tendency vector is determined directly from the observed values of the mixed layer mean potential temperature and the mixing ratio at a later time. Finally, the direction of the entrainment flux vector is determined from rawinsonde observations as the sonde penetrates the inversion layer. The magnitude of the entrainment flux vector is calculated using the assumption that the surface and entrainment fluxes together determine the evolution of the mixed layer, such that the sum of

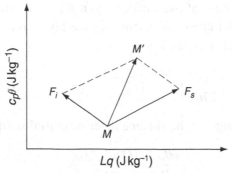

Figure 5.12. Vector diagrams for the mixed layer budget, with M representing the initial mixed layer, M' representing the mixed layer at a later time, F_s representing the surface flux vector, F_i representing the entrainment flux vector. After Betts (1992).

the surface flux and entrainment flux vectors equals the mixed layer tendency vector (Fig. 5.12). The entrainment coefficient can then be determined by projecting the vectors $\vec{F}_s$ and $\vec{F}_i$ onto a dry virtual adiabat (see Betts 1992). This technique provides a good conceptual picture of boundary layer evolution, and is another way in which mixed layer schemes continue to be valuable in meteorology. Mixed layer schemes have even been used to study cirrus outflow dynamics (Lilly 1988).

5.4.2 *Penetrative convection scheme*

Estoque (1968) visualizes the exchange of heat in the boundary layer as taking place between the ground surface and each level within the mixed layer, a clear example of a non-local viewpoint. This assumption leads to a heat flux distribution that depends upon the entire temperature distribution within the boundary layer, such that countergradient fluxes are a natural outcome of this approach. Blackadar (1978) develops a one-dimensional boundary layer model based upon this idea, as outlined and tested in Zhang and Anthes (1982). This scheme is often referred to as the Blackadar scheme.

Assume that the intensity of mixing under free convection depends upon the surface layer temperature (θ_{sl}) and the heat flux at the top of the surface layer ($Q_{H_{sl}}$). Priestley (1954) derives an expression for the heat flux at the top of the surface layer, in which

$$Q_{H_{sl}} = \rho c_p z_{sl} b (\theta_{sl} - \theta_{sl+\Delta z})^{3/2}, \qquad (5.31)$$

where z_{sl} is the thickness of the surface layer, $\theta_{sl+\Delta z}$ is the potential temperature of the first model layer above the surface layer, ρ is the density, g is the acceleration due to gravity, and

$$b = \left(\frac{2g}{27\theta_{sl}}\right)^{1/2} \frac{1}{z_{sl}} \left[z_{sl}^{-1/3} - 2(z_{sl+\Delta z})^{-1/3}\right]^{-3/2}. \qquad (5.32)$$

The time rate of change of the surface layer potential temperature is then

$$\frac{\partial \theta_{sl}}{\partial t} = \frac{(Q_H - Q_{H_{sl}})}{\rho c_p z_{sl}}, \qquad (5.33)$$

which is simply proportional to the difference of the fluxes into and out of the surface layer. This is another example of how the depth of the atmospheric layers influences parameterization scheme behavior, as a thinner surface layer leads to a faster response in the surface layer potential temperature.

Within the convective boundary layer, the time rate of change of potential temperature is given by

$$\frac{\partial \theta(z)}{\partial t} = \overline{m}w(z)[\theta_{sl} - \theta(z)]. \qquad (5.34)$$

Here $\overline{m}$ is the fraction of mass exchanged between the boundary layer and the free atmosphere above, and $w(z)$ is a weighting function used to account for variations in the exchange rate. Typically, $w(z) = 1$ at all heights, although Estoque (1968) proposed that $w(z)$ should decrease linearly as a function of boundary layer height.

To determine $\overline{m}$, it is assumed that no flux occurs across the top of the boundary layer, such that

$$Q_{H_{sl}} - \overline{m}\rho c_p \int_{z_{sl}}^{z_{top}} [\theta_{sl} - \theta(z)]\, dz = 0, \qquad (5.35)$$

and therefore

$$\overline{m} = Q_{H_{sl}} \left[\rho c_p (1 - k_e) \int_{z_{sl}}^{z_{top}} [\theta_{sl} - \theta(z)]\, dz\right]^{-1}, \qquad (5.36)$$

where k_e again is the entrainment coefficient. The value of k_e can be viewed schematically as the ratio of negative area to positive area within the planetary boundary layer in Fig. 5.13.

Results from Zhang and Anthes (1982) suggest that this non-local scheme typically does very well in non-saturated boundary layers. Bright and Mullen (2002) also find that it does very well in simulating the boundary layers over

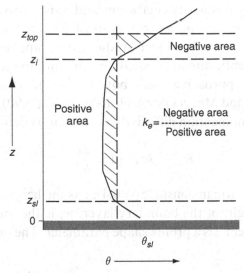

Figure 5.13. Illustration of potential temperature versus height within the boundary layer, highlighting the positive and negative areas on the diagram that are used to define the boundary layer top (z_{top}). The entrainment coefficient is k_e and z_i is the inversion level. After Zhang and Anthes (1982).

Arizona during the summertime, when boundary layer depths can exceed 3 km. As applied in numerical models, this scheme reverts to a local closure approach during neutral and stable boundary layer conditions. An extension of this scheme that retains the direct, non-local upward flux from rapidly rising plumes while also including a gradual downward flux due to compensatory subsidence is found in Pleim and Chang (1992).

5.4.3 *Non-local diffusion scheme*

Another non-local closure approach to modeling the boundary layer is used in the National Centers for Environmental Prediction (NCEP) Global Forecast System (GFS). This scheme is described by Hong and Pan (1996), based upon Troen and Mahrt (1986), and the description of it here largely follows their derivation. This scheme uses results from large-eddy simulations in its formulation. Deardorff (1972), Troen and Mahrt (1986), Holtslag and Moeng (1991) and Holtslag and Boville (1993) show that the turbulence diffusion equations for prognostic variables (u, v, θ, q) can be expressed as

$$\frac{\partial C}{\partial t} = \frac{\partial}{\partial z}\left[K_c\left(\frac{\partial C}{\partial z} - \gamma_c\right)\right], \tag{5.37}$$

where K_c is the eddy diffusivity coefficient and γ_c is a correction to the local gradient that incorporates the contributions of the large-scale eddies. In this scheme, the correction to the local gradient only applies to the potential temperature and specific humidity equations, and not to momentum. In addition, a local closure approach is used above the boundary layer.

Following Troen and Mahrt (1986), Holtslag *et al.* (1990), and Holtslag and Boville (1993), the momentum diffusivity coefficient is defined as

$$K_{zm} = kzw_s\left(1 - \frac{z}{h}\right)^2, \tag{5.38}$$

where k is the von Karman constant ($=0.4$), z is the height above the ground surface, h is the height of the boundary layer, w_s is the mixed layer velocity scale, and the exponent 2 is a profile shape parameter. The mixed layer velocity scale is defined as

$$w_s = u_*\phi_m^{-1}, \tag{5.39}$$

where u_* is the surface friction velocity and ϕ_m is the wind profile function evaluated at the top of the surface layer. The correction to the local gradient from the large-scale eddies for θ and q is given by

$$\gamma_c = 7.8\frac{\overline{(w'c')}_{sfc}}{w_s}. \tag{5.40}$$

Here the constant 7.8 is a proportionality constant defined following the derivation in Troen and Mahrt (1986). To ensure that the top of the surface layer and the bottom of the boundary layer are compatible, identical profile functions are used for both layers. For unstable and neutral conditions, defined as having positive buoyancy flux, define

$$\phi_m = \left(1 - 16\frac{0.1h}{L}\right)^{-1/4}, \tag{5.41}$$

$$\phi_t = \left(1 - 16\frac{0.1h}{L}\right)^{-1/2}, \tag{5.42}$$

where L is the Monin–Obukhov length, ϕ_m is used for momentum variables, and ϕ_t is used for mass variables. For the stable regime with buoyancy fluxes less than zero, the profile functions are defined as

$$\phi_m = \phi_t = \left(1 + 5\frac{0.1h}{L}\right). \tag{5.43}$$

Finally, to close the system the boundary layer height is calculated from

$$h = Ri_c \frac{\theta_{v_a} |U(h)|^2}{g(\theta(h) - \theta_s)},$$ (5.44)

where Ri_c is the critical Richardson number ($=0.25$), $U(h)$ is the horizontal wind speed at the top of the boundary layer (h), θ_{v_a} is the virtual potential temperature at the first model level above the ground surface, and θ_s is a near-surface temperature defined as

$$\theta_s = \theta_{v_a} + \theta_T = \theta_{v_a} + 7.8 \frac{\overline{(w'\theta'_v)}_{sfc}}{w_s h}.$$ (5.45)

In the early testing of this scheme, it was found that the value of θ_T could become too large when the winds were weak, yielding an unrealistically large boundary layer depth, so an upper value of $\theta_T = 3.0$ was chosen. In addition, the thermal diffusivity coefficient K_{zt} is proportional to the value of K_{zm} through the Prandtl number (Pr), such that

$$Pr = \frac{K_{zm}}{K_{zt}} = \left(\frac{\phi_t}{\phi_m} + 7.8k \frac{0.1h}{h} \right).$$ (5.46)

The value for boundary layer height (h) is determined iteratively. First, h is estimated without considering the thermal excess θ_T. This value of h is then used to compute all other variables needed, and then to compute the thermal excess. Then h is calculated again using the value for θ_T. This is done until a value for h is found that is stable.

The scheme perhaps is easier to understand if one rewrites the equation for boundary layer height in terms of the virtual potential temperature at the boundary layer top. This manipulation yields

$$\theta_v(h) = \theta_{v_a} + \theta_T + \frac{Ri_c \theta_{v_a} |U(h)|^2}{gh}.$$ (5.47)

From here it is easier to see that as the buoyancy flux from the ground surface increases, leading to a larger value of θ_T, the boundary layer deepens since $\theta_v(h)$ increases (see Fig. 5.14). The third term shows that $\theta_v(h)$ increases as the wind speed at the top of the boundary layer increases and is larger for smaller boundary layer depths. These behaviors are consistent with many observations of boundary layer development.

Results from Hong and Pan (1996) indicate that this non-local diffusion scheme produces better boundary layer structures during the First ISLSCP (International Satellite Land Surface Climatology Project) Field Experiment

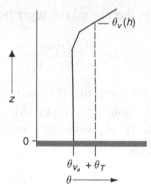

Figure 5.14. Boundary layer potential temperature versus height, showing the relationship between surface layer $\theta_{v_a} + \theta_T$ and the temperature at the top of the boundary layer $\theta_v(h)$.

(FIFE) than a local closure scheme (Fig. 5.15). The local closure scheme for this case produces a boundary layer that is too shallow, which is a well-known problem when countergradient fluxes are important. The non-local diffusion scheme produces an adiabatic lapse rate in the boundary layer, very much in agreement with observations. In addition, while the non-local scheme underestimates the specific humidity near local noon, it produces a very reasonable moisture profile by late afternoon (Fig. 5.15).

There are many other non-local closure schemes available in the literature. One of the more unique schemes is from Stull (1984, 1988, 1993) who develops a transilient turbulence scheme that allows for a range of eddy sizes to influence the turbulent mixing. This theory is based upon a discrete view of the non-local mixing that occurs in the boundary layer. Consider a boundary layer that is divided into a finite set of discrete layers, which for simplicity are assumed to have equal depth. Since large eddies dominate the vertical mixing in many boundary layers, it is possible that the evolution of a given vertical layer is influenced by mixing with any or all of the other vertical layers. Thus, there is a need to represent how each layer mixes with each other layer. This is accomplished by defining a transilient matrix $c(i, j)$ to represent the fraction of air mixed into vertical layer i from vertical layer j.

The transilient matrix has a number of constraints in order to satisfy mass and state conservation. The sum of all the values of $c(i, j)$ for a fixed value of j must equal 1, and the sum of all the mass-weighted values of $c(i, j)$ for a fixed value of i must equal 1. In addition, no element can have a value less than 0 or greater than 1. As an example, if no vertical mixing occurs, then the diagonal elements of $c(i, j)$ are equal to 1 and all the other elements are equal to zero,

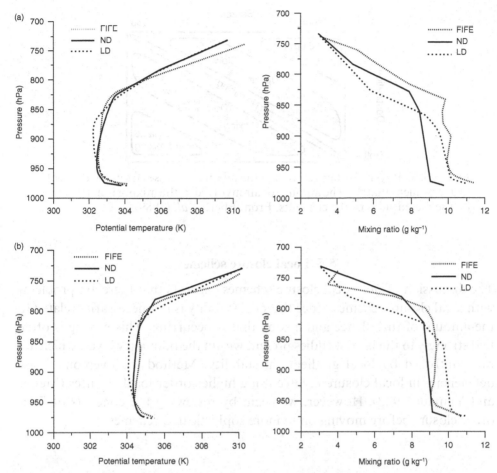

Figure 5.15. Comparisons of boundary layer profiles of potential temperature (K) and mixing ratio (g kg^{-1}) for the 9–10 August sonde averages (shaded lines) with averages from the non-local (solid lines) and local (dotted lines) schemes for (a) 1845 UTC and (b) 2145 UTC. Potential temperature on the left side, and mixing ratio on the right-hand side of the figure. From Hong and Pan (1996).

indicating that no mixing occurs. In contrast, if all the elements of $c(i, j)$ are non-zero, then the boundary layer is mixing throughout its depth.

A variety of different physical processes can be simulated with this approach depending on the form of the transilient matrix. The mixing processes active at any given time can be discerned from the location within the transilient matrix and the amount of air involved as indicated by the size of the matrix elements (Fig. 5.16). This scheme unifies the boundary layer and turbulence parameterizations under a single approach. Raymond and Stull (1990) apply this unified turbulence scheme to several case studics and show good results.

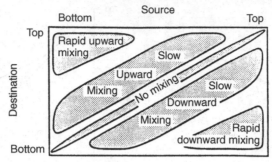

Figure 5.16. Physical interpretation of the mixing processes from the elements in a transilient matrix. The amount of air involved in the mixing is determined by the magnitudes of the elements. From Ebert *et al.* (1989).

5.5 Local closure schemes

The discussion of non-local closure schemes suggests that there are problems with local closure schemes, since the local stability is not necessarily related to the amount of turbulence and mixing that is occurring. This mixing is often tied strongly to the largest eddies present within the boundary layer, which are not controlled by local gradients in stability. Methods to overcome these deficiencies in local closure involve using higher-order local closures (Mellor and Yamada 1982). However, we begin by reviewing the concepts of first-order closure before moving on to more sophisticated schemes.

5.5.1 First-order closure scheme

One of the most commonly used forms of turbulent closure is first-order closure, commonly referred to as *K*-theory. In first-order closure, only the prognostic equations for the means of the variables are retained and the turbulent fluxes are parameterized (Stull 1988). Thus, we begin with

$$\frac{\partial \bar{u}_i}{\partial t} + \bar{u}_j \frac{\partial \bar{u}_i}{\partial x_j} = -\delta_{i3} g + f \varepsilon_{ijk} \bar{u}_j - \frac{1}{\bar{\rho}} \frac{\partial \bar{p}}{\partial x_i} + \nu \frac{\partial^2 \bar{u}_i}{\partial x_j^2} - \frac{\partial \overline{(u_i' u_j')}}{\partial x_j}, \qquad (5.48)$$

$$\frac{\partial \bar{\theta}}{\partial t} + \bar{u}_j \frac{\partial \bar{\theta}}{\partial x_j} = -\frac{\partial \overline{(u_j' \theta')}}{\partial x_j} + \text{diabatic terms}, \qquad (5.49)$$

$$\frac{\partial \bar{q}}{\partial t} + \bar{u}_j \frac{\partial \bar{q}}{\partial x_j} = -\frac{\partial \overline{(u_j' q')}}{\partial x_j} + \text{source/sink terms}, \qquad (5.50)$$

where the covariance terms are those that need to be parameterized. Note that on the right-hand side of (5.48), the second to last term represents molecular effects and is so small that it can be neglected. The covariance terms are parameterized using

$$\overline{u_j'\zeta'} = -K\frac{\partial\bar\zeta}{\partial x_j},$$ (5.51)

where K has units of $m^2\,s^{-1}$. For $K>0$, the flux $\overline{u_j'\zeta'}$ flows down the local gradient of $\bar\zeta$. This relationship is called K-theory. If we use this relationship in the equations for the mean quantities, then we obtain

$$\frac{\partial\bar u_i}{\partial t} + \bar u_j\frac{\partial\bar u_i}{\partial x_j} = -\delta_{i3}g + f\varepsilon_{ijk}\bar u_j - \frac{1}{\bar\rho}\frac{\partial\bar p}{\partial x_i} + K_m\frac{\partial^2\bar u_i}{\partial x_j^2},$$ (5.52)

$$\frac{\partial\bar\theta}{\partial t} + \bar u_j\frac{\partial\bar\theta}{\partial x_j} = K_H\frac{\partial^2\bar\theta}{\partial x_j^2} + \text{diabatic terms},$$ (5.53)

$$\frac{\partial\bar q}{\partial t} + \bar u_j\frac{\partial\bar q}{\partial x_j} = K_H\frac{\partial^2\bar q}{\partial x_j^2} + \text{source/sink terms},$$ (5.54)

where K_m and K_H are the different coefficients for the mixing of momentum and mass, respectively. Typically, when one refers to K-theory they only mean the vertical derivatives and the horizontal derivatives are referred to as horizontal diffusion.

It is often helpful to view this relationship schematically. To begin, look at the $\overline{w'\theta'}$ term since we have some idea how this flux behaves in the planetary boundary layer. As shown in Fig. 5.17, if θ decreases with height as occurs in the lowest portion of the boundary layer, then $\overline{w'\theta'}$ is positive and turbulence

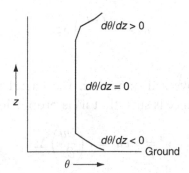

Figure 5.17. Illustration of the boundary layer potential temperature versus height, with the lapse rates for three very different layers indicated.

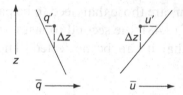

Figure 5.18. Profiles of and $\bar{q}$ and $\bar{u}$ (solid lines) as a function of z. The perturbations q' and u' are created by moving a parcel upward within this environment an amount Δz.

moves warm air upward. This is consistent with expectations from rising thermals moving hot air near the ground upward. If θ increases with height, as occurs at the very top of the boundary layer, then $\overline{w'\theta'}$ is negative and turbulence moves warm air downward. This is what one would expect from the entrainment of warmer air from aloft into the boundary layer. However, a problem occurs when θ is constant with height as occurs in the middle of a very well-mixed layer during the daytime. In this case, K-theory indicates that no turbulence is occurring, whereas observations would probably show very strong thermals moving upward through the boundary layer and penetrating the capping inversion.

A simple thought experiment can be used to illustrate a variation on K-theory. Assume that an idealized boundary layer exists with linear profiles of q and u as in Fig. 5.18, where only small eddies are present (i.e., no non-local transport). A turbulent eddy moves a parcel upward a distance of Δz and the parcel does not mix with the surrounding environment. Thus, the parcel differs from its environment by

$$q' = -\left(\frac{\partial \bar{q}}{\partial z}\right)\Delta z, \tag{5.55}$$

$$u' = -\left(\frac{\partial \bar{u}}{\partial z}\right)\Delta z. \tag{5.56}$$

Now, in order to move a distance Δz, the parcel requires some upward motion w'. If the turbulence is such that w' is proportional to u', then

$$w' = -cu' = c\left(\frac{\partial \bar{u}}{\partial z}\right)\Delta z. \tag{5.57}$$

From the equations above, the product $w'q'$ caused by that one idealized eddy can be determined. Averaging over many such eddies to find $\overline{w'q'}$ yields

$$\overline{w'q'} = -\left(\frac{\partial \overline{q}}{\partial z}\right)\Delta z c \left(\frac{\partial \overline{u}}{\partial z}\right)\Delta z = -c\overline{(\Delta z)^2}\left(\frac{\partial \overline{q}}{\partial z}\right)\left(\frac{\partial \overline{u}}{\partial z}\right), \qquad (5.58)$$

where $\overline{(\Delta z)^2}$ is the variance of the parcel displacement. If a mixing-length l is defined, such that $l^2 = c\overline{(\Delta z)^2}$ then

$$\overline{w'q'} = -l^2 \left|\frac{\partial \overline{u}}{\partial z}\right| \frac{\partial \overline{q}}{\partial z}. \qquad (5.59)$$

Recalling from K-theory that

$$\overline{w'q'} = -K_H \frac{\partial \overline{q}}{\partial z}, \qquad (5.60)$$

then it follows that

$$K_H = l^2 \left|\frac{\partial \overline{u}}{\partial z}\right|, \qquad (5.61)$$

which is called mixing-length theory (Stull 1988). This theory suggests that K increases as the shear increases, and also that K increases as the variance of the parcel displacement increases. This allows for K-theory to apply in a more realistic way to boundary layer development. The limitations of mixing-length theory are that it is valid only in boundary layers not having large eddies, and that it only allows for downgradient transport (unless $K < 0$ is allowed). Also note that the vertical transport of perturbation velocities is neglected, which may be large.

Let us return again to examine observations and compare against mixing-length theory. It is not unusual for the winds within the boundary layer to be relatively constant with height within a well-mixed layer during the daytime (Fig. 5.19). This situation often occurs when large eddies are present, which implies mixing-length theory will have difficulty. Zhang and Anthes (1982) show the development of such a mixed layer over Marfa, Texas, at both 1800 UTC 10 April and 0000 UTC 11 April 1979. Using mixing-length theory, an examination of Fig. 5.19 suggests that the value of K_H is again near zero throughout the boundary layer, and then changes to a large value at the top of the boundary layer where the wind shear is greatest. So, K_H from mixing-length theory is influenced by dU/dz, but not by the stability of the layer. This is not realistic, and indicates that more reasonable parameterizations for K are needed. Blackadar (1979) suggests that one such parameterization valid for heights less than 200 m is

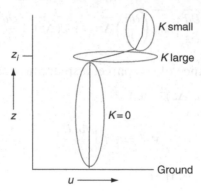

Figure 5.19. Mean wind speed u versus height z within a daytime boundary layer from the ground surface to above the inversion level (z_i). Values of K indicated within three different layers.

$$K_H = 1.1 \left[\frac{(Ri_c - Ri)l^2}{Ri} \right] \left| \frac{\partial \bar{u}}{\partial z} \right|, \qquad (5.62)$$

where Ri is the Richardson number defined as

$$Ri = \frac{(g/\theta_v)(\partial \bar{\theta}_v / \partial z)}{(\partial \bar{u} / \partial z)^2}, \qquad (5.63)$$

and $l = kz = 0.4z$. For heights greater than 200 m, one might instead use

$$K_H = (1 - 18Ri)^{-1/2} l^2 \left| \frac{\partial \bar{u}}{\partial z} \right|, \qquad (5.64)$$

where $l = 70$ m and K_H cannot be negative (Stull 1988).

One of the common threads in these various definitions of K is that they depend only upon the local values of the model variables, such that K-theory is a local closure approach to turbulence closure. Because of the local closure assumption, when surface heating is strong, K-theory can produce very deep superadiabatic layers near the surface that are unrealistic. In addition, not all turbulent transport is downgradient. This last point can easily be illustrated by examining the boundary layer during a typical summer morning. A super-adiabatic layer exists near the ground surface, with the remaining stable nocturnal inversion layer above it. Large eddies associated with the rise of warm air parcels transport heat from hot to cold regardless of the local gradient of the background environment. So in this case warm thermals move heat upward, even though the background environment would suggest from K-theory that warm air is being transported downward. Thus, when large eddies exist, K-theory often fails owing to upgradient, or countergradient, fluxes. This is perhaps the most important criticism of K-theory – that the mass and momentum transport in the boundary layer during the daytime is mostly accomplished by the largest eddies, and that these eddies are more representative of the properties of the entire boundary layer than the local conditions at one vertical level (Deardorff 1972; Troen and Mahrt 1986; Holtslag and Moeng 1991; Stull 1991). In addition, Ayotte *et al.* (1996) show that this type of scheme tends to underestimate entrainment when the capping inversion is strong.

5.5.2 *1.5-order local closure scheme*

As the order of the closure increases, the parameterizations include more equations for the higher moments. Thus, for 1.5-order closure the parameterizations typically include equations not only for the standard prognostic variables ($\bar{u}, \bar{v}, \bar{\theta}, \bar{q}$), but also the potential temperature variance ($\overline{\theta'^2}$) and the turbulent kinetic energy ($\bar{e}$). Also recall that most boundary layer schemes in use today are one-dimensional and only consider the vertical derivatives. The equations for a typical 1.5-order closure scheme are (Stull 1988)

$$\frac{\partial \bar{u}_i}{\partial t} + \bar{u}_j \frac{\partial \bar{u}_i}{\partial x_j} = -\delta_{i3}g + f\varepsilon_{ijk}\bar{u}_j - \frac{1}{\bar{\rho}}\frac{\partial \bar{p}}{\partial x_i} - \frac{\partial \overline{(u'_i w')}}{\partial z}, \tag{5.65}$$

$$\frac{\partial \bar{\theta}}{\partial t} + \bar{u}_j \frac{\partial \bar{\theta}}{\partial x_j} = -\frac{\partial \overline{(w'\theta')}}{\partial z} + \text{diabatic terms}, \tag{5.66}$$

$$\frac{\partial \bar{q}}{\partial t} + \bar{u}_j \frac{\partial \bar{q}}{\partial x_j} = -\frac{\partial \overline{(w'q')}}{\partial z} + \text{source/sink terms}, \tag{5.67}$$

$$\frac{\partial \bar{e}}{\partial t} = -\overline{u'w'}\frac{\partial \bar{u}}{\partial z} - \overline{v'w'}\frac{\partial \bar{v}}{\partial z} + \frac{g}{\bar{\theta}}\overline{w'\theta'} - \frac{\partial}{\partial z}\overline{\left(\frac{w'p'}{\rho} + ew'\right)} - \varepsilon, \tag{5.68}$$

$$\frac{\partial \overline{\theta'^2}}{\partial t} = -2\overline{w'\theta'}\frac{\partial \bar{\theta}}{\partial z} - \frac{\partial}{\partial z}\left(\overline{w'\theta'^2}\right) - 2\varepsilon_\theta - \varepsilon_R, \tag{5.69}$$

where dissipation is indicated by the ε, ε_θ, and ε_R terms.

The unknown variables in this set of equations include the fluxes $\overline{u'w'}, \overline{v'w'}, \overline{w'\theta'}, \overline{w'p'}/\bar{\rho}$, and $\overline{w'q'}$, the third moments $\overline{w'e}, \overline{w'\theta'^2}$, and the three dissipation terms. Thus, as mentioned by Stull (1988), it initially appears that the addition of the variance equations has created havoc instead of producing conceptual stability. With first-order closure there are four unknowns to specify, whereas here there are six more unknowns for a total of 10! The reason behind this apparent madness is that the additional equations for the potential temperature variance and the turbulent kinetic energy provide information on the intensity and effectiveness of the turbulence within the boundary layer. This information is used to develop improved parameterizations for the eddy diffusivities K that can now be functions of $\overline{\theta'^2}$ and $\bar{e}$ instead of just functions of wind shear and stability as developed for first-order closure.

Following Yamada and Mellor (1975), one suggested set of parameterizations for the 10 unknowns is as follows:

$$\overline{u'w'} = -K_m\left(\bar{e}, \overline{\theta'^2}\right) \frac{\partial \bar{u}}{\partial z},$$ (5.70)

$$\overline{v'w'} = -K_m\left(\bar{e}, \overline{\theta'^2}\right) \frac{\partial \bar{v}}{\partial z},$$ (5.71)

$$\overline{w'\theta'} = -K_H\left(\bar{e}, \overline{\theta'^2}\right) \frac{\partial \bar{\theta}}{\partial z} - \gamma_c\left(\bar{e}, \overline{\theta'^2}\right),$$ (5.72)

$$\overline{w'q'} = -K_H\left(\bar{e}, \overline{\theta'^2}\right) \frac{\partial \bar{q}}{\partial z} - \gamma_c\left(\bar{e}, \overline{\theta'^2}\right),$$ (5.73)

$$\overline{w'\left(\frac{\rho'}{\bar{\rho}} + e\right)} = \frac{5}{3} L_4 e^{-1/2} \frac{\partial \bar{e}}{\partial z},$$ (5.74)

$$\overline{w'\theta'^2} = -L_3 e^{1/2} \frac{\partial \overline{\theta'^2}}{\partial z},$$ (5.75)

$$\varepsilon = \frac{\bar{e}^{3/2}}{L_1},$$ (5.76)

$$\varepsilon_R = 0,$$ (5.77)

$$\varepsilon_\theta = \frac{\bar{e}^{1/2} \overline{\theta'^2}}{L_2}.$$ (5.78)

Here L_x are empirical length-scale parameters that are often chosen by trial and error to make the simulated flow best match the observed flow for a given set of cases. Often the values of L approach kz for small z, and approach a specified constant L_0 for large z. Typically, values for L_0 are between 50 and 100 m. Also, the γ_c term is used so that flux occurs even when there is no mean vertical gradient in the variable, thereby allowing countergradient fluxes.

The expressions for the eddy diffusivities K are complex and include terms related to the environmental wind shear and stability, but can be represented in conceptual form as

$$K = L\sqrt{\bar{e}}, \qquad (5.79)$$

where L is again one of the empirical length scales. Further information on the details of these types of closure schemes can be found in Mellor and Yamada (1982) and Janjic (1994). Another approach to determining the empirical length scale is to include a separate prognostic equation for the dissipation rate ε that is then used with the turbulent kinetic energy to calculate the length scales (Stull 1988). However, these e–ε closure schemes are not very prevalent in the meteorological literature.

Note that in the parameterizations of the unknowns a consistency is seen that applies for almost all of the definitions – downgradient diffusion. The parameterizations for the covariance terms are specified as functions of the vertical gradients of the mean variables, i.e. the $\overline{w'\theta'}$ term is related to the vertical gradient of $\bar{\theta}$, and the value of K_H depends upon the potential temperature variance and the turbulent kinetic energy. Similarly, the triple correlation terms are parameterized as flowing down the vertical gradients of the covariances, i.e. the $\overline{w'\theta'^2}$ term is related to the vertical gradient of $\overline{\theta'^2}$ and the magnitude of the turbulent kinetic energy. Note also that the dissipation terms are parameterized as being proportional to their respective variables. Thus, as the turbulent kinetic energy increases, its dissipation also increases.

The ability of 1.5-order and higher closure schemes to account for countergradient fluxes is shown by Deardorff (1966). In the thermal variance equation (5.69), the first term on the right-hand side is a production term that increases the thermal variance when the heat flux is downgradient, while the dissipation terms act to smooth the flow. It is the second term on the right-hand side that allows for countergradient flux. Deardorff (1966) notes that $\overline{w'\theta'}$ is generally large when $\theta' > 0$ in the boundary layer and small otherwise, so $\overline{w'\theta'^2}$ is positive but decreases with height in the countergradient region. This situation leads to countergradient fluxes as the thermal variance is increased.

One would expect that countergradient fluxes would be particularly important in dry convective boundary layers. Teixeira *et al.* (2004) develop and test a 1.5-order closure scheme that shows improvements in simulating dry

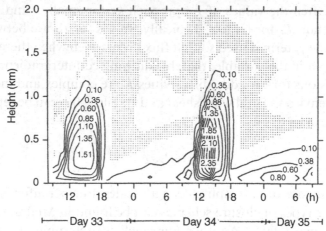

Figure 5.20. Time and vertical variation of the computed $\bar{e}^2$ (twice the turbulent kinetic energy in $m^2 s^{-2}$) for days 33–35 of the Wangara experiment. From Yamada and Mellor (1975).

convective boundary layers over southern Portugal. Results suggest that the improved boundary layer predictions are due to more realistic representations of entrainment.

One of the benefits of 1.5-order or higher closure schemes is that they explicitly predict the intensity of the turbulent kinetic energy (Fig. 5.20). The diurnal cycle of the boundary layer is clearly seen in the turbulent kinetic energy, and this type of information may be useful in studies of air pollution dispersion. As discussed by Yamada and Mellor (1975), the computed turbulence field helps to explain the behavior of the mean quantities, even if it is difficult to use in comparisons with observations of turbulence.

The NCEP Eta model uses a 1.5-order closure scheme (Janjic 1994) for the boundary layer that is slightly different from that discussed above. In the Eta model scheme there is no equation for the potential temperature variance, so the eddy diffusivities are defined only in terms of the turbulent kinetic energy. This highlights that one must be careful when discussing schemes with non-integer closures, since the exact application of the intermediate closure assumptions is uncertain.

5.5.3 Second-order closure scheme

The many details involved in a full second-order closure scheme can be found in Mellor and Yamada (1982) and Stull (1988), which are both excellent sources for information on local closure schemes. In addition to the equations listed for 1.5-order closure, second-order closure also has predictive equations for all the remaining covariance terms $\overline{u_i'u_j'}$, $\overline{u_j'\theta'}$, and $\overline{u_j'q'}$. The number of unknowns is large and includes some very complex and rather daunting-looking terms. For example, if we look at the time rate of change of the momentum covariance term we find

$$\frac{\partial \overline{(u_i'u_j')}}{\partial t} = - \overline{u_i'w'}\frac{\partial \bar{u}_j}{\partial z} - \overline{u_j'w'}\frac{\partial \bar{u}_i}{\partial z} - \frac{\partial \overline{(u_i'u_j'w')}}{\partial z} + \frac{g}{\theta}\left(\delta_{i3}\overline{u_j'\theta'} + \delta_{j3}\overline{u_i'\theta'}\right)$$
$$+ \overline{\left(\frac{p'}{\rho}\right)\left(\frac{\partial u_i'}{\partial x_j} + \frac{\partial u_j'}{\partial x_i}\right)} - 2\varepsilon_{u_i u_j}. \tag{5.80}$$

This equation has a triple correlation term and a pressure correlation term. The pressure correlation term

$$\overline{\left(\frac{p'}{\rho}\right)\left(\frac{\partial u_i'}{\partial x_j} + \frac{\partial u_j'}{\partial x_i}\right)}, \tag{5.81}$$

partitions energy among the three components while not contributing to the total energy. Rotta (1951) provides a key suggestion for the parameterization of this term by calling it an "energy redistribution term" and specifying it as being

$$\overline{\left(\frac{p'}{\rho}\right)\left(\frac{\partial u_i'}{\partial x_j} + \frac{\partial u_j'}{\partial x_i}\right)} = c\overline{u_i'u_j'} + d\frac{\partial \bar{u}_i}{\partial x_j}, \tag{5.82}$$

where c and d are constants. There are many proposed ways to parameterize the various terms in higher-order closure schemes. Most use some variant of mixing-length theory and Monin–Obukhov similarity theory as a foundation. Examples can be found in Launder *et al.* (1975), Lumley and Khajeh-Nouri (1974), Mellor and Yamada (1974), Rotta (1951), Zeman (1981), and Wichmann and Schaller (1986). Others are undoubtedly given in the literature. While it is nearly impossible to summarize all the proposed parameterizations, a brief overview of some commonly used closures is helpful.

The downgradient diffusion model suggests that

$$\frac{\partial \overline{(w'e)}}{\partial z} = -\frac{\partial}{\partial z}\left(L_1\sqrt{\bar{e}}\frac{\partial \bar{e}}{\partial z}\right), \tag{5.83}$$

where L_1 is another empirical length scale. As mentioned previously, downgradient diffusion models have difficulty in simulating convective boundary layers where large eddies are important.

The diffusion term ε can be parameterized as

$$\varepsilon = \frac{\bar{e}^{3/2}}{L_2},\tag{5.84}$$

which is obtained through scaling arguments (Mellor and Yamada 1982). L_2 is another empirical length scale.

The triple correlation terms can also be parameterized using the down-gradient diffusion model, such that

$$\frac{\partial \overline{w'^2\theta'}}{\partial z} = -\frac{\partial}{\partial z}\left[L_3\sqrt{\bar{e}}\frac{\partial(\overline{w'\theta'})}{\partial z}\right],\tag{5.85}$$

$$\frac{\partial \overline{w'\theta'^2}}{\partial z} = -\frac{\partial}{\partial z}\left[L_4\sqrt{\bar{e}}\frac{\partial(\overline{\theta'^2})}{\partial z}\right],\tag{5.86}$$

where L_3 and L_4 are two more empirical length scales. Typically, one finds that

$$(L_1, L_2, L_3, L_4, \ldots) = (\alpha_1, \alpha_2, \alpha_3, \alpha_4, \ldots)L,\tag{5.87}$$

where

$$L = \frac{kz}{1 + kz/L_0},\tag{5.88}$$

and

$$L_0 = \frac{0.1\int_0^{z_i} \sqrt{\bar{e}}z\,dz}{\int_0^{z_i} \sqrt{\bar{e}}\,dz}.\tag{5.89}$$

In their review of local closure schemes, Mellor and Yamada (1982) state that "the major weakness of all the models probably relates to the turbulent master length scale (or turbulent macroscale, or turbulent inertial scale), and, most important, to the fact that one sets all process scales proportional to a single scale." A comparison of a second-order closure simulation to observations during 2 days of the Wangara experiment shows that the boundary layer scheme reproduces the general evolution of the boundary layer very well, yet some important differences between the simulation and observations are also seen (Fig. 5.21). At 12 local time the simulated boundary layer shows the mean virtual potential temperature decreasing with height up to just below 1 km, whereas observations indicate a nearly constant value up to above 1 km. The boundary layer depth at 18 local time remains near 1 km in the simulation, but is closer to 1.25 km in the observations. The simulated and observed profiles at 3 and 9 local time agree fairly well, indicating that the local closure approach does well at reproducing the nocturnal boundary layer structures for this case.

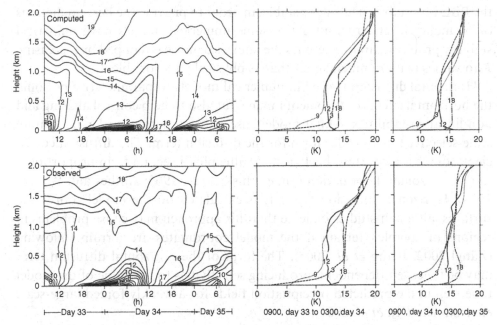

Figure 5.21. Simulated (top) and observed (bottom) boundary layer virtual potential temperature (θ_v) and vertical and temporal variations taken during the Wangara experiment over Hays, Australia. Isolines on the left-hand plot are $\theta_v - 273$ and shown in units of kelvin. Times indicated on the right-hand plots are local. From Yamada and Mellor (1975).

Thus, as mentioned previously, the local closure approach may have difficulties in predicting the daytime convective boundary layer, although it does a reasonable job in reproducing the nocturnal boundary layer for this case.

5.6 Turbulence and horizontal diffusion

Above the planetary boundary layer, turbulence and vertical mixing still occur and need to be represented in numerical models. Local closure schemes typically are used to calculate mixing throughout the vertical extent of the model domain from the surface through the PBL and upwards to the model top. Transilient turbulence theory also handles turbulence both in and above the boundary layer, in addition to predicting boundary layer depth. Other non-local closure schemes often are developed only to determine the PBL depth, and another closure approach is used above the PBL to account for turbulent mixing. For example, Zhang and Anthes (1982) use a non-local closure scheme to represent daytime penetrative convection, but a first-order local closure scheme is used above the PBL and under stable conditions at night. Thus, some models mix both non-local and local closure schemes depending upon whether the PBL is unstable, neutral, or stable, in order to provide the best predictions of boundary layer structure. Also, some numerical models have numerical stability problems, and sometimes the numerical noise is damped by increasing K to values that are much larger than is physically reasonable.

Horizontal diffusion is used in numerical models to parameterize not only the horizontal effects of turbulent mixing but also to help control aliasing and non-linear instabilities. Many models use either a second-order or a more scale-selective fourth-order form for the diffusion term. The diffusion coefficient often has a constant background value plus a second term proportional to the horizontal deformation (Smagorinsky *et al.* 1965; Anthes and Warner 1978). Horizontal diffusion typically is calculated on the model coordinate surfaces, although studies indicate that this approach may cause problems in regions of complex terrain if the model coordinates are terrain following (Zängl 2002; Juang *et al.* 2005). The form of the horizontal diffusion term may be time-step-dependent, producing sensitivities to the value of the model time step in the predicted precipitation fields for a weakly forced large-scale environment (Xu *et al.* 2001).

5.7 Discussion

The evolution of the planetary boundary layer is very important, as it sets the stage for many of the sensible weather phenomena that can occur in the atmosphere, such as deep convection. Yet predicting the development and evolution of the boundary layer is quite challenging, since the dominant mechanism for boundary layer development is turbulence. During the daytime, turbulence is often dominated by buoyancy gradients produced by surface forcing. However, at night and in certain environments, turbulence is created from the shear of the mean wind profile and is intermittent. Boundary layer schemes, of course, have to account for all of these mechanisms by which turbulence is generated, which is not a trivial task.

In contrast, it is easy to see that the magnitudes of the surface sensible and latent heat fluxes are important to the boundary layer scheme during the daytime when solar insolation dominates the surface energy budget. At night, the balance between the incoming and outgoing longwave radiation plays a large role in determining the evolution of the boundary layer, as does the mean wind profile. The connections between the boundary layer and the soil–vegetation–atmosphere schemes discussed in Chapter 3 are many and multifaceted. It is not difficult to imagine the complexities and sensitivities one may encounter in boundary layer evolutions around the world that could lead any scheme to reproduce the observed boundary layer poorly on a given

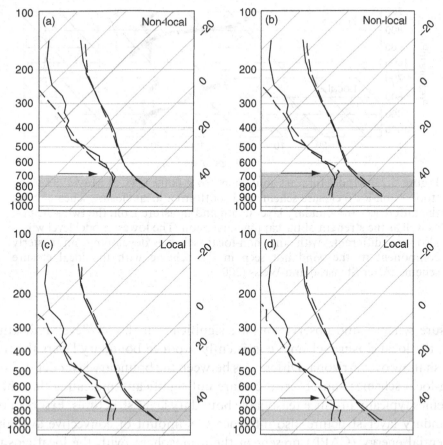

Figure 5.22. Skew-T log p plot of 0000 UTC observed (solid) and 12 h forecast (dashed) composite soundings for Tucson, Arizona, for (a, b) non-local closure schemes and (c, d) local closure schemes. Gray shading indicates model-predicted PBL depths, while the arrow indicates the estimated PBL depth from the observed composite sounding. Notice how both local closure schemes underpredict the depth of the boundary layer by roughly half. After Bright and Mullen (2002).

day. However, it also appears that certain boundary layer scheme behaviors are consistent over time. For example, Bright and Mullen (2002) show that the local closure schemes consistently underpredict the boundary layer depth over Arizona during the summertime (Fig. 5.22). This results in boundary layers that are too cool and moist. During this time of year, boundary layer depths are typically greater than 2 km and can approach and even exceed 3 km. However, the local closure schemes consistently underpredict the boundary layer depth, often by a factor of 2. Comparisons against observed soundings indicate that the differences between the boundary layer depth from the local

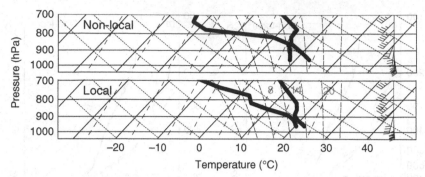

Figure 5.23. Model forecast soundings from 0000 UTC 4 May 1999 using (top) a non-local closure scheme and (bottom) a local closure scheme. Note the differences in boundary layer depth and moisture from the two schemes, as well as the strength of the capping inversion. The lowest model level winds also are different, with the non-local scheme developing an easterly component to the wind not seen in the scheme with the local closure scheme. After Stensrud and Weiss (2002).

closure schemes and observations are significant at the 5% level, indicating that the local closure schemes consistently produce boundary layers that are too shallow over Arizona. Differences between the boundary layer depth from non-local schemes and observations are not significant, indicating that these schemes typically produce reasonable boundary layer structures. Errors in the boundary layer structure also influence the amount of convective available potential energy (CAPE) present in the atmosphere, with the local closure schemes overestimating CAPE by a factor of 2 on average (Bright and Mullen 2002). This result emphasizes the connections between the boundary layer evolution and the potential for deep convection.

Unfortunately, these differences in boundary layer structure from local and non-local closure schemes are not limited to Arizona in the summertime. Results from a mesoscale model forecast of the 3 May 1999 tornado outbreak over Oklahoma also reveal distinct differences between the boundary layers produced by local and non-local closure schemes (Fig. 5.23). The boundary layer depth, structure, and wind profile produced by the non-local scheme compares better with observations than does the local scheme (Stensrud and Weiss 2002). However, there also are occasions when a local closure scheme provides a better forecast than a non-local closure scheme (Deng and Stauffer 2006). Selecting which type of scheme will produce the best forecast for a given location in advance is not easy.

While the difference in boundary layer structure from local and non-local closure is often noticed at observation times, these structures differ throughout the model forecasts (Fig. 5.24). Similar differences are seen in Alpaty *et al.*

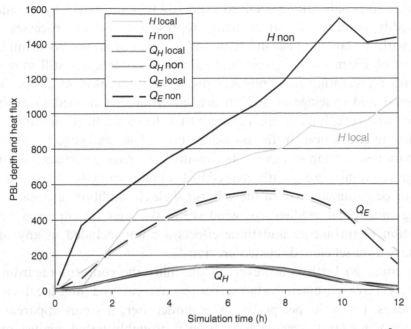

Figure 5.24. Boundary layer height H (m), sensible heat flux Q_H (W m^{-2}) and latent heat flux Q_E (W m^{-2}) for grid point in southeastern South Dakota from simulation starting 1200 UTC 28 June 1997. Model initial conditions and parameterizations are identical, except for the boundary layer. Either a non-local (non) or local scheme is used. While the surface fluxes generally are within 20 W m^{-2} throughout the first 10 h of simulation time, the differences in boundary layer depth are large with the local scheme always producing a shallower boundary layer than the non-local scheme for this case.

(1997) who compared four different boundary layer schemes with observations from both Wangara and FIFE. Since boundary layer schemes influence the wind profiles, not only the potential for deep convection, but the type of deep convection that may be expected to develop (e.g., a multi-cell versus a supercell thunderstorm) is influenced by the boundary layer scheme.

The performance of five boundary layer schemes in the prediction of the diurnal cycles of surface temperature and wind speed is evaluated by Zhang and Zheng (2004), who find that while the diurnal cycle of surface temperature is predicted well by all schemes there are large differences in the diurnal cycle of surface wind speed. In general, all five schemes underestimate the surface wind speed during the daytime and several of the schemes overestimate the surface wind speed at night. Again, the feedbacks between parameterization schemes are widespread and are often difficult to understand for a given forecast situation a priori.

As model grid spacing decreases to the point where the horizontal grid spacing approaches the vertical depth of the boundary layer, then some of the largest thermals can be resolved explicitly to some extent by the model. Large-eddy simulations that explicitly represent turbulent processes associated with the larger eddies in the boundary layer often use horizontal grid spacings of 100 m or less (Agee and Gluhovsky 1999), yet still may have problems representing some physical processes (Stevens *et al.* 2005). Thus, horizontal grid spacings of 1–2 km are far from that needed to explicitly resolve turbulence, but are small enough that the model likely develops some overturning circulations in the boundary layer. One challenge to boundary layer parameterization schemes as horizontal grid spacing decreases is how to distinguish the mixing explicitly created by the model from the mixing that still needs to be parameterized on the subgrid scale. In addition, at smaller grid spacings horizontal gradients of wind shear also may be important to the generation of turbulence and these effects are not included in any of the boundary layer schemes discussed previously.

One could go further, however, in outlining the concerns regarding all boundary layer parameterizations, for both local and non-local closure approaches. From the perspective of a model user, it seems apparent that most boundary layer schemes have been thoroughly tested against only a handful of data sets. Most of these data sets are from detailed boundary layer observational studies that focused on areas of relatively flat and well-behaved terrain regions with consistent vegetation. While these locales offered good test grounds for examining boundary layer evolution in the best of circumstances, the surface of the Earth is often far from these conditions. Mountains, streams, rolling hills, and vegetation patchiness abound! When observations are taken in regions near complex terrain, the resulting behavior of the boundary layer does not always correspond well with our expectations. Schneider and Lilly (1999) indicate that a number of common simplifying parameterizations are not appropriate for the boundary layer behavior they observed (Fig. 5.25). In addition, boundary layer roll circulations can develop in clear air and influence boundary layer development. The effects of these rolls are not included in most boundary layer schemes.

Few boundary layer schemes have been rigorously evaluated in complex terrain or under highly variable surface conditions, so what we are doing is applying schemes that have been tuned to a handful of data sets (often in pristine surface and environmental conditions) to all conditions across the globe. Amazingly, the success of this application can be seen daily in the consistent utility of numerical model weather forecasts, suggesting strongly that some of the signals must be correct. But this does not mean that these forecasts of boundary layer development should be expected to be correct, or nearly correct, in all circumstances. The boundary layer certainly is not

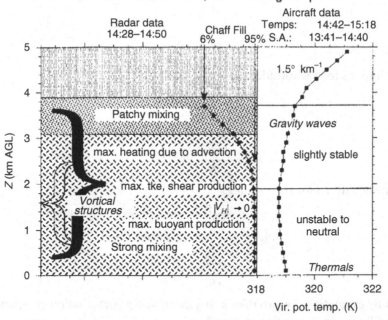

Figure 5.25. A structure schematic of the daytime boundary layer characteristics observed near Boulder, Colorado, during the Phoenix II experiment. Strong mixing occurs near the ground surface, with vortical structures in the middle of the boundary layer, and patchy mixing from gravity waves and shear turbulence production near the top of the boundary layer. From Schneider and Lilly (1999).

observed everywhere, nor do we really know what to expect regarding boundary layer evolution over some regions. There is no guarantee that the predictions from boundary layer schemes will always be successful. Furthermore, none of the schemes summarized previously allows for the direct influence of clouds on boundary layer development, except when saturation occurs on the grid scale. Instead, shallow convective parameterizations and cloud cover parameterizations have been developed to account for the effects of clouds within boundary layers (see Chapters 6 and 9). As this discussion suggests, there are many unsolved issues surrounding boundary layer parameterization that deserve attention.

5.8 Questions

1. Following the outline provided by the section on Reynolds averaging and turbulence closure, derive the equations for the perturbation velocity and potential temperature

$$\frac{\partial u_i'}{\partial t} = \cdots$$

$$\frac{\partial \theta'}{\partial t} = \cdots$$

the equations for the turbulent fluxes

$$\frac{\partial (\overline{u_i' u_j'})}{\partial t} = \cdots$$

$$\frac{\partial (\overline{u_i' \theta'})}{\partial t} = \cdots$$

and the equations for the variances

$$\frac{\partial \overline{u_i'^2}}{\partial t} = \cdots$$

$$\frac{\partial \overline{\theta'^2}}{\partial t} = \cdots$$

Begin by expanding the variables into a mean and a perturbation component and use the Boussinesq equation (5.3) and

$$\frac{\partial \theta}{\partial t} + u_j \frac{\partial \theta}{\partial x_j} = \nu \frac{\partial^2 \theta}{\partial x_j^2}.$$

Show all work.

2. Using knowledge of turbulence closure, discuss the pros and cons to going to higher-order closure schemes (i.e., third-order closure).

3. One set of equations for a mixed layer model is

$$\frac{d\bar{\theta}}{dt} = \frac{(1 + k_e) C_T V_S (\theta_S - \bar{\theta})}{H},$$

$$\frac{dH}{dt} = k_e C_T V_S \frac{(\theta_S - \bar{\theta})}{\Delta\theta},$$

$$\frac{d\bar{q}}{dt} = C_T V_S \frac{M(q_S - \bar{q}) + [k_e(\theta_S - \bar{\theta})\Delta q / \Delta\theta]}{H},$$

where M denotes the moisture availability, C_T is a transfer coefficient, and V_S is the surface wind speed. Let us now examine the consequences of changing both the entrainment parameter k_e and the moisture availability. Assume the following: $C_T = 0.015$; $V_S = 10 \, \text{m s}^{-1}$; $M = 0.5$; $\theta_S = 311 \, \text{K} + 10\Delta t \, (3 \, \text{h})^{-1}$ (the potential temperature increases linearly with time to 321 K at 3 hours and 331 K at 6 hours); and $q_S = 17 \, \text{g kg}^{-1} - 2.5\Delta t \, (3 \, \text{hours})^{-1}$ (the mixing ratio decreases by 2.5 g kg^{-1} over 3 hours). Also assume that the initial environmental potential temperature and mixing ratio profiles look like

$\theta = 310\,\text{K} + 5\,\text{K} \times z\,(1000\,\text{m})^{-1}$

$q = 11\,\text{g}\,\text{kg}^{-1}$ at and below $z = 1000\,\text{m}$

$q = 3\,\text{g}\,\text{kg}^{-1}$ above $z = 1000\,\text{m}$,

where $H = 30\,\text{m}$ at time zero (if H is zero at time zero, then the model blows up). Develop a finite-difference version of these mixed layer equations and integrate the equations for 3 and 6 hours using three different values of k_e: 0.1, 0.3, and 0.5. Then answer the following questions. What are the final values of θ and q in the mixed layer? What are the final values of the mixed layer depth H? What do these results say about the importance of choosing an appropriate value for the entrainment coefficient when using a PBL model? Do these differences increase or decrease over longer time periods?

One simple integration scheme is to define $x'_{i,n} = f(x_{i,n})$, where i denotes the variable ($i = 1, 2, 3$) and n is the time level, and then integrate forward one time step to $n+1$ using $x_{i,n+1} = 0.5\left[x_{i,n} + \tilde{x}_{i,n+2}\right]$, where one first determines $\tilde{x}_{i,n+1} = x_{i,n} + f(x_{i,n}) \times \Delta t$ and then determines $\tilde{x}_{i,n+2} = \tilde{x}_{i,n+1} + f(\tilde{x}_{i,n+1}) \times \Delta t$. This simple time integration scheme is stable for the mixed layer model, but requires a time step of about 1 s.

4. Using the program developed in Question 3, assume that the land surface scheme has a bias of $+1\,\text{K}$ in the temperature forecast over a 3 h period. Thus, $\theta_S = 311\,\text{K} + 11\Delta t / (3\ \text{hours})$. Rerun the mixed layer scheme with this surface temperature evolution and with $k_e = 0.3$. What are the values of the mixed layer depth H at 3 and 6 hours? How do these changes compare with changes in the entrainment coefficient?

5. Using the same initial potential temperature and mixing ratio profiles as in Question 3, and with the same time rate of change of the surface potential temperature and the mixing ratio, develop a simple finite-difference model of the lowest 10 000 m of the atmosphere based only upon K-theory mixing. Use

$$\frac{\partial \bar{\theta}(z)}{\partial t} = K_H \frac{\partial^2 \bar{\theta}(z)}{\partial z^2}$$

and

$$\frac{\partial \bar{q}(z)}{\partial t} = K_H \frac{\partial^2 \bar{q}(z)}{\partial z^2},$$

and set up a vertical grid starting at the surface ($z = 0$ so that $\theta_s = \bar{\theta}(0)$) with 50 m vertical increments. Here $\bar{\theta}$ is the mean potential temperature of a given layer and it is allowed to vary with height. Assume that K_H is a constant. The time evolution of potential temperature and the mixing ratio at $z = 0$ is then specified. Try the finite-difference approximation

$$F_{m,n+1} = F_{m,n} + \frac{K_H \Delta t}{(\Delta x)^2} \left(F_{m+1,n} + F_{m-1,n} - 2F_{m,n}\right),$$

where m is the vertical grid level and n is the time level. Try time steps of 1 s. Integrate the model out to 6 h and examine the potential temperature and mixing ratio profiles. Try using several values of K_H between 10 and 200. What behaviors are seen? Is this result realistic?

6. Obtain an observed sounding from near local noon or late afternoon (00:00 UTC in the USA is fine) and a corresponding 2 m temperature and mixing ratio. Determine the positive and negative areas of this sounding, as defined from the penetrative convection non-local closure scheme, and the boundary layer top. Using the diffusion scheme, calculate the surface value of $\theta_v + \theta_T$ and from this value also the boundary layer top. Finally, using the observed wind and thermodynamic profiles, estimate values for K_H within this boundary layer using mixing-length theory as in (5.59) and (5.61). How does the implied mixing from the local closure scheme compare to the implied mixing from the non-local closure schemes? Explain.

7. Choose a single observed sounding location site and compare the observed soundings against model forecast soundings at all available observation times for a 7 day period. Construct a table comparing the boundary layer depth, the mean boundary layer potential temperature, and the mean boundary layer mixing ratio from both the model and the observational data. Separate the data based upon the observation time. Summarize the results.

6

Convective parameterizations

6.1 Introduction

Moist convection comes in many shapes and sizes. For some people, memories of severe thunderstorms with strong and gusty winds, heavy rainfall and lightning probably come to mind when thinking about convection. Sometimes these thunderstorms merge over time and form organized lines of deep convective storms with trailing stratiform rainfall regions. However, others may remember stratocumulus clouds that often form near coastlines when thinking of convection, or the clouds and attendant rainfall that often occur with the passage of a cold front. Of course, all of these examples describe moist convection – widely varying in shape and size, but similar in that the cloud motions generally are vertical and driven by buoyancy forces.

Moist convection is important to the prediction of the atmospheric circulation for a number of reasons. Large-scale horizontal gradients of latent heating produced by deep moist convection help to drive vertical circulations that are large scale, such as the Hadley and Walker cells. Deep convection also is a key component in the El Niño Southern Oscillation (ENSO), which can strongly influence seasonal climate in the northern hemisphere. During ENSO events, the sea surface temperatures in the tropical eastern Pacific are warmer than normal. Deep convection develops in association with this warm surface temperature anomaly, releasing latent heating in a deep atmospheric column and producing upper-level divergence. The upper-level divergence produced by the convection then helps to excite a train of Rossby waves that alter the hemispheric flow patterns (Tribbia 1991). In contrast to vertically deep convective clouds, shallow cumulus clouds are the most frequently observed tropical cloud (Johnson *et al.* 1999). Shallow convection modifies the surface radiation budget, influences the structure and turbulence of the planetary boundary layer and thereby also influences global climate (Randall *et al.* 1985).

In the midlatitudes, organized regions of deep convection also can modify the large-scale circulation patterns over continental-sized regions for several days (Stensrud 1996; Stensrud and Anderson 2001). On smaller scales, upper-level meso-α-scale anticyclones and mid-level convectively generated vortices are associated with mesoscale convective systems (MCSs) in the midlatitudes (Ninomiya 1971a, b; Maddox 1980; Fritsch and Maddox 1981; Bartels and Maddox 1991; Trier *et al.* 2000). Some of the mid-level vortices persist for days and influence the development of new convection. Organized convective systems are observed across much of the globe (Laing and Fritsch 1997) and are known to enhance upper-level jet streaks (Keyser and Johnson 1984; Wolf and Johnson 1995a, b) and modify the large-scale environment to be more favorable for cyclogenesis (Zhang and Harvey 1995).

Shallow convection also is common outside of the tropics, occurring in isolation and across broad regions. It is found in association with midlatitude cyclones and subtropical anticyclones and across broad areas in the Arctic (Houze 1993). When cold air moves over warm water, as often occurs with the passage of frontal boundaries during the cold season, shallow cumulus clouds develop over the water and commonly align themselves in well-defined bands or streets (Kuettner 1959). The importance of convection to both local and global processes suggests that without a reasonable representation of all types of moist convection in numerical models it is impossible to predict either the small- or the large-scale atmospheric circulations correctly.

In addition to its role in modifying the local and large-scale circulation patterns, moist convection is a key component of the hydrological cycle. Large quantities of water are on the move in the climate system (Peixoto and Oort 1992) and precipitation is what links the atmospheric and terrestrial components of the hydrological cycle. Moist convection also is a key process in regulating the water vapor in the atmosphere, which provides the largest feedback for climate change. Arguably the most important component of the weather and climate system is precipitation, which provides the water to grow plants, sustain human beings and other animals, feed lakes and rivers, and build glaciers and snow packs. Precipitation from deep convection also is associated with strong and damaging winds, hail, tornadoes, lightning, and flash floods. Thus, the effects of latent heat release, the transport of heat, moisture, and momentum, and the precipitation that deep moist convection produces need to be accounted for within numerical weather prediction models.

As suggested above, it is convenient to divide the various forms of moist convection into just two major categories: deep convection and shallow convection. As the category names aptly imply, deep convection refers to convective elements that vertically span much of the troposphere, while shallow convection refers to convective elements that vertically span only a small portion of the troposphere. Owing to its smaller depth and weaker vertical motions, shallow convection sometimes is assumed to not produce precipitation. When categorized in this simple yet efficient approach, moist convection alters the environment in two very different ways. Deep convection, associated with strong updrafts and precipitation, acts to warm and dry the environment as gravity waves initiated by the convection warm the environment via subsidence and precipitation removes water vapor from the atmosphere. Conversely, non-precipitating shallow convection produces no net warming or drying because water is not removed from the atmosphere. Instead, vertical dipole effects occur as the convection acts to cool and moisten the upper half of the cloud layer, as cloud water detrains from the cloud into the environment and evaporates, and to warm and dry the lower half of the cloud layer, as moisture condenses to produce cloud water and releases latent heat.

A difficulty arises in that deep convection can be further subdivided into convective and stratiform components (Houze 1997; see Fig. 6.1). The convective component refers to convection associated with individual cells, horizontally small regions of more intense updrafts and downdrafts associated with young, active convection. When sampled with a radar, cells appear as vertically oriented concentrated peaked regions of high reflectivity. The stratiform component refers to convection associated with older, less active convection with vertical motions generally less than $1 \, \mathrm{m \, s^{-1}}$. Radar echoes from

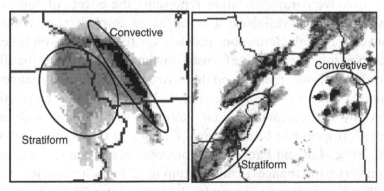

Figure 6.1. Radar depiction of a mesoscale convective system with both convective and stratiform precipitation regions (left) and isolated convective cells and stratiform rain along a frontal boundary (right). The convective region is associated with stronger updrafts, higher reflectivities, and more vertically oriented cells. The stratiform region is associated with weak updrafts, lower reflectivities, and sometimes is composed of remnants from previous active cells from the convective region.

stratiform convection have weak horizontal gradients and/or a bright band (Houze 1997). The convective and stratiform components are often related to each other, such as when intense convective cells organize and develop an adjacent stratiform region. However, the convective and stratiform components also can be distinct, such as with isolated convective cells and stratiform regions associated with baroclinic cyclones and frontal boundaries.

Since moist convection comprises very small-scale mixtures of updrafts and downdrafts, it is computationally impossible to represent these processes directly on the grids of most numerical weather prediction models that are used operationally today. Indeed, the grid spacing needed to resolve individual convective elements is likely to be between 25 and 1000 m, such that the computational requirements to run true convection resolving forecasts still lie a few years in the future. However, there are those who believe that convective parameterization is near the end of its useful lifetime, since explicit bulk representations of the microphysical variables on the model grids have been developed and are found to produce reasonable simulations of many convective events when the grid spacing is small. This is an arguable point, but it certainly is not true for global models that produce extended range forecasts or that produce seasonal or climate predictions and simulations. For seasonal or climate predictions, decades may pass before computers become fast enough to explicitly resolve convective motions. Thus, convective parameterization, which has been a staple of numerical weather prediction models for the past 40 years, will continue to be a necessary component of many numerical models for years to come.

Since convective parameterization represents the effects of subgrid scale processes on the grid variables, it is called an implicit parameterization. Some convective parameterizations account for the effects of both deep and shallow convection, but many are focused only upon reproducing the effects of either deep or shallow convection on the environment. Only the changes to the environmental temperature and the specific humidity due to convection are determined by many schemes, even though several studies indicate that momentum transport can be important (Zhang and Cho 1991; Grubisic and Moncrieff 2000; Han and Pan 2006). However, there are a few schemes that also parameterize the changes in momentum (e.g., Fritsch and Chappell 1980; Donner 1993; Kershaw and Gregory 1997; Han and Pan 2006). Passive tracers and chemistry also can be included and offer new observational opportunities.

One common feature of convective parameterization schemes is that they evaluate the convective available potential energy (CAPE) of the environment – the maximum energy available to an ascending parcel as determined from simple parcel theory (see Emanuel 1994). The CAPE is calculated by taking an imaginary parcel representative of the environment at a given height and lifting it adiabatically from its starting height to its level of neutral buoyancy, or equilibrium level (EL), while holding the environment constant and allowing no mixing between the parcel and the environment. Assuming that the parcel initially is unsaturated and the environment is conditionally unstable, the lifting at first is dry adiabatic and the parcel is negatively buoyant (the parcel density is greater than that of its environment). If the parcel becomes saturated as lifting continues, it thereafter rises moist-adiabatically and may at some height reach its level of free convection (LFC) above which the parcel is positively buoyant (the parcel density is less than that of its environment) and can rise freely. As the parcel continues to rise it eventually reaches the EL where the parcel and environmental temperatures are again equal. Above the EL the parcel is negatively buoyant. Thus, CAPE can be defined as

$$\text{CAPE} \equiv g \int_{\text{LFC}}^{\text{EL}} \frac{\theta(z) - \bar{\theta}(z)}{\bar{\theta}(z)} \, dz, \tag{6.1}$$

where g is the acceleration due to gravity, θ is the potential temperature of a parcel rising adiabatically from its starting point, $\bar{\theta}$ is the environmental potential temperature, and z is the height. In between the LFC and the EL the work done by the buoyancy force acting on the ascending parcel can be converted to parcel kinetic energy.

The importance of CAPE can be explained simply by recognizing that deep convection cannot occur unless CAPE > 0. Although CAPE can have both positive and negative values, only positive values are important for determining if convection is possible. While positive CAPE must be present for deep convection, CAPE can be created and stored for long time periods. On a thermodynamic diagram, such as a skew-T or tephigram, CAPE is the positive area of a sounding. CAPE values can exceed $6000 \, \text{J} \, \text{kg}^{-1}$ in severe storm environments.

A second term common in discussions of convection and its parameterization is convective inhibition (CIN). The CIN of a parcel is defined as the energy needed to lift the parcel vertically and pseudoadiabatically from its starting level to its LFC. For a parcel with positive CAPE, the CIN is represented by the negative area on a thermodynamic diagram and is calculated as

$$\text{CIN} \equiv -g \int_{\text{SL}}^{\text{LFC}} \frac{\theta(z) - \bar{\theta}(z)}{\bar{\theta}(z)} \, dz, \tag{6.2}$$

where SL is the parcel starting level used to calculate the CAPE. The minus sign is used so that CIN > 0 for parcels that require some lift to reach their

LFC. Even in environments with large CAPE, convection may not develop if the values of CIN are large enough, since the energy needed to lift parcels to their LFC in order to initiate convection is too large. Thus, both CAPE and CIN are important environmental parameters when considering deep convection.

Convective parameterization can be conceptualized in many different ways and the schemes separated into a number of basic types. Here the literature is challenging and many points of dispute remain, as is true in most active and vigorous areas of research. One scheme may make assumptions about how a unit of convection influences the environment, while another may make assumptions about the amount of convection allowed in a given environment. Most schemes determine a single end result of the convective process, while others take a stochastic approach to determine how convection influences the environment. The closure assumptions used in convective parameterization also could be subdivided into those that place constraints on large-scale states, moist-convective processes, or both as done by Arakawa (1993). It is important to recognize that the framework through which the process of convection is viewed influences how the convective parameterization scheme operates and behaves.

One of the ways to distinguish between the types of closures used in convective parameterization schemes is outlined by Mapes (1997). He separates convective schemes into two general types based upon the vertical extent of the atmospheric forcing that controls the convection. Thus, convective parameterization schemes are grouped into deep-layer control schemes and low-level control schemes. Deep-layer control schemes tie the creation of CAPE by large-scale processes to the development of convection. Mapes (1997) suggests that these schemes could be termed "supply-side" approaches, as it often is assumed that convection consumes the CAPE that is created. Low-level control schemes tie the development of convection to the initiation processes by which CIN is removed. In these schemes CAPE can be generated and stored for long periods before it is consumed by the scheme. Many schemes incorporate elements of both deep-layer and low-level controls.

Another way to view convective parameterization is based upon how the environmental changes due to convection are defined. A static scheme determines the final environmental state after convection is done working its magic and adjusts the model fields toward this final state. It is not burdened with the details that produce this state, and the only remaining question is how long it takes the atmosphere to reach this final state, i.e. the adjustment time period. The final state often is one that is neutral to convective overturning. In contrast, a dynamic scheme assumes that the myriad physical processes involved in convection are important and should influence how the scheme functions. Some of these schemes use entraining plumes to approximate the effects of convection and calculate the transfers of mass in updrafts and downdrafts from one vertical level to the next (mass flux schemes). The assumptions used in dynamic schemes lead to them being very sensitive to the local environment in which convection develops, in stark contrast to static schemes. However, to account for all the physical processes involved in convection using simple methods, dynamic schemes use a number of parameters whose values are poorly known yet can influence the final environmental state.

Convection also can be viewed conceptually as being driven by the buoyancy realized by parcels being lifted past their LFC and rising freely into the local environment. From this instablity viewpoint, local buoyancy is the key variable needed to determine the convective response, with the scheme acting to remove the local buoyancy. Simple plume models often are used to evaluate the amount of buoyancy available to a parcel and are crucial components of many convective parameterization schemes. Yet it can be argued that moisture also is a key variable for convection, since convective parameterization is simply a method to account for the effects of subgrid-scale saturation. From this perspective, the moisture content should drive the behavior of the convective scheme, controlling the amount of convection produced in an unstable environment based upon the available moisture that can be removed from the atmosphere. While these different conceptual perspectives may not seem important, the environmental states created after activating an instability-control and a moisture-control convective scheme starting from the same initial environmental state could be quite different.

The closure assumptions used within a scheme sometimes define where and when convection is activated in the model. However, in other schemes the closure assumptions only determine the amount or intensity of the convection and a separate set of criteria are used to determine where and when convection develops. These criteria that determine convective development are called "trigger functions." Every scheme can be said to have a trigger function, whether it is part of the closure assumption or a separate entity. Trigger functions are very important to convective parameterization, as they determine how convection evolves over time.

As a further complication, the distinction between implicit and explicit parameterization of cloud processes is blurred in many numerical models. Many models run at 10–40 km grid spacing use both a convective parameterization scheme and an explicit cloud microphysical parameterization. This is because some convective parameterization schemes are designed to activate deep convection prior to the occurrence of water vapor saturation at a grid point, but then the resolvable-scale model equations are assumed to develop the appropriate mesoscale mass, heat, moisture, and momentum transports of any larger than convective-scale precipitation processes, such as the stratiform regions of mesoscale convective systems or precipitation within frontal zones. In this way, the implicit and explicit representations of convection act in concert to produce the resulting convective activity. For some combinations of implicit and explicit schemes, the convective parameterization ends up mimicking the leading line of deep convection in a mesoscale convective system, with the explicit microphysics acting to help support the leading line but also representing the trailing stratiform precipitation region. Hybrid schemes have been developed that deposit a fraction of the parameterized precipitation from the convective scheme directly to the model grid as inputs into the microphysical variables (Frank and Cohen 1987; Molinari and Dudek 1992; Molinari 1993).

Many demands are placed upon convective parameterization schemes. The public pays the most attention to precipitation forecasts, since this often has the most dramatic influence on their plans for a given day. However, precipitation arguably is the most difficult model output to predict correctly, since it depends upon predicting all the other forecast variables very accurately. Thus, one of the biggest hopes of our ever-changing forecast models is that precipitation forecasts will become better over time. There is a strong desire for convective parameterizations to develop convection at the right place and time, evolve it correctly, and then produce the correct changes to the larger-scale environment so that any subsequent convection also will be forecast correctly. This is a pretty demanding set of requirements, and one that may never be met. Yet it is the goal that all the creators of convective schemes attempt to meet. Indications are that when the large-scale forcing for upward motion is strong, the convective schemes do a better job than when the large-scale forcing for upward motion is weak. Reasons for this behavior are discussed later, but one always needs to realize that any given parameterization scheme is just part of the whole. Errors in precipitation forecasts are due to more than just errors in the convective scheme. The links between the various physical parameterization schemes used in any model are large and reducing the errors attributed to one scheme often requires improvements across the board to all facets of the model.

We begin by examining the heat and moisture budgets over various regions from several experiments in order to illustrate the influences that convection has on the environment. This is followed by a discussion of the deep-layer control schemes, which were developed first, and then the low-level control schemes. For those who are interested in more of the details, a good overview is provided in Emanuel and Raymond (1993). Many other important and interesting issues involving convective parameterization are discussed by Mapes (1997).

6.2 Influences of deep convection on the environment

Deep convective clouds have sizes much smaller than can be sampled by most observational networks, and so the effects of deep convection on the environmental temperature and moisture fields cannot be measured directly. Using such observations, the effects of deep convection are inferred as a residual from large-scale heat and moisture budgets. These budgets typically are calculated using special observational data taken during field experiments, in which rawinsonde observations are taken at least every 6 h. As part of this process, it often is convenient to capture the entire region of convective activity within the experimental domain. Owing to the generally stronger tropospheric wind speeds and more rapid motion of convection in midlatitudes, many of the field experiments that examine the heat and moisture budgets were conducted in the tropical regions (Fig. 6.2). Alternatively, Mapes and Lin (2005) use single Doppler radar data to derive the wind divergence profiles associated with convection that can be related to the latent heating produced by convection. However, the following discussion draws heavily upon the earlier studies of the larger-scale heat and moisture budget by Reed and Recker (1971), Yanai *et al.* (1973), Johnson and Young (1983), Johnson (1984), and Yanai and Johnson (1993).

We begin by defining the dry (s) and moist (h) static energy as

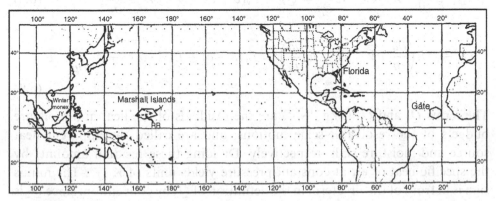

Figure 6.2. Locations of observational networks used in calculating the large-scale heat and moisture budgets. From Johnson (1984).

$$s = c_p T + gz, \tag{6.3}$$

$$h = c_p T + gz + L_v q, \tag{6.4}$$

where c_p is the specific heat of air at constant pressure, T is temperature (K), z is height, L_v is the latent heat of vaporization, and q is the mixing ratio of water vapor. Dry static energy is conserved for dry adiabatic processes, while moist static energy is conserved for both dry and moist adiabatic processes for non-precipitating systems. All quantities are averaged horizontally over an observing area. The equations for mass continuity, heat energy, and moisture continuity are then

$$\overline{\nabla \cdot \vec{V}} + \frac{\partial \overline{\omega}}{\partial p} = 0, \tag{6.5}$$

$$\frac{\partial \overline{s}}{\partial t} + \overline{\nabla \cdot s\vec{V}} + \frac{\overline{\partial s\omega}}{\partial p} = Q_R + L_v(\overline{c} - \overline{e}), \tag{6.6}$$

$$\frac{\partial \overline{q}}{\partial t} + \overline{\nabla \cdot q\vec{V}} + \frac{\overline{\partial q\omega}}{\partial p} = \overline{e} - \overline{c}. \tag{6.7}$$

Here, ω is the vertical motion in pressure coordinates, p is pressure, Q_R is the heating rate due to radiation, $\overline{c}$ is the rate of condensation per unit mass, and $\overline{e}$ is the rate of evaporation per unit mass.

The variables are expanded into terms representing the horizontal mean and perturbations from the mean (e.g., $\omega = \overline{\omega} + \omega'$), and the above equations, (6.6) and (6.7), are Reynolds averaged (see Chapter 5) to obtain

$$Q_1 \equiv \frac{\partial \overline{s}}{\partial t} + \overline{\vec{V}} \cdot \nabla \overline{s} + \overline{\omega} \frac{\partial \overline{s}}{\partial p} = Q_R + L_v(\overline{c} - \overline{e}) - \nabla \cdot \overline{s'\vec{V}'} - \frac{\partial}{\partial p}(\overline{s'\omega'}), \tag{6.8}$$

$$Q_2 \equiv -L\left(\frac{\partial \overline{q}}{\partial t} + \overline{\vec{V}} \cdot \nabla \overline{q} + \overline{\omega} \frac{\partial \overline{q}}{\partial p}\right) = L_v(\overline{c} - \overline{e}) + L_v\nabla \cdot \overline{q'\vec{V}'} + L_v\frac{\partial}{\partial p}(\overline{q'\omega'}). \tag{6.9}$$

The terms Q_1 and Q_2 ($\mathrm{J\,kg^{-1}\,s^{-1}}$) are the residuals of the heat and moisture budgets of the resolvable-scale motions, and are called the apparent heat source and moisture sink, respectively. Dividing these residual amounts by the specific heat of air at constant pressure yields units of $\mathrm{K\,s^{-1}}$. The minus sign is in front of the first term in the definition of Q_2 by convention, since Q_2 is then positive when drying occurs (or when condensation is greater than evaporation). Both equations have terms that are the product of perturbations, representing the effects of subgrid-scale motions on the large-scale budgets.

If the horizontal transports by eddies are ignored (a common assumption), then one finds that

$$Q_1 - Q_2 - Q_R = -\frac{\partial}{\partial p}(\overline{h'\omega'}).$$ (6.10)

If there are no subgrid-scale transports, then $Q_1 - Q_2 - Q_R = 0$. Conversely, if $Q_1 - Q_2 - Q_R \neq 0$, then the subgrid-scale motions associated with convection influence the large-scale environment. It also is possible for $Q_1 - Q_2 - Q_R = 0$ when the profiles of Q_1 and Q_2 are approximately equal, which may be an indicator of stratiform rainfall (Luo and Yanai 1984; Gallus and Johnson 1991).

This budget approach has been widely used to measure the activity of cumulus convection, but it is important to be able to assess how well it works. By defining

$$\langle \; \rangle \equiv \frac{1}{g} \int_{p_{tropopause}}^{p_{sfc}} (\;) \, dp,$$ (6.11)

and assuming that the horizontal eddy flux terms still can be ignored, then one can integrate Q_1 and Q_2 over the depth of the troposphere from the surface (*sfc*) to the tropopause to yield

$$\langle Q_1 \rangle = \langle Q_R \rangle + L_v \langle \overline{c} - \overline{e} \rangle - \frac{1}{g}(\overline{s'\omega'})_{sfc}$$

$$\approx \langle Q_R \rangle + L_v P + \rho_s c_p (\overline{w'T'})_{sfc}$$

$$= \langle Q_R \rangle + L_v P + Q_H,$$ (6.12)

$$\langle Q_2 \rangle = L_v \langle \overline{c} - \overline{e} \rangle + \frac{L_v}{g}(\overline{q'\omega'})_{sfc}$$

$$\approx L_v P - \rho_s L_v (\overline{q'w'})_{sfc}$$

$$= L_v (P - E),$$ (6.13)

where P is the precipitation rate per unit area at the surface, Q_H is the sensible heat flux per unit area at the surface, E is the evaporation rate per unit area at the surface ($L_v E = Q_E$), and zero fluxes are assumed at the tropopause level. A simple subtraction yields

$$\langle Q_1 \rangle - \langle Q_2 \rangle = \langle Q_R \rangle + Q_H + Q_E, \tag{6.14}$$

which allows one to examine the accuracy of the estimates of Q_1 and Q_2 using estimates of the radiative flux cooling profile and the surface heat and moisture fluxes. Results suggest that the estimates of Q_1 and Q_2 are more accurate at larger scales (see Mapes *et al.* 2003). The neglect of the horizontal eddy transports becomes more problematic as the perimeter of the observing area is sampled more sparsely, such that eddies of intermediate scales may transport anomalies of temperature and moisture between soundings unsampled. When this situation occurs, the interpretation of the results is much more difficult.

Typical time-averaged vertical profiles of Q_1 and Q_2 show that the apparent heat source has a maximum at lower pressures than the apparent moisture sink. For example, using data from the Marshall Islands, Yanai *et al.* (1973) show that the average maximum of Q_1 is located in the upper troposphere near 475 hPa, whereas the average maximum of Q_2 is located in the lower tropo-sphere near 775 hPa (Fig. 6.3). This difference suggests via (6.10) the presence of an eddy vertical flux of moist static energy (h) from cumulus convection. The Q_1 profile suggests that as deep convection removes CAPE and establishes vertical temperature profiles that are close to being moist adiabatic, the maximum heating occurs in the mid-troposphere above layers with the highest lapse rates (or in the layers where parcel buoyancy is maximized).

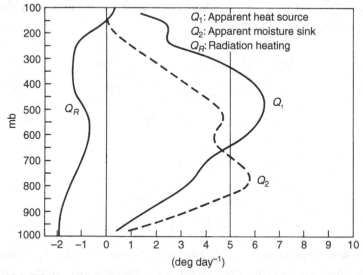

Figure 6.3. The vertical profiles of Q_1 (solid, the mean apparent heat source) and Q_2 (dashed, the mean apparent moisture sink) calculated over the Marshall Islands (from Yanai *et al.* 1973). The vertical profile of Q_R, the radiational heating rate, is on the left as given by Dopplick (1972).

The Q_2 profile has two maximums, one at 775 hPa and one at 525 hPa, that are more difficult to explain. If one assumes that $Q_2 \propto -\overline{\omega}(\partial\overline{q}/\partial p)$ from (6.9) and that $\overline{\omega}$ has a simple D-shaped vertical profile, then the surprising feature of the Q_2 profile is the layer in between the two maximum values. Since $\overline{\omega}$ likely is decreasing within this layer, the decrease in Q_2 implies that the vertical gradient of $\overline{q}$ must be smaller within this layer than elsewhere and reaches a minimum near 600 hPa. This is supportive of enhanced detrainment from cumulus congestus and cumulonimbus clouds near the 0 °C stable layer, leading to a moist layer at this level as discussed by Johnson *et al.* (1999).

Since the 1980s it has been recognized that tropical and midlatitude precipitation systems tend to organize in a very similar fashion. These MCSs often are a combination of a leading line of deep convective cells, followed by a region of stratiform precipitation that is clearly organized on the mesoscale. The total rainfall produced by the stratiform region is quite large, and can represent over 40% of the system total (Cheng and Houze 1979). The heat and moisture budgets calculated using the techniques described above, in which the convective system is entirely contained within the observation network, provide information on the total influence of convection on the large-scale environment. However, while some convective parameterizations appear to account for this total amount, a few schemes are designed to only account for the influence of the most intense convective portion of the precipitation system. For these schemes, in which it is assumed that the explicit microphysical parameterizations develop the needed heating and moistening profiles within the stratiform precipitation region, another type of analysis is needed.

The MCSs observed during the Winter Monsoon Experiment (Winter MONEX) have the characteristics of typical MCSs observed in both the tropics and midlatitudes, containing both a leading convective line and a trailing stratiform precipitation region (Johnson and Young 1983). These MCSs are seen to undergo a very typical cycle, in which the systems initiate with clusters of cumulonimbus clouds, develop upscale into a mesoscale, stratiform cloud system with some cumulonimbus clouds remaining on the peripheries, and then dissipate. Fortuitously, on seven days in a single month, reports indicate that the MCSs develop, mature, and decay at nearly the same times each day, such that the 1400 UTC observations all correspond to the mature stage of the system when the stratiform cloud system is the dominant feature. Thus, with the benefit of radar data and 6 hourly rawinsonde data launched from a triangular array of research ships, the possibility exists to tease out the contributions of the stratiform rain region to the total heating source and moisture sink. Results similar to those presented below are also found in Mapes and Lin (2005) using single Doppler radar data.

The same Q_1 and Q_2 equations as derived previously, (6.8) and (6.9), are used by Johnson and Young (1983) to calculate the individual terms involving storage $(\partial/\partial t)$, horizontal advection $(\vec{V} \cdot \nabla)$ and vertical advection $(\overline{\omega}\partial/\partial p)$ of $\bar{s}$ and $\bar{q}$. Interpolations in the vertical direction and in time are needed, and some data gaps need to be filled. Results indicate that the vertical advection term dominates in the heat and moisture budgets, as seen in many other studies, so that errors in the storage and horizontal advection terms are not important. While errors can arise owing to the calculation of vertical velocity, these errors are not sensitive to the methods used.

The results of Johnson and Young (1983) are partitioned by Johnson (1984) into the relative contributions of the mesoscale stratiform and convective regions. Johnson assumes that a fraction F of the total accumulated rainfall (R_0) is ascribed to the stratiform portion of the MCS. If the fraction of the total area A covered by the mesoscale stratiform region is $\sigma_m \equiv A_m/A$, and the fraction of A covered by cumulus clouds is $\sigma_c \equiv A_c/A$, then one can define the area-averaged precipitation rate (P_0) as

$$P_0 = \sigma_m P_m + \sigma_c P_c, \tag{6.15}$$

where P_m and P_c are the average mesoscale and cumulus precipitation rates, respectively. The total accumulated rainfall amounts $(R_m$ and $R_c)$ over a time period Δt can then be defined as

$$R_m = A_m P_m \Delta t, \tag{6.16}$$

$$R_c = A_c P_c \Delta t. \tag{6.17}$$

From the definition of $F = R_m/R_0$ one can define the following relationships:

$$F = \sigma_m \frac{P_m}{P_0} = 1 - \sigma_c \frac{P_c}{P_0}, \tag{6.18}$$

$$\sigma_m = F\frac{P_0}{P_m}, \tag{6.19}$$

$$\sigma_c = (1 - F)\frac{P_0}{P_c}. \tag{6.20}$$

The apparent heat source Q_1 can then be written as

$$Q_1 = \sigma_m Q_{1m} + \sigma_c Q_{1c} + (1 - \sigma_m - \sigma_c)Q_{1e}, \tag{6.21}$$

where Q_{1m} is the heating in the mesoscale stratiform cloud region, Q_{1c} is the heating in the cumulus cloud region (both shallow and deep clouds), and Q_{1e} is the heating in the environment of the convective system. Q_{1e} is assumed to be the radiative heating rate of the environment, or $Q_{1e} = Q_{Re}$.

If the apparent heat source Q_1 is normalized by the total precipitation P_0, then a good approximation for the apparent heat source is

$$\frac{Q_1}{P_0} \approx F\frac{Q_{1m}}{P_m} + (1 - F)\frac{Q_{1c}}{P_c} + \frac{Q_R}{P_0}. \tag{6.22}$$

Thus, with good information on the precipitation rates (amounts per day), the apparent heat source Q_1, the apparent heat source for the mesoscale stratiform region Q_{1m}, as derived by Johnson and Young (1983), and the radiative heating of the environment Q_{Re}, it is possible to choose a reasonable value for F and determine the apparent heat source for the cumulus region Q_{1c} as a residual. A similar expression and method of solution is also found for Q_2 in order to calculate Q_{2c}. These calculations also provide information on how much the stratiform regions of MCSs contribute to the total heating and moistening of the environment.

Results from Johnson (1984) indicate that the mesoscale stratiform region and the deep cumulus convective cells are both important components of the total heat source and moisture sink (Figs. 6.4 and 6.5). The single normalized $\hat{Q}_1 (= Q_1/P_0)$ peak near 500 hPa (Fig. 6.4) actually is a reflection of the contributions from two very different vertical heating profiles from two very different convective phenomena. The cumulus or convective-scale heating peaks in the lower troposphere near 600 hPa, whereas the mesoscale, stratiform heating peaks in the upper troposphere near 350 hPa. Positive heating below 500 hPa is due to the convective-scale heating being larger than the mesoscale cooling beneath the stratiform cloud region. This mesoscale cooling generally is ascribed to the evaporation of rain below the stratiform clouds. Although the value of F is assumed to be 0.2, the results are qualitatively not sensitive to this value or to the choice of Q_{Re}. In addition, the results of Johnson (1984) are very similar to those described by Houze (1982), giving greater confidence in the profiles shown here. Divergence profiles from Mapes and Lin (2005) also are supportive of these heating profile shapes.

The partitioning of total drying $\hat{Q}_2$ (Fig. 6.5) also indicates that both the mesoscale stratiform component and the convective-scale component have similar magnitudes. The two local maximums in the total $\hat{Q}_2$ vertical profile are seen to be from the widely separated peaks in drying of the two components. The convective-scale component produces strong drying in the lowest

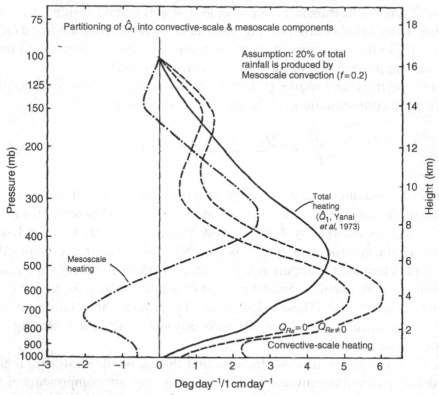

Figure 6.4. Partitioning of $\hat{Q}_1$, the normalized apparent heat source, into convective-scale and mesoscale stratiform components for $F = 0.2$. Curves for $Q_{Re} = 0$ and $Q_{Re} \neq 0$ are shown, where the non-zero radiative heating profile in the environment of the convective clouds is equivalent to that given by Cox and Griffith (1979). From Johnson (1984).

levels, with a peak near 800 hPa, while the mesoscale stratiform component produces a smaller maximum of drying around 350 hPa and strong moistening near 775 hPa due to evaporation of falling raindrops and cloud water.

These results indicate that the vertical heating distributions produced by the convective-scale and mesoscale stratiform components are quite different from each other and from the total heating distribution. It is often unclear or debatable whether the convective parameterization being used in a given model should produce the total heating profile, which by definition then includes both the convective-scale and mesoscale stratiform components, or only the convective-scale component. The answer also may depend upon the model grid spacing. If a scheme reproduces the total heating distribution, then the addition of explicit microphysical variables may produce unrealistic results if additional heating is provided by the phase changes of water as represented in these variables. In this case, it is important to examine the model output and

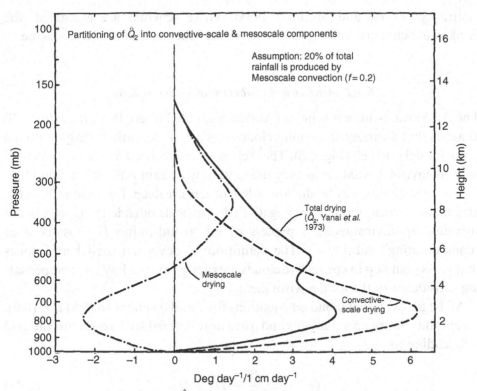

Figure 6.5. Partitioning of $\hat{Q}_2$, the normalized apparent moisture sink, into convective-scale and mesoscale stratiform components for $F = 0.2$. From Johnson (1984).

to see how and where the rainfall is divided between the convective parameterization and the resolvable-scale microphysics. Alternatively, one could examine the heating profiles directly to determine if they are top or bottom heavy, although these diagnostics may not be readily available.

6.3 Deep-layer control convective schemes

Now that we have seen how convection influences the thermodynamic properties of the large-scale environment, we can turn our attention to the various ways in which convective parameterization schemes have been created. The schemes summarized in this section are deep-layer control schemes. Often these schemes assume that the large-scale environment supplies CAPE and that convection is an efficient and speedy consumer of the CAPE produced. These types of schemes may be called equilibrium control, quasi-equilibrium, or CAPE adjustment schemes, since convection is assumed to maintain the instability of the large-scale environment in a state of equilibrium near neutrality (Frank and Molinari 1993). Three schemes are examined: the Arakawa–Schubert scheme, the Betts–Miller scheme, and the Kuo scheme.

6.3.1 *Arakawa–Schubert convective scheme*

The Arakawa–Schubert scheme (Arakawa and Schubert 1974; hereafter AS) assumes that a variety of cumulus cloud sizes are all potentially active within a given model grid cell (Fig. 6.6). These clouds are idealized as plumes which all entrain environmental air as they rise, and they detrain only at the cloud top. Some of the clouds may be shallow, while others are deep. Deviations from the grid-mean vertical motion are negative (downward) outside the clouds in the surrounding environment, a process called "cloud-induced" subsidence or "compensating" subsidence. The definition of deviation used here implies that mass going up in convective clouds is exactly balanced by this compensating subsidence in the local environment.

AS begin by deriving budget equations for $\bar{s}$ and $\bar{q}$ similar to (6.8) and (6.9), except that the mean values of s and q are now defined with respect to the grid cell, leading to

$$Q_{1c} = M_c \frac{\partial \bar{s}}{\partial z} + D_S + \bar{Q}_R, \qquad (6.23)$$

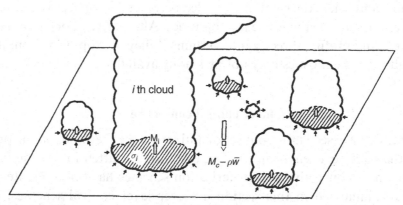

Figure 6.6. Schematic showing a unit horizontal area (e.g. a grid cell) at some vertical level between the cloud base and the highest cloud top. All the clouds are shown to entrain environmental air at each level, with detrainment occurring at the cloud top. The smallest cloud pictured has lost its buoyancy at this particular level and is detraining cloud air into the environment as shown by the outwardly directed arrows. From Arakawa and Schubert (1974).

$$Q_{2c} = M_c \frac{\partial \overline{q}}{\partial z} + D_q, \tag{6.24}$$

where

$$M_c = \sum_i \rho w_i \sigma_i. \tag{6.25}$$

Here σ_i is the fractional area of the ith cloud, w_i is the updraft velocity of the ith cloud, D_S and D_q are the detrainment effects at the cloud top, and $\overline{Q}_R$ is the radiative heating of the environment. Thus, M_c $(\mathrm{kg\,m^{-2}\,s^{-1}})$ is the mass flux in the vertical, while D represents the effects of cloud detrainment on the grid cell mean values. Since this scheme is concerned with the transfer of mass from one vertical level to another it also can be categorized as a mass flux scheme. Comparing (6.24) and (6.8) it is evident that

$$M_c \frac{\partial \overline{s}}{\partial z} + D_s \approx L_v(\overline{c} - \overline{e}) - \frac{\partial}{\partial z}(\rho \overline{s'w'}). \tag{6.26}$$

Thus, it is assumed that Q_1 and Q_2 are due to convective processes only, and that these terms can be estimated accurately with a simple cloud model: infinitesimally narrow and intense updrafts within a uniform environment, so that the environmental properties are equal to the large-scale mean. The total effects of convection are defined to be whatever is implied by having the mass flux M_c crossing the isopleths of $\overline{s}(p)$ and $\overline{q}(p)$ in the mean (grid cell) stratification.

While the original AS scheme describes a spectrum of entrainment rates to obtain a variety of cloud depths, most applications have altered this (owing to practical concerns) to define simply a spectrum of cloud depths. Thus, knowing the cloud base for a given grid cell, clouds are assumed to have depths that correspond to the difference in depth between the cloud base and each of the model levels above the cloud base (Fig. 6.7). Entrainment coefficients λ are backed out, defined such that the cloud top occurs precisely at model grid levels with one λ per model level. The normalized cumulus mass flux $\eta(\lambda, z)$ is assumed to have the form

$$\eta(\lambda, z) = e^{\lambda(z - z_{cloud\_base})}, \tag{6.27}$$

where again the value of λ controls the entrainment rate. Shallower clouds correspond to a larger value of λ, since they must entrain more environmental air in order to lose their buoyancy at lower levels. Note that $\eta(\lambda, z)$ is normalized to a value of 1 at the cloud base, and that the cumulus mass flux of the ith cloud type is defined to be $M_i(\lambda, z) = M_B(\lambda)\eta(\lambda, z)$, where $M_B(\lambda)$ is the cumulus mass flux at the cloud base and can be different for each cloud depth.

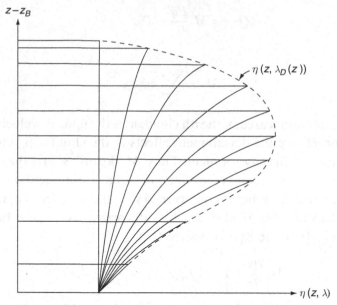

Figure 6.7. Schematic profiles of the normalized mass flux η for various values of the entrainment rate λ that lose buoyancy and detrain at model level D. Note that the cloud tops are at specific height levels and correspond to the heights of the model levels. From Arakawa and Schubert (1974).

The key to the AS scheme is the definition of a cloud work function, which for clouds with a λ entrainment rate is

$$A(\lambda) = \int_{z_{cloud\_base}}^{z_{cloud\_top(\lambda)}} g\eta(\lambda, z) \frac{T_c(\lambda, z) - \overline{T}(z)}{\overline{T}(z)} \, dz, \qquad (6.28)$$

where η is the cumulus mass flux normalized to a value of 1 at the cloud base, T_c is the cloud temperature (a mixture of updraft and environmental air), and $\overline{T}$ is the environmental temperature. This expression represents the work done by the buoyancy force. Note that when $\lambda = 0$, $A(\lambda)$ is just the CAPE of the cloud base parcel. In contrast, for a given λ, $A(\lambda)$ is the mass flux-weighted buoyancy integral.

Lord *et al.* (1982) show that in the tropics $A(\lambda)$ can be approximated by characteristic values based upon observations from the Marshall Islands (the location is shown in Fig. 6.2). They define

$$A(\lambda) = A_N(\lambda) \left[p_{cloud\_base} - p(\lambda)_{cloud\_top} \right], \qquad (6.29)$$

where pressures at the cloud base and the cloud top are in hPa, and the values of $A_N(\lambda)$ are derived from observations and vary from 1.7 for clouds with tops at 150 hPa to 0.04 for clouds with tops at 800 hPa. Generally, the values of $A_N(\lambda)$ increase with increasing cloud depth.

To close the scheme, AS assume a quasi-equilibrium state in which any creation of potential buoyant energy (as defined using A) by large-scale processes is quickly consumed by convection. They write the tendency of the cloud work function as

$$\left(\frac{dA(\lambda)}{dt}\right)_{convection} = \int_0^{\lambda_{max}} K(\lambda, \lambda')M_B(\lambda')\, d\lambda', \tag{6.30}$$

where $K(\lambda, \lambda')$ is the mass flux kernel which defines the stabilization of the λ cloud type due to changes in the environment from the λ' cloud type, and $M_B(\lambda')$ is the mass flux at the cloud base for the λ' cloud type. If $F(\lambda)$ is the large-scale forcing, defined as

$$F(\lambda) = \left[\frac{dA(\lambda)}{dt}\right]_{large-scale}, \tag{6.31}$$

such that it represents the changes in $A(\lambda)$ due to large-scale processes, then the closure assumption requires that

$$\int_0^{\lambda_{max}} K(\lambda, \lambda')M_B(\lambda')\, d\lambda' + F(\lambda) = 0, \tag{6.32}$$

where $M_B(\lambda')$ must be greater than zero.

The kernel $K(i, j)$ is calculated by modifying the mass flux $M_B(\lambda_j)$ for the λ_j cloud type and evaluating the change to the cloud work function $A(\lambda)$ for the λ_i cloud type. In essence, an initial small value of $M_B(\lambda)$ is chosen for all values of λ. Smaller perturbations are made one at a time to these initially specified values of $M_B(\lambda_i)$ and the subsequent changes to $A(\lambda)$ for all values of λ are determined. This yields a matrix K of values representing how changes to the cloud base mass fluxes of all the different cloud types influence all the other cloud types.

The $F(\lambda)$ term is calculated using knowledge of the time tendencies of the large-scale (grid resolved) forcing terms from the equations of motion. These tendencies are integrated forward in time to provide estimates of the environmental variables $\overline{T}$ and $\overline{q}$ at the next time step. Once these future values of $\overline{T}$ and $\overline{q}$ are known, the new cloud work function $A(\lambda)$ is calculated. Note that since the definition $A(\lambda)$ includes only the normalized cumulus mass flux, it can be calculated easily with knowledge of the environmental conditions, the cloud base conditions, and the entrainment rate parameter λ.

With knowledge of the large-scale forcing $F(\lambda)$ and the kernel $K(i, j)$, then the only unknown is the cloud base mass flux $M_B(\lambda)$ for all values of λ. The resulting equation is a Friedholm integral equation of the first kind, and can be solved in several different ways (see Lord *et al.* 1982), but has problems involving non-existence or non-uniqueness of solutions. The resulting distribution of $M_B(\lambda)$ defines a spectrum of cloud sizes and types that precisely consume the buoyant energy created by the large-scale forcing rate $F(\lambda)$.

The AS scheme assumes that detrainment occurs only at the cloud top. Thus, for each cloud top an amount of suspended liquid water is detrained at the cloud top level. The rest of the water condensed during plume ascent is converted to precipitation, which is assumed to fall to the ground. The summation of these precipitation rates over all clouds yields the total precipitation rate at the surface.

Since the AS scheme relates convective activity to changes in the cloud work function, which is influenced by the structure of the entire vertical column, the AS scheme is a deep-layer control scheme. It also is an instability-control scheme, since moisture is of secondary importance compared to the creation of instability. Finally, since the scheme is concerned with the details of how convection produces the final profiles it can be categorized as a dynamic scheme.

Owing to the computational demands of the AS scheme, numerous alternatives and simplifications have been investigated. Ding and Randall (1998) assume that the convective mass flux η grows linearly with height, instead of exponentially, in order to simplify the iterative computations needed to determine the value of λ for each cloud depth. Grell *et al.* (1991) develop a simplified version in which only a single cloud type is allowed, and they also add downdrafts to the scheme. Pan and Randall (1998) develop a prognostic cumulus kinetic energy as a different closure assumption. Grell and Devenyi (2002) develop an ensemble convective parameterization in which a variety of closures and parameters are varied and the ensemble mean tendencies given back to the numerical model.

Evidence in support of the quasi-equilibrium assumption that forms the basis of the AS scheme is provided by Arakawa and Schubert (1974) by using the Marshall Islands rawinsonde array data at 6 h intervals to calculate the rate of changes in the cloud work function A (called $\dot{A}$) and the rate of expected changes to A from large-scale forcing ($\dot{A}_{LS}$). Recall that if quasi-equilibrium is true, then convection should respond very quickly to changes in the environment due to large-scale forcing. Thus, regardless of the values of $\dot{A}_{LS}$ one expects to see very little change in $\dot{A}$ over time. Results support this expectation as there is little observed change in $\dot{A}$ over time even though there are some large values of $\dot{A}_{LS}$ (Fig. 6.8a). However, Mapes (1997) illustrates that this relationship between $\dot{A}$ and $\dot{A}_{LS}$ holds even if a heating process, resembling

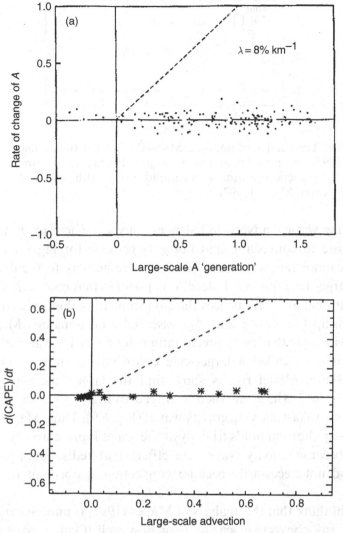

Figure 6.8. Tests of the quasi-equilibrium closure hypothesis using (a) Marshall Islands rawinsonde data (from Arakawa and Schubert 1974) and (b) an imposed-heating model from Mapes (1997). From Mapes (1997).

that of a tropical MCS, fluctuates arbitrarily within a similarly sized rawinsonde array (Fig. 6.8b). In this scenario, convection is defined to be unresponsive to the large-scale flow yet the results appear on the surface to support the quasi-equilibrium assumption. Mapes argues that the large-scale flow quickly redistributes convective heating over large areas, so that quasi-equilibrium holds for reasons opposite to the causality presumptions that are inherent in the Arakawa–Schubert scheme. Thus, any convection that develops within the rawinsonde array is capable of driving large-scale divergent flows where the

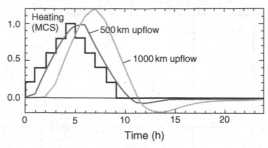

Figure 6.9. Time series of imposed MCS-like heating (black line) from the Mapes (1997) linear model, and area-integrated values of upward motion as sampled by synthetic low-latitude rawinsonde arrays with ∼500 and ∼1000 km diameters. From Mapes (1997).

corresponding vertical advection balances convective heating, but as a large-scale response to convection and not a large-scale forcing for convection. Thus, one cannot tie the amount of convective activity to the diagnosed or predicted large-scale forcing. Indeed, it is possible that convection outside of the observational array influences the environment within the array, making the relationship between $\dot{A}$ and $\dot{A}_{LS}$ essentially meaningless. Mapes (1997) argues convincingly that every precipitating cloud has a "large-scale essence," or what might be called a large-scale contribution, since precipitation is positive definite. Model results show that the large-scale vertical motion responds within 2–3 h, even when averaged over 500 or 1000 km diameter regions, owing to fast-moving gravity waves (Fig. 6.9). Thus, Mapes concludes that "quasi-equilibrium holds trivially at the scales represented by radiosonde networks, because gravity waves are efficient at redistributing convective heating, and not necessarily because convection is obedient to large-scale forcing."

One might think that the analysis of Mapes (1997) is unnecessary, since the real test of any convective scheme is in how well it can predict convective activity and its associated precipitation. While this is a valid and practical viewpoint, the arguments regarding the truthfulness of quasi-equilibrium closure are important to consider. These arguments provide a viewpoint from which one can learn about the process of convection and how convection interacts with the large-scale flow. In many atmospheric conditions it may not matter if the underlying closure is quasi-equilibrium or something else, since the model forecasts are all similar regardless of these details. It is in those really sensitive, marginal events where the assumptions underlying the behaviors of the parameterization schemes can play a large role in how the model forecast evolves. One would hate to have the wrong assumption rise up and produce a poor model forecast in these most crucial forecast events.

6.3.2 Betts–Miller convective scheme

The Betts–Miller convective scheme (Betts and Miller 1986, 1993) is based upon observations within tropical regions that show soundings in the vicinity of convective systems have similar temperature structures. This structure is very close to a moist adiabat through the cloud base equivalent potential temperature (θ_e) in regions of deep convection, although it departs from the moist adiabat by several degrees in the middle troposphere. Compared to the computational complexity of the AS scheme, the Betts–Miller scheme is a relatively straightforward lagged static scheme. Betts and Miller (1993) state (p. 107) that the scheme "was designed to represent directly the quasi-equilibrium state established by deep convection, so as to avoid the uncertainties involved in attempting to determine this state indirectly using increasingly complex cloud models, whose closure parameters can themselves ultimately be determined only by comparison with atmospheric variables." Results presented later clearly show that it is a deep-layer control scheme and so, even though it is quite different from the AS scheme, it is correctly placed in this section. Owing to its sensitivity to moisture, it also could be called a moisture-control scheme.

Instead of working in the typical temperature (T), mixing ratio (q), and pressure (p) space that most meteorologists use, the Betts–Miller scheme uses a transformed set of thermodynamic variables based upon the concept of saturation points (SPs). To gain a good appreciation for the Betts–Miller scheme, it is important to understand this saturation point concept. Saturation points are defined at the lifting condensation level of a parcel, and the values of the variables at their saturation point are indicated by a *. At a parcel's SP, the parcel's conserved parameters (θ^*, q^*) are known from just the saturation level (T^*, p^*), (θ^*, p^*), or (q^*, p^*), whereas for unsaturated parcels the values of p, T, and q are all needed to calculate both θ and q. Thus, on any thermodynamic diagram an SP is defined uniquely by only two variables, instead of the typical three variables required to define a parcel.

The concept of SPs can be very helpful for diagnosing mixing. Betts (1985, 1986) finds that if mixing occurs in a vertical layer, the SPs align along what is called a "mixing line." This behavior arises because q^*, θ^*, and θ_e^* are all approximately conserved in isobaric mixing as well as in adiabatic motion. Convective mixing, or vertical mixing, can be viewed as the sum of adiabatic motion followed by isobaric mixing.

Mixing line structures are seen in many soundings. A schematic of a typical sounding that passes through a cloud (Fig. 6.10) illustrates many of the important features that can be diagnosed. Even though the temperature and

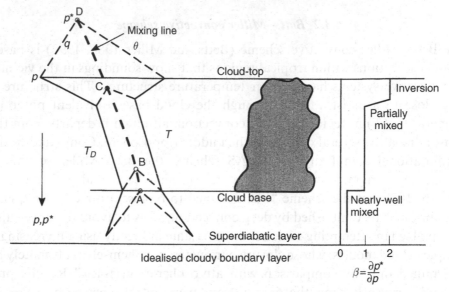

Figure 6.10. Schematic environmental sounding showing the relationship between the temperature and dewpoint temperature profiles, the mixing line, and the mixing parameter β for an idealized convective boundary layer. Thin dashed lines represent lines of constant θ and q. From Betts (1986).

dewpoint temperature profiles of the sounding have a rather erratic behavior, moving together and apart as the sounding characteristics change, the mixing line connecting the SPs of the parcels has a very smooth variation with height. A convenient way to look at mixing is through the parameter

$$\beta = \frac{\partial p^*}{\partial p}, \tag{6.33}$$

where p^* is the pressure at the SPs and p is the pressure at the parcel's originating level (i.e., the pressure from which the parcel is originally lifted to reach the SP). As shown in Fig. 6.10, the values of β provide information about the amount of vertical mixing that occurs within layers. A layer is thoroughly well mixed when $\beta = 0$. In this case, all the SPs occur at the same point on the thermodynamic diagram. For $\beta = 1$, the layer is partially mixed and the temperature and dewpoint temperature profiles are parallel both to each other and to the mixing line. For values of $\beta > 1$, the temperature and dewpoint temperature profiles are diverging from one another and from the mixing line.

Average soundings from within typhoons, slow and fast moving tropical squall lines, and tropical convection over land all show characteristic SP

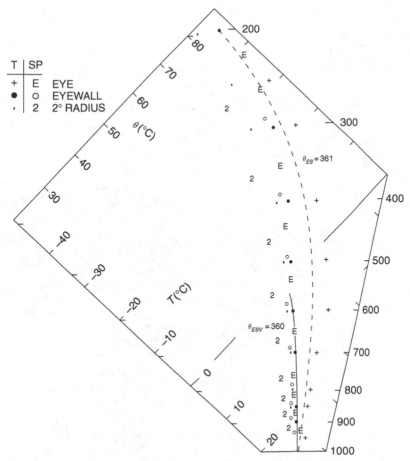

Figure 6.11. Composite typhoon sounding from inside the eye (E), the eyewall (○), and at 2° radius (2) showing temperature and SP structures. The solid line represents a moist virtual adiabat and the dashed line denotes a regular moist adiabat. From Betts (1986).

mixing lines that largely fall along a moist virtual adiabat (θ^*_{vc}; Betts 1986). Moist virtual adiabats join the SPs of cloudy parcels that have the same density when brought to the same pressure level. Betts finds that the SPs from these tropical convective systems fall along a moist virtual adiabat associated with cloud base conditions until the freezing level is reached, at which point the SPs transition to following a regular moist adiabat (Figs. 6.11 and 6.12). Thus, even though some of these average soundings are from the typhoon eyewall, and others are from tropical convective systems over land, they all have this characteristic and repeatable behavior. The Betts–Miller convective scheme adjusts the atmosphere towards these characteristic SP profiles of deep convection.

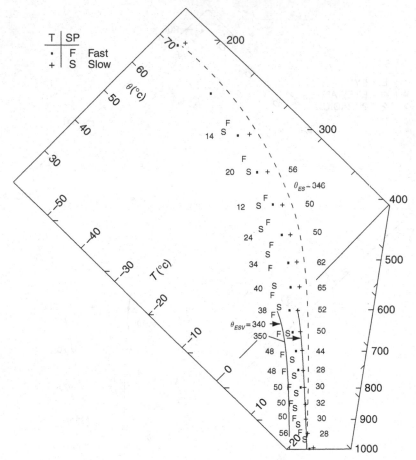

Figure 6.12. Composite wake soundings for slow (S) and fast (F) moving squall lines observed in the eastern Atlantic showing temperature (■ +), SP (S, F), and listing the $|P|$ values (numbers). Solid lines are two representative moist virtual adiabats and the dashed line is a regular moist adiabat. From Betts (1986).

Two distinct reference profiles, that are characteristic of shallow and deep convection, are used in the Betts–Miller scheme. These are partly specified and partly internally determined by the scheme. The formal argument for the development of this convective scheme starts with the large-scale thermo-dynamic tendency equation in SP form (Betts and Miller 1993). If one denotes an SP (θ^*, q^*) by a two-dimensional vector $\vec{S}$ (Betts 1983), then the large-scale thermodynamic tendency equation can be written as

$$\frac{\partial \vec{S}}{\partial t} = -\vec{V} \cdot \nabla \vec{S} - \varpi \frac{\partial \vec{S}}{\partial p} - g \frac{\partial \vec{N}}{\partial p} - g \frac{\partial \vec{F}}{\partial p}, \qquad (6.34)$$

where $\vec{N}$ is the net radiative flux vector and $\vec{F}$ is the net convective flux vector in (θ^*, q^*) space. These vectors point in the direction of the changes in the SP.

The net convective flux vector is defined in Betts and Miller (1993) as

$$-g\frac{\partial \vec{F}}{\partial p} = \frac{\vec{R} - \vec{S}}{\tau},$$

(6.35)

where $\vec{R}$ is the reference (characteristic) thermodynamic structure suggested by the observations and τ is the adjustment timescale. Thus, $\vec{R}$ is the reference SP structure seen in Figs. 6.11 and 6.12 from a given cloud base moist virtual adiabat. So $\vec{S}$ moves toward $\vec{R}$ during the adjustment time period. Betts and Miller (1993) further simplify the large-scale forcing to the vertical advection term only, assuming that the large-scale horizontal advection and radiative effects are small in comparison. This yields

$$\frac{\partial \vec{S}}{\partial t} = -\varpi\frac{\partial \vec{S}}{\partial p} + \frac{\vec{R} - \vec{S}}{\tau}.$$

(6.36)

If the large-scale forcing is steady on timescales much greater than the adjustment timescale, then the atmosphere eventually reaches a quasi-equilibrium condition with the local change of $\vec{S}$ approximately zero, such that

$$\vec{R} - \vec{S} \approx \varpi\left(\frac{\partial \vec{S}}{\partial p}\right)\tau.$$

(6.37)

This relationship states that $\vec{R} - \vec{S}$ is related to the large-scale forcing directly, i.e. to the mean vertical motion (ϖ) and the change in mean SP vector with pressure ($\partial \vec{S}/\partial p$). For small adjustment times, the atmosphere approaches $\vec{R}$ and so one can substitute $\vec{S} \approx \vec{R}$ in the vertical advection term. This yields

$$\vec{R} - \vec{S} \approx \varpi\tau\frac{\partial \vec{R}}{\partial p}.$$

(6.38)

From this expression, the convective fluxes (F) over the domain can be calculated, with

$$\vec{F} = \int \frac{\vec{R} - \vec{S}}{\tau}\frac{dp}{g} \approx \int \varpi\frac{\partial \vec{R}}{\partial p}\frac{dp}{g}.$$

(6.39)

This expression shows that the convective fluxes are tied to the structure of the reference profile $\vec{R}$. Therefore, if a good reference profile is chosen the convective flux vectors are constrained to be realistic and have a structure similar to those derived diagnostically.

Now that the foundation for the Betts–Miller convective scheme is in place, one can examine the details of how the scheme works in practice. The scheme begins by finding the cloud base and the cloud top. In Betts and Miller (1986) this is done by sequentially lifting parcels to their lifting condensation level (LCL), starting from the model levels closest to the ground surface, and finding the lowest parcel that is buoyant at the first model level above its LCL. In the National Centers for Environmental Prediction (NCEP) Eta model, the cloud base is defined as the LCL of the parcel in the lowest 200 hPa of the model that produces the largest CAPE. The cloud top is defined as the equilibrium level found using the moist adiabat of the cloud base parcel. The application of this scheme in the Eta model is called the Betts–Miller–Janjic (BMJ) convective scheme, owing to modifications made by Janjic (1994).

Once the cloud base and the cloud top have been determined, the Betts–Miller scheme constructs reference profiles of the SPs. If the moist adiabatic lapse rate is defined as

$$\Gamma_m = \frac{\partial \theta^*}{\partial p^*}, \tag{6.40}$$

then the reference profile of SPs θ_R^* is defined as

$$\theta_R^* = \theta_{cb}^* + 0.85\Gamma_m(p - p_{cb}), \tag{6.41}$$

where θ_{cb}^* is the cloud base potential temperature, p_{cb} is the cloud base pressure, and p is the pressure of the model levels. The coefficient 0.85 is chosen by Betts and Miller (1993) to better match the observed reference profiles that fall along a moist virtual adiabat (see Figs. 6.11 and 6.12). Equation (6.41) applies only for pressures above those of the freezing level. For pressures below the freezing level pressure (i.e., when the model temperatures are below freezing), the θ_R^* profile returns to the moist adiabat representative of the cloud base parcel using a quadratic function such that θ_R^* asymptotes to the moist adiabat at the cloud top. This approach thus reproduces the typical profile of SPs that is observed in association with tropical convection (Figs. 6.11 and 6.12).

After the reference SP profile has been defined, it must be converted into actual temperature and mixing ratio profiles. To accomplish this, a saturation pressure departure parameter P is defined as $P = p^* - p$. Observations of P from the same tropical systems used to construct the reference SP profile show a range of P values from -10 to -75 hPa (Fig. 6.13). There is a large variability

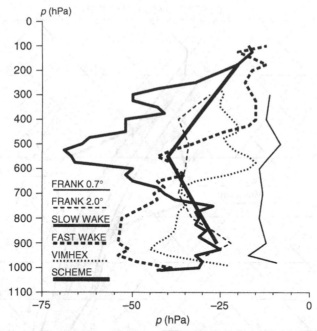

Figure 6.13. Profile of the saturation pressure departure P (hPa) with respect to pressure showing the averages from five different tropical convective types and the profile used by the Betts–Miller scheme. From Betts and Miller (1993).

in the values of P from different convective systems, in comparison to the persistent SP structure of the soundings. The Betts–Miller scheme uses the mean values of P at the cloud base, the freezing level, and the cloud top to define the P values associated with the reference profiles (Fig. 6.13). Once the values of P are known, then it is easy to calculate the reference profiles of temperature (T_R) and mixing ratio (q_R) that are needed by the model. On a thermodynamic diagram, the reference profiles of T_R and q_R are calculated by starting at the SP reference profile and then descending dry adiabatically for T, and along a line of constant q for q, until the defined saturation pressure deficit P is reached. This process is exactly opposite to the graphical approach to calculating the lifting condensation level of a parcel. An example of the typical first-guess reference profiles is shown in Fig. 6.14.

The reference profiles of T_R and q_R are shifted jointly, keeping the same shape and separation distance, to either the right or left to conserve the total enthalpy, such that

$$\int_{p_{cb}}^{p_{ct}} \left[(c_p T_R + L_v q_R) - (c_p \overline{T} + L_v \overline{q}) \right] dp = 0, \qquad (6.42)$$

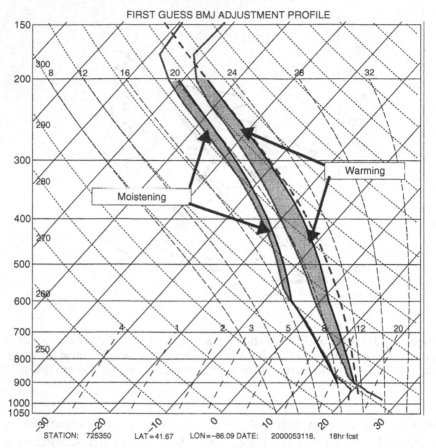

Figure 6.14. Model forecast sounding (solid lines) and preliminary first-guess Betts–Miller–Janjic reference profiles (dark solid lines). Thick, dashed line shows the reference moist adiabat. Note that in the BMJ application of the scheme, the value of T_R at the cloud base is anchored to the environmental temperature at that level and q_R is specified accordingly. The shading highlights the differences between the forecast environment and the reference profiles. From Baldwin *et al.* (2002).

where the overbar denotes a grid cell (environment) value, c_p is the specific heat of air at constant pressure, p_{cb} is the cloud base pressure, and p_{ct} is the cloud top pressure. This conservation equation requires that the release of latent heat from convection is balanced by the removal of water from the column. Suggested enthalpy changes to the environment both before and after enthalpy adjustment (Fig. 6.15) show that non-trivial adjustments are often needed. Since the profiles are shifted together either to the left or the right until enthalpy is conserved, the corrections to the reference profiles are applied equally throughout the convective layer (Fig. 6.16). Once the final T_R and q_R

(a)

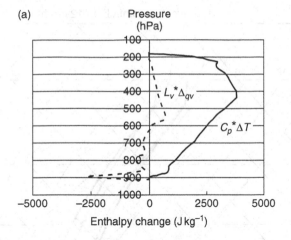

(b)

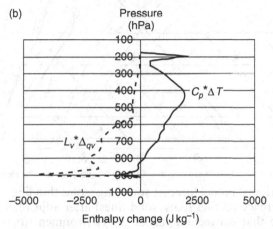

Figure 6.15. Vertical profiles of enthalpy change ($\mathrm{J\,kg^{-1}}$) suggested by the reference temperature (solid line) and mixing ratio (dashed line) both before (a) and after (b) enthalpy conservation is applied by the BMJ application of the scheme. From Baldwin *et al.* (2002).

profiles are known, the convective tendencies that are applied to the model can be calculated using an adjustment time, typically assumed to be 1 h.

Finally, the precipitation rate (PR) is defined as

$$PR = \int_{P_{cb}}^{P_{ct}} \left(\frac{q_R - \bar{q}}{\tau} \right) \frac{dp}{g} = -\frac{c_p}{L_v} \int_{P_{cb}}^{P_{ct}} \left(\frac{T_R - \bar{T}}{\tau} \right) \frac{dp}{g}. \tag{6.43}$$

No liquid water is stored in the Betts–Miller scheme, and the deep scheme is not activated if the value of PR after adjusting the profiles to conserve

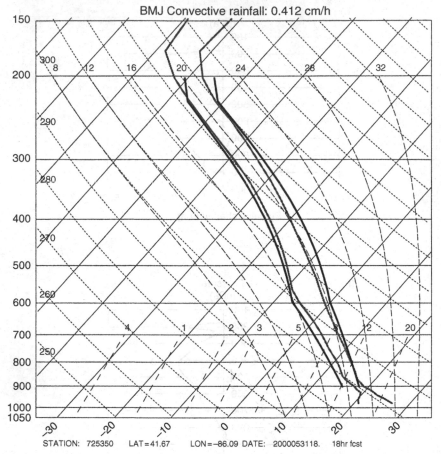

Figure 6.16. Original model forecast sounding (solid thin lines) and the final T_R and q_R reference profiles (dark solid lines) after adjustment to conserve enthalpy. Note that convection causes the environment to warm and dry throughout most of the layer where the scheme is acting, as opposed to the warming and moistening seen from the first-guess profiles shown in Fig. 6.14. From Baldwin *et al.* (2002).

enthalpy is less than zero. Quite reasonably, it is assumed that deep convection must both produce heat and remove water from the atmosphere.

As mentioned earlier, the Betts–Miller scheme also has a shallow convection component. If the cloud depth is determined to be less than 200 hPa, or if the *PR* is negative, then the shallow scheme is used. The cloud base for the shallow scheme is the same as for the deep scheme, but the cloud top is defined as the layer within 200 hPa of the cloud base that has the fastest decrease in relative humidity (Betts and Miller 1993). The shallow scheme uses a mixing line structure of SPs, as suggested by observations, from the cloud base to the cloud top. A value of $\beta = 1.2$ is assigned to determine the reference T_R and q_R

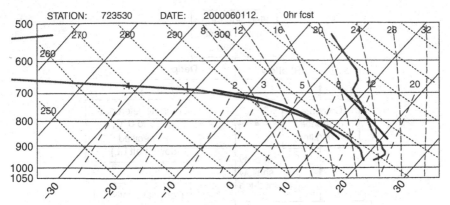

Figure 6.17. Model sounding (solid lines) and final T_R and q_R from the BMJ shallow convective scheme (thick solid lines). Note how the scheme warms the lower portion of the cloud layer and cools the upper portion of the cloud layer. From Baldwin *et al.* (2002).

profiles in the original Betts–Miller approach, although changes to the humidity profile specified are made in some operational versions of the scheme. Then the constraints

$$\int_{p_{cb}}^{p_{ct}} c_p(T_R - \overline{T}) \, dp = 0, \tag{6.44}$$

$$\int_{p_{cb}}^{p_{ct}} L_v(q_R - \overline{q}) \, dp = 0, \tag{6.45}$$

are used to adjust the profiles. Note that these adjustments yield no precipitation from the shallow scheme. As expected, the shallow scheme acts to warm and dry the lower half of the cloud layer and to cool and moisten the upper half of the cloud layer (Fig. 6.17). A flowchart of the scheme summarizes how it operates in a numerical weather prediction model, and chooses between deep and shallow components (Fig. 6.18).

Betts and Miller (1993) outline a simple approach to include downdrafts in the scheme that injects downdrafts into the lowest three model levels. However, there are no indications that downdrafts are included in any of the versions of the Betts–Miller scheme that are used routinely for numerical weather prediction. In addition, momentum transport is not included in the Betts–Miller scheme.

The Betts–Miller scheme is a deep-layer control scheme because, except for the specification of the cloud base, the scheme does not evaluate or consider any conditions in the low levels of the atmosphere. It also is a static scheme,

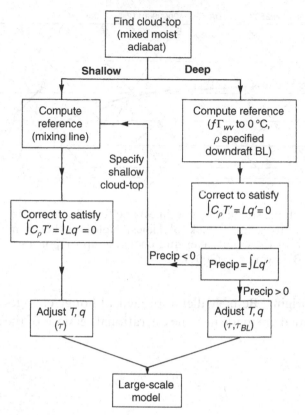

Figure 6.18. Flowchart for the Betts–Miller convective parameterization
scheme. From Betts and Miller (1993).

since it is not concerned with the details of the convective processes that
produce the final state. Finally, the Betts–Miller scheme is concerned more
with moisture than instability. Indeed, examination of the reference profiles
and the changes that are required to conserve enthalpy strongly suggests that
the Betts–Miller scheme is very sensitive to cloud-layer moisture. Changes in
cloud-layer relative humidity are shown by Baldwin *et al.* (2002) to produce
very noticeable changes in scheme precipitation values (Fig. 6.19). Even with
large values of CIN, if the environment has CAPE and is moist throughout the
cloud layer, the Betts–Miller scheme can activate. Forecasters at the Storm
Prediction Center (SPC) have also noticed that the BMJ scheme does not
activate in conditions that are prime for tornadic supercell thunderstorms,
i.e. environments with high CAPE owing to very moist boundary layers, but
with very dry mid levels representative of elevated mixed layers (i.e., there is no
positive precipitation solution). Thus, even though strong low-level forcing
may be able to remove the CIN on these days, the dry mid levels control

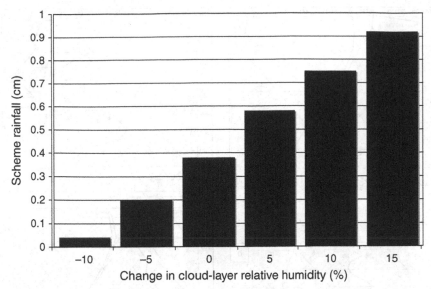

Figure 6.19. Parameterized convective rainfall from the BMJ scheme as a function of relative humidity changes in the cloud layer. After Baldwin *et al.* (2002).

whether or not the BMJ scheme will activiate and for this typical scenario deep convection fails to develop.

The operation of the BMJ scheme in the NCEP Eta model is summarized by Baldwin *et al.* (2002). They highlight that while the quantitative precipitation forecasts from this model have been very good, in no small part because of the performance of the BMJ scheme, artificial structures are sometimes produced that influence other forecast processes. For example, both CAPE and CIN can be greatly changed by the BMJ scheme and both of these parameters are important to forecasting the potential for severe convective activity.

Two examples of typical cases of modifications to the environment due to the activation of the BMJ scheme are shown in Baldwin *et al.* (2002). The first case involves only the activation of the shallow scheme near Birmingham, Alabama from 1200, 11 May to 0000 UTC, 12 May 2000 (Fig. 6.20). Diagnostics indicate that the shallow scheme activates almost immediately upon model startup (see the profiles in Figs. 6.20b, c). Over time the shallow scheme acts to remove much of the warm layer between 900 and 700 hPa (Fig. 6.20d), and also acts to warm the lower part of the cloud layer and thus lead to the entrainment of warmer air into the boundary layer. After 12 h of on and off modification to this environment by the shallow scheme, the resulting sounding (Fig. 6.20f) shows fairly significant differences when compared to the observed sounding. These changes result in operationally

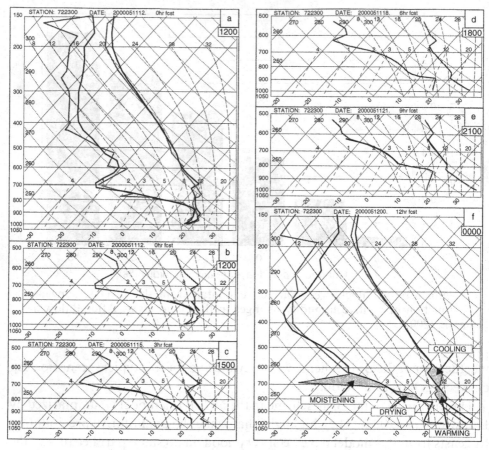

Figure 6.20. Sequence of Eta model forecast soundings from 1200 UTC 11 May to 0000 UTC 12 May 2000 over Birmingham, Alabama. Model initial condition (thick lines) and observed sounding (thin lines) (a), BMJ shallow convection reference profiles (thick lines) at 1200 with model sounding (b), BMJ shallow convection reference profiles with model sounding at 1500 UTC (c), model sounding at 1800 UTC (d), BMJ shallow convection reference profiles and model sounding at 2100 (e), and model (thick line) and observed (thin line) sounding at 0000 UTC (f). The shading in (f) denotes differences in the model and observed soundings associated with the shallow convection scheme. From Baldwin *et al.* (2002).

important alterations of CIN and CAPE owing only to the shallow convective scheme.

The second case is an example of shallow convection transitioning to deep convection within the BMJ scheme on 24 April 2001 over Greensboro, North Carolina (Fig. 6.21). While the shallow scheme does not activate immediately, after a few hours the boundary layer evolution creates some instability and the scheme activates, although the mid levels are too dry for deep convection

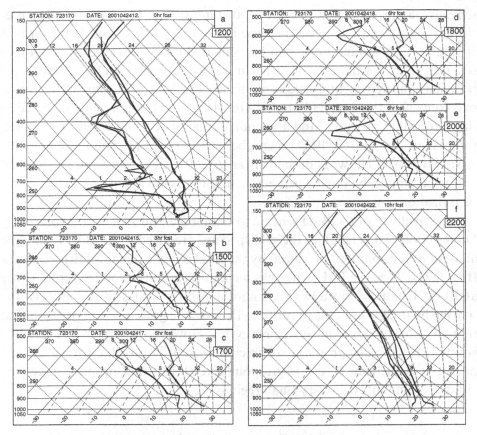

Figure 6.21. Sequence of Eta model forecast soundings from 1200 to 2200 UTC 24 April 2001 over Greensboro, North Carolina. Model initial condition (thick lines) and observed sounding (thin lines) (a), BMJ shallow convection reference profiles (thick lines) at 1500 with model sounding (b), BMJ shallow convection reference profiles with model sounding at 1700 UTC (c), BMJ shallow convection reference profiles with model sounding at 1800 UTC (d), BMJ shallow convection reference profiles and model sounding at 2000 (e), and model deep convection (thick line) and model sounding (thin line) at 2200 UTC (f). From Baldwin *et al.* (2002).

(Fig. 6.21b). The shallow scheme continues to modify the environment over the next hours as the boundary layer deepens, which raises the cloud base and thus the cloud top as well (Figs. 6.21b–d). This process acts to remove much of the dry air between 850 and 700 hPa, since the shallow scheme also is efficient at transporting moisture upward. This eventually leads to the activation of deep convection within the BMJ scheme (Fig. 6.21f).

As emphasized by Baldwin *et al.* (2002), this discussion is not intended to denigrate the BMJ scheme. It is a very robust convective parameterization scheme that has helped to provide needed improvements in quantitative precipitation forecasts. However, because it has a very discernable behavior that is seen in model soundings it can be evaluated much more easily than other schemes. These evaluations lead to an improved understanding of both the beneficial and detrimental aspects of this scheme that can be improved upon or overcome with the intervention of a knowledgeable human being involved in the forecasting process. Model sounding analysis can be a key part of the forecast process, and the BMJ scheme behaviors, once understood, can be used advantageously. The work of Baldwin *et al.* (2002) highlights what can be done and what should be done when evaluating the behavior of convective parameterization schemes.

The sensitivity of the BMJ scheme to changes in grid spacing is explored by Gallus (1999). In general, as the grid spacing is reduced from 78 to 12 km, the amount of precipitation produced by the BMJ scheme remains fairly constant (Fig. 6.22). Since the BMJ scheme is most influenced by cloud-layer moisture and CAPE, which typically vary smoothly on the scales of the models, the scheme by itself is not able to produce the extreme gradations in precipitation that are often observed in convective events. However, when the BMJ scheme is not the dominant producer of precipitation, it is possible for an explicit microphysics scheme to create gridpoint-like storms that produce large precipitation amounts over very small regions.

6.3.3 Kuo convective scheme

The Kuo convective scheme (Kuo 1965) is one of the earliest proposed convective schemes, and as stated by Raymond and Emanuel (1993), remains one of the (p. 145) "most enduringly popular schemes for parameterizing cumulus convection." It is computationally less demanding than either the Arakawa–Schubert or Betts–Miller schemes, which likely is one reason for its continued usage. As discussed below, it relates convective activity to total column moisture convergence, and so is aptly categorized as a deep-layer control scheme. It also can be categorized as a static scheme, since it is not concerned with the details of the convective processes, and a moisture-control scheme since its behavior is closely tied to the available moisture.

The derivation of the Kuo scheme begins from the large-scale equations in pressure coordinates (x, y, p) for the potential temperature and the water vapor mixing ratio, with

$$\frac{\partial \bar{\theta}}{\partial t} + \nabla \cdot (\vec{V}\bar{\theta}) + \frac{\partial}{\partial p}(\bar{\omega}\bar{\theta}) = \frac{L_v(\bar{c} - \bar{e}) + Q_R}{\pi} - \frac{\partial}{\partial p}(\overline{\omega'\theta'}), \qquad (6.46)$$

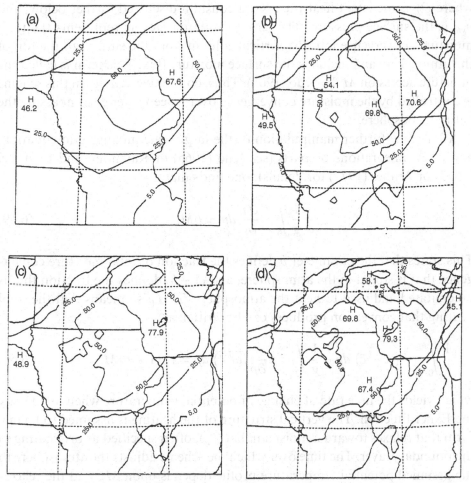

Figure 6.22. Accumulated precipitation (mm) over the 24 h period from 1200 UTC 16 June to 1200 UTC 17 June 1996 from the NCEP Eta model with the BMJ scheme at grid spacings of (a) 78, (b) 39, (c) 22, and (d) 12 km. The first isoline is at 5 mm; otherwise the isoline interval is 25 mm. From Gallus (1999).

$$\frac{\partial \overline{q}}{\partial t} + \nabla \cdot (\overrightarrow{V}\overline{q}) + \frac{\partial}{\partial p}(\overline{\omega}\overline{q}) = -(\overline{c} - \overline{e}) - \frac{\partial}{\partial p}(\overline{\omega'q'}), \qquad (6.47)$$

where π is the Exner function and the horizontal bar again denotes grid cell mean values.

The key assumption in the Kuo scheme is that the precipitation rate PR is

$$PR = (1-b)\left[-\frac{1}{g}\int_0^{p_{sfc}} \nabla \cdot (\overrightarrow{V}\overline{q})\, dp + \frac{1}{L_v}Q_E\right] = (1-b)M_t, \qquad (6.48)$$

where Q_E is the latent heat flux, b is a constant discussed further below, and p_{sfc} is the surface pressure. The term in square brackets represents the total moisture convergence into the vertical column from the surface to the top of the atmosphere and includes the surface moisture flux. This term is called the moisture accession M_t by Kuo (1965). Thus, convective activity in this scheme is controlled by the moisture convergence over a deep layer encompassing the entire troposphere.

After some further manipulation of the large-scale mixing ratio equation, and a few integrations by parts (see Kuo (1965) or Raymond and Emanuel (1993) or Anthes (1977) for details), one finds that

$$\frac{1}{g} \int_0^{p_{sfc}} \frac{\partial \overline{q}}{\partial t} \, dp = bM_t. \tag{6.49}$$

This states that the parameter b defines the fraction of total moisture convergence that is stored in the atmosphere, and $(1 - b)$ defines the fraction that is precipitated and used to heat the atmosphere. This also means that the total heating that convection produces can be written as

$$\frac{1}{g} \int_0^{p_{sfc}} L_v(\overline{c} - \overline{e}) \, dp - \frac{1}{g} \int_0^{p_{sfc}} \pi \frac{\partial}{\partial p}(\overline{\omega'\theta'}) \, dp = L_v(1 - b)M_t + Q_H, \tag{6.50}$$

which yields the time rate of change of potential temperature when divided by the Exner function. The vertical structure of the heating is assumed to be in a form that relaxes toward a moist adiabat (θ_a), often specified as originating in the boundary layer. The time over which the scheme adjusts the atmosphere to the assumed potential temperature profile depends upon M_t, and the relaxation time decreases as M_t increases.

Numerous modifications have been made to the Kuo scheme over the ensuing decades. Many of these involve definitions for the b parameter (Anthes 1977; Molinari 1982), the vertical heating profile (Anthes 1977; Molinari 1985), or the moisture accession parameter (Krishnamurti *et al.* 1976). A particularly interesting extension is the development of a slantwise convective scheme by Nordeng (1987, 1993) to represent precipitation bands produced by conditional symmetric instability in numerical models that cannot resolve this instability. Results from Lindstrom and Nordeng (1992) illustrate the potential utility of such a scheme. While all of these studies have their advantages, the main concern about the Kuo scheme is the fundamental assumption that convective development is directly tied to moisture convergence.

Raymond and Emanuel (1993) offer a strong and well-argued criticism of the Kuo scheme. The Kuo scheme assumes a statistical equilibrium of water

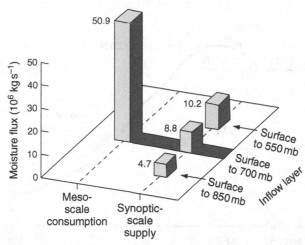

Figure 6.23. Comparison of synoptic-scale moisture flux into a squall line with the mesoscale moisture flux of the squall line for a grid-sized atmospheric column in northern Oklahoma. These consumption and supply terms should roughly balance if the Kuo closure is reasonable. From Fritsch *et al.* (1976).

substance, such that convection is assumed to consume water (not energy) at the rate at which water is supplied by the atmosphere. For a convective event over Oklahoma, Fritsch *et al.* (1976) clearly show that the moisture fluxes due to a mesoscale convective system are 10 times larger than the low-level synoptic-scale supply and five times larger than the deep-layer synoptic-scale supply (Fig. 6.23). This result indicates that convection is not caused or limited by the large-scale water supply. Thus, the Kuo scheme closure fundamentally violates causality. Raymond and Emanuel (1993) conclude that "the illposedness of this closure underlies many of the problems encountered with the Kuo scheme in practice."

6.4 Low-level control convective schemes

The convective schemes summarized in this section are low-level control schemes. These schemes also have been called activation control schemes, since they assume that the deep convection is determined by the physical processes that control convective initiation. This assumption is very consistent with an ingredients-based approach to forecasting deep convection (Johns and Doswell 1992), in which instability, moisture, and lift are needed to create deep convection. One often observes broad regions in which the large-scale forcing creates CAPE (basically a combination of instability and low-level moisture) over many days, but deep convection only occurs in specific areas where

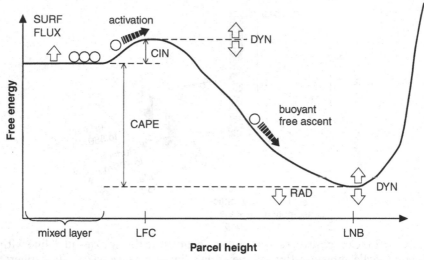

Figure 6.24. A free-energy diagram of a parcel model for deep convection. The vertical height of a parcel is indicated by the horizontal axis, with the level of free convection (LFC) and level of neutral buoyancy (LNB) indicated. The cumulative integral of work done in lifting the parcel is indicated by the heavy solid line, as a function of the parcel height. Parcels are indicated by circles, and must overcome an energy barrier (CIN) to activate the CAPE. Dynamical processes (DYN) can vary CIN and CAPE, while surface fluxes (SURF FLUX) and radiation (RAD) change the amount of free energy. After Mapes (1997).

mesoscale and small-scale forcing are able to lift low-level parcels to their level of free convection. Low-level control schemes attempt, in various ways, to incorporate this kind of behavior into how the scheme operates.

Another illustration of the differences between low-level and deep-layer control schemes is shown in Fig. 6.24. Assuming that parcels originate in the mixed layer (at the far left of the diagram), as often occurs with deep convection, it takes an increase in energy to overcome or remove the region of convective inhibition before the parcels can activate the CAPE and rise quickly to the level of neutral buoyancy (LNB) or equilibrium level. Low-level control schemes are concerned with how the parcel is able to overcome the convective inhibition and activate CAPE, while deep-layer control schemes constrain the amount of deep convection to changes in CAPE. The low-level control schemes discussed in this chapter are the Tiedtke, Gregory–Rowntree, Kain–Fritsch and Emanuel schemes. Since these schemes are concerned with the details of the convective process, they also are dynamic schemes and can be further classified as mass flux schemes, since they calculate the transfer of mass from one vertical level to the next.

6.4.1 Mass flux convective schemes

The Tiedtke (1989), Gregory and Rowntree (1990), Kain and Fritsch (1993), and Emanuel (1991) convective schemes are all mass flux schemes and share similar foundations and assumptions. As such, it is perhaps easier to summarize the basic ideas behind these mass flux schemes and then to delve into each scheme separately to discuss some important details and closure assumptions. These schemes also can be categorized as dynamic schemes, since they are concerned with the details of convective processes. For this discussion, the approach of Tiedtke (1989) is followed, although the approaches outlined in the other two schemes are nearly identical. As always, we begin with the large-scale equations for heat and moisture as in (6.4) and (6.5), but now in height coordinates

$$Q_{1c} = -\frac{1}{\rho}\frac{\partial}{\partial z}(\overline{\rho w's'}) + L_v(\overline{c} - \overline{e}) + \overline{Q}_R, \tag{6.51}$$

$$Q_{2c} = -\frac{1}{\rho}\frac{\partial}{\partial z}(\overline{\rho w'q'}) - (\overline{c} - \overline{e}), \tag{6.52}$$

where s is the dry static energy, q is the mixing ratio, ρ is the atmospheric density, c is the rate of condensation, e is the rate of evaporation, and w is the vertical velocity.

The vertical eddy transports are subdivided into contributions from convective updrafts, convective downdrafts, and the "compensating" subsidence of the environmental air needed to balance the mass flux going up from convection, such that

$$\overline{\rho(w's')}_{cu} = \overline{\rho}\sum_i a_{ui}(w_{ui} - \overline{w})(s_{ui} - \overline{s}) + \overline{\rho}\sum_i a_{di}(w_{di} - \overline{w})(s_{ui} - \overline{s})$$

$$+ \overline{\rho}\left[1 - \sum_i(a_{ui} + a_{di})\right](\tilde{w} - \overline{w})(\tilde{s} - \overline{s}), \tag{6.53}$$

where the subscript cu denotes the contributions from convection, the subscript i denotes the cloud type, subscripts u and d denote convective updrafts and downdrafts, respectively, tilde denotes the environmental air, and a is the fractional area coverage. The updrafts, downdrafts, and environmental subsidence are all average values, as it is assumed that a one-dimensional model is capable of adequately modeling their effects. It is further reasonable to assume that $\tilde{s} = \overline{s}$ and $\tilde{q} = \overline{q}$. To simplify the expressions, it is convenient to define

$$M_{ui} = \bar{\rho} a_{ui}(w_{ui} - \overline{w}), \tag{6.54}$$

$$M_{di} = \bar{\rho} a_{di}(w_{di} - \overline{w}), \tag{6.55}$$

where M_{ui} and M_{di} are the updraft and downdraft mass flux, respectively, for cloud type i.

Using (6.53)–(6.55), the large-scale budget equations become

$$Q_{1c} = -\frac{1}{\bar{\rho}}\frac{\partial}{\partial z}[M_u s_u + M_d s_d - (M_u + M_d)\bar{s}]$$

$$+ L_v(c_u - e_d - \tilde{e}_l - \tilde{e}_p) - \frac{1}{\bar{\rho}}\frac{\partial}{\partial z}(\overline{\rho w' s'})_{pbl} + \overline{Q}_R, \tag{6.56}$$

$$Q_{2c} = -\frac{1}{\bar{\rho}}\frac{\partial}{\partial z}[M_u q_u + M_d q_d - (M_u + M_d)\bar{q}]$$

$$- (c_u - e_d - \tilde{e}_l - \tilde{e}_p) - \frac{1}{\bar{\rho}}\frac{\partial}{\partial z}(\overline{\rho w' q'})_{pbl}, \tag{6.57}$$

where M_u, M_d, c_u, and e_d are the net contributions from all clouds to the upward mass flux, the downward mass flux, the condensation, and the evaporation, respectively. Variables s_u, s_d, q_u, and q_d are the weighted averages of s and q from all updrafts and downdrafts, $\tilde{e}_l$ is the evaporation of cloud air that is detrained into the environment, and $\tilde{e}_p$ is the evaporation of precipitation into the unsaturated cloud air. The subscript *pbl* refers to a turbulence contribution from the planetary boundary layer, which is parameterized by another scheme in the model. Lastly, there is a final equation for the mass continuity for rain water, defined as

$$P(z) = \int_z^\infty (G_p - e_d - \tilde{e}_p)\bar{\rho}\, dz, \tag{6.58}$$

where $P(z)$ is the rain water flux at height z and G_p is the conversion from cloud water into precipitation.

A simple bulk model based on parcel theory is used to represent an ensemble of convective clouds. Some evidence that a bulk model approach is sufficient is provided by Yanai *et al.* (1976), who show that a bulk model and a spectral model yield nearly identical vertical cloud mass flux for tropical convection. The updrafts of the cloud are assumed to be in a steady state, which then leads to the bulk equations for mass, heat, moisture, and cloud water content

$$\frac{\partial M_u}{\partial z} = E_u - D_u, \tag{6.59}$$

$$\frac{\partial(M_u s_u)}{\partial z} = E_u \bar{s} - D_u s_u + L_v \bar{\rho} c_u, \tag{6.60}$$

$$\frac{\partial(M_u q_u)}{\partial z} = E_u \bar{q} - D_u q_u - \bar{\rho} c_u, \tag{6.61}$$

$$\frac{\partial(M_u l)}{\partial z} = -D_u l + \bar{\rho} c_u - \bar{\rho} G_p, \tag{6.62}$$

where E and D are the rates of mass entrainment and detrainment per unit length, respectively, l is the cloud liquid water content, and c_u is the net condensation in the convective updrafts. These equations simply state that the convective mass, heat, moisture, and cloud water fluxes change owing to the effects of entrainment, detrainment, condensation, and precipitation. Generally, the precipitation rate is determined by the amount of liquid water content left over after the effects of condensation and evaporation are determined, using a mass continuity equation for liquid water.

Similarly, expressions for the downdraft mass flux, the dry static energy, and the moisture are defined as

$$\frac{\partial M_d}{\partial z} = E_d - D_d, \tag{6.63}$$

$$\frac{\partial(M_d s_d)}{\partial z} = E_d \bar{s} - D_d s_d + L_v \bar{\rho} e_d, \tag{6.64}$$

$$\frac{\partial(m_d q_d)}{\partial z} = E_d \bar{q} - D_d q_d - \bar{\rho} e_d, \tag{6.65}$$

below the downdraft origination level. Similar expressions are found in Gregory and Rowntree (1990) and Kain and Fritsch (1993). At this point, the schemes start diverging and are discussed separately. However, note that all of these schemes include a number of assumptions that are difficult to verify with observations, such as the relationships between the updraft and the downdraft mass fluxes. These assumptions clearly influence the behaviors of the schemes.

6.4.2 Tiedtke convective scheme

Tiedtke (1989) assumes that the convection is dependent upon the moisture supply from large-scale moisture convergence and boundary layer turbulence, making this scheme also a moisture-control scheme. The entrainment rate of the plume model used to represent the convective clouds depends upon the amount of large-scale convergence with smaller values of entrainment when large-scale convergence is present. This assumption leads to the partitioning of the updraft entrainment into a term that represents the turbulent exchange of mass through the cloud edges (E_u^1), and a second term that represents organized inflow into the cloud from large-scale convergence (E_u^2). A similar assumption is made for detrainment (D_u), leading to the relationships

$$E_u = E_u^1 + E_u^2, \tag{6.66}$$

$$D_u = D_u^1 + D_u^2. \tag{6.67}$$

The turbulent components of updraft entrainment and detrainment are defined as $E_u^1 = \varepsilon_u M_u$ and $D_u^1 = \delta_u M_u$. These fractional entrainment rates are set to a smaller value ($1 \times 10^{-4}\,\mathrm{m}^{-1}$) in the presence of large-scale flow convergence, and to a larger value ($3 \times 10^{-4}\,\mathrm{m}^{-1}$) otherwise. A smaller entrainment rate generally results in deeper clouds, such that the deep convection in this scheme is more likely in the presence of large-scale flow convergence. The large-scale convergence component of the updraft entrainment rate is defined as

$$E_u^2 = -\frac{\bar{\rho}}{\bar{q}}\left(\overrightarrow{V} \cdot \nabla \bar{q} + \bar{w}\frac{\partial \bar{q}}{\partial z}\right), \tag{6.68}$$

such that the entrainment rate is directly proportional to the large-scale moisture convergence. The large-scale component of the detrainment rate is set to zero, although a large-scale component of detrainment is used for shallow convection. For downdrafts, the downdraft entrainment and the detrainment rates are both set to $2 \times 10^{-4}\,\mathrm{m}^{-1}$.

Following Fritsch and Chappell (1980), downdrafts are assumed to begin at the level of free sink (LFS), the highest model level where the temperature of a saturated mixture of equal amounts of saturated environmental air at the wet-bulb temperature and updraft air becomes cooler than the environment. The downdraft mass flux is assumed to be directly proportional to the upward mass flux, so once the updraft mass flux is defined the downdraft mass flux at the LFS is defined to be 20% of the updraft mass flux at the cloud base.

Tiedtke (1989) closes the scheme by referring back to the study of Kuo (1965) linking convective activity to large-scale moisture convergence. This assumption leads to imposing a moisture balance for the *low-level* subcloud layer so that the cloud mass flux is tied directly to the low-level, large-scale moisture convergence. This relationship is defined as

$$[M_u(q_u - \bar{q}) + M_d(q_d - \bar{q})]_{cb} = -\int_0^{cb} \left(\vec{V} \cdot \nabla \bar{q} + \overline{w} \frac{\partial \bar{q}}{\partial z} + \frac{1}{\bar{\rho}} \frac{\partial}{\partial z} (\overline{\rho w' q'})_{pbl} \right) \bar{\rho} \, dz,$$

$$(6.69)$$

where the subscript *cb* indicates cloud base. From a small initial cloud base updraft mass flux, one can determine the relative downdraft mass flux at the cloud base, and thereby iterate to find the required amount of updraft mass flux at the cloud base needed to match the large-scale moisture convergence in the subcloud layer. Shallow convection is included as part of the scheme with shallow clouds detraining both at their level of neutral buoyancy and the model level above it.

Comparisons of single column tests against observational data from the GATE field experiment indicate that this closure works well for tropical systems (Fig. 6.25). However, it has already been argued in the discussion of the Kuo scheme that the idea of moisture convergence causing convection is illposed (Raymond and Emanuel 1993) and is not supported well by observations (Fritsch *et al.* 1976).

Modifications to the Tiedtke convective scheme are discussed in Gregory *et al.* (2000) for which the discrimination between shallow versus deep convective clouds depends upon whether or not the cloud depth is deeper or shallower than 200 hPa. In addition, the cloud base updraft mass flux is determined based upon the removal of CAPE over an adjustment time period. This approach is more like an instability-based scheme than the original moisture-based approach of Tiedtke.

6.4.3 Gregory–Rowntree convective scheme

Gregory and Rowntree (1990) assume that the initial mass flux is proportional to the excess buoyancy of the parcel within the initial convecting layer (which also could be called the updraft source layer). Thus, this scheme also is clearly a low-level control scheme, as convection can occur without the presence of large-scale forcing. However, the scheme appears to come into some semblance of balance with large-scale and boundary layer changes in the thermodynamic structure (Gregory and Rowntree 1990). While this scheme also is a mass flux scheme, it has several differences in comparison to the Tiedtke (1989) scheme. Since the scheme is concerned with parcel buoyancy, it also could be classified as an instability-control scheme. Like the Tiedtke scheme, this scheme also allows for shallow convection.

In contrast with the Tiedtke scheme, the Gregory–Rowntree scheme does not partition the entrainment rate into turbulent and large-scale components.

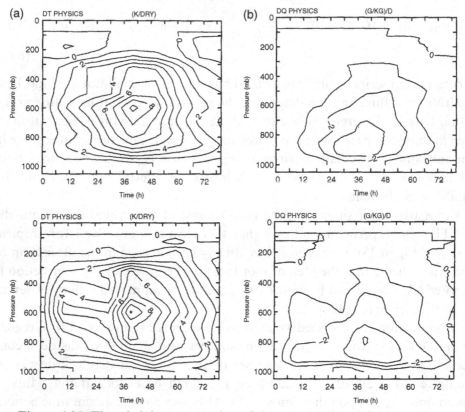

Figure 6.25. Time–height cross-section of the computed and diagnosed heat source Q_1 (K day^{-1}) and moisture sink Q_2 (g kg^{-1} day^{-1}) from GATE. Left-hand panels show the Q_1 values, whereas the right-hand panels show the Q_2 values. The top panels are the diagnosed values from Thompson *et al.* (2004). The bottom panels are the computed values from Tiedtke (1989). From Tiedtke (1989).

Only a turbulent component is assumed, such that the updraft entrainment rate (E_u) is

$$E_u = 3A_E\sigma_k, \tag{6.70}$$

where $A_E = 1.0$ at the lowest near-surface model level and 1.5 elsewhere, and σ_k is the value of the vertical coordinate σ at model level k (σ varies from 1.0 at the surface to 0 at the model top). This allows the updraft entrainment rate to vary with height as suggested by observations. The turbulent updraft detrainment rate (D_u) is defined as a fraction of the entrainment rate (E_u), with D_u generally being roughly one-third of the entrainment rate. Calculated mass entrainment rates using this approach agree reasonably well with those determined from observations (Fig. 6.26).

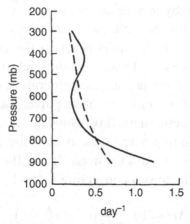

Figure 6.26. Comparison of the mass entrainment rates from the Gregory and Rowntree scheme (dashed) calculated using the convective mass flux of Yanai *et al.* (1973) with the rates diagnosed (solid) by Yanai *et al.* (1973). From Gregory and Rowntree (1990).

Besides the differences in how the entrainment and detrainment rates are determined, there are three other major differences between the Tiedtke and Gregory–Rowntree schemes. First, the Gregory–Rowntree scheme allows for an ensemble of parcels within the cloud model, such that a single cloud type is not specified. Second, the closure assumption is very different between the two schemes. Third, the Gregory–Rowntree scheme does not include the effects of downdrafts as part of the cloud model. However, non-frozen precipitation is allowed to evaporate below the updraft cloud base, and so low-level cooling can occur. These three differences are now explained in more detail.

The Gregory–Rowntree scheme makes use of the concept of forced detrainment. This approach allows a single bulk updraft parcel to be composed of a number of smaller parcels that undergo a variety of mixing profiles with the environment. The computational simplicity of a bulk scheme is thereby maintained, but with a representation of the complexity of an ensemble of parcels. Forced detrainment occurs as a convecting parcel ascends from one level to the next level and fails to retrain a minimum value of excess buoyancy (b). If the difference between the virtual potential temperature of the convecting parcel (θ_v^P) and the environment (θ_v^E) is such that

$$\theta_v^P - \theta_v^E < b, \tag{6.71}$$

then forced detrainment occurs. Generally, b is set to 0.2 K, although it is argued that larger values also may be reasonable. Forced detrainment indicates that less buoyant, smaller members of the total bulk parcel are detrained at this model level, thereby increasing the buoyancy of the remaining parcel members and allowing for the bulk parcel to continue rising. The potential temperature of the rising bulk parcel is set to the environmental potential temperature plus the value of b. Two constraints act to bound the temperature of this rising parcel. First, the maximum temperature of the bulk parcel must be no more than that of an initial undilute parcel ascending from the cloud base. Second, if the convective mass flux is smaller than a minimum value, then detrainment of the remaining parcel mass occurs and the parcel stops rising.

The closure assumption relates the mass flux in the initial convective layer (M_I) to the stability of the lowest convecting layers by

$$M_I = 10^{-3} \frac{c}{\Delta\sigma} \left(\theta_v^P - \theta_v^E - b \right), \tag{6.72}$$

where $c = 1.33 \times 10^{-4}$, the potential temperature differences are determined at the first model level above cloud base, and $\Delta\sigma$ is the difference in the vertical coordinate σ between the cloud base and the first model level above the cloud base. Thus, parcel buoyancy is the only criterion that determines whether or not convection develops, and the initial mass flux is proportional to the excess buoyancy of the parcel starting from one layer and rising to the next layer. The value of c is selected based upon tests that yield realistic atmospheric thermodynamic structures and rainfall rates. An enthalpy check is made to ensure that enthalpy is conserved by the scheme and temperatures are adjusted if needed.

While a downdraft is not included as part of the cloud model, the Gregory–Rowntree scheme does allow for the effects of precipitation. First, precipitation only occurs if the cloud depth is at least 4.0 km over land, 1.5 km over sea, or 1.0 km if the temperature of the level above the cloud base is less than 263 K. Until these depths are reached, the condensed water is stored within the parcel. After these depths have been reached, the precipitation (P_k) produced by the parcel rising over one model layer (from level k to level $k-1$) is defined as

$$P_k = (l_{k-1}^P - l_{min}) M_{k-1} \frac{p^*}{g}, \tag{6.73}$$

where l is the parcel condensate, l_{min} is set to $1 \, \text{g kg}^{-1}$, M_{k-1} is the convective mass flux at level $k-1$, and p^* is the surface pressure (Pa). The total precipitation falling from the cloud base is defined by summing the values of P_k from the cloud base to the cloud top. Freezing and melting occur at the freezing level of the environment and influence the model temperature fields accordingly. Falling snow does not sublimate, but evaporation of falling rain is allowed below the cloud base with the rate being proportional to the subsaturation of the environment. Thus, low-level cooling due to the effects of precipitation occurs in this scheme, even though downdrafts are not included explicitly as part of the cloud model.

6.4.4 Kain–Fritsch convective scheme

Kain and Fritsch (1990, 1993) also develop a mass flux scheme in which the cloud base mass flux is determined by the amount of CAPE in the environment that needs to be removed. This might lead one to believe that the scheme is a form of deep-layer control, but the activation of convection is based upon low-level forcing. Thus, the Kain–Fritsch scheme also is a low-level control scheme. Since the scheme focuses upon CAPE it also is an instability-control scheme. This scheme also uses one of the more complex trigger functions. Shallow convection is not considered in this scheme.

The Kain–Fritsch convective scheme differs from the schemes of Tiedtke (1989) and Gregory and Rowntree (1990) in several important ways, although the basic concepts of all the mass flux schemes are very similar. First, the Kain–Fritsch scheme uses variable entrainment and detrainment rates based upon the properties of updraft parcels that mix with the environment. Second, the closure assumption specifies the amount of updraft mass flux based upon that needed to remove CAPE. Third, the scheme includes a downdraft in the cloud model, but the downdraft is initiated 150–200 hPa above the cloud base and is fed by the evaporation of condensate. Lastly, the determination of when convection activates (the so-called trigger function) is contingent upon the lifting of a parcel from its LCL to its LFC in agreement with parcel theory and an ingredients-based approach to understanding convection. These differences are now summarized.

The Kain–Fritsch convective scheme uses the traditional concept for the entrainment rate, namely that the fractional increase in cloud-mass flux per unit height is inversely proportional to the updraft radius, but also incorporates a linear approximation to this relationship and modifies this expression to be a function of pressure. These changes lead to a rate δM_e (kg s^{-1}) at which the environmental air mixes into an updraft over a pressure interval δp, such that

$$\delta M_e = -\frac{0.03\delta p M_{u0}}{R}, \tag{6.74}$$

where R and M_{u0} are the updraft radius (m) and the mass flux (kg s^{-1}) at the cloud base, respectively. From this expression, an entraining updraft plume with radius 1500 m and no detrainment grows to double its mass flux as it ascends to 500 hPa. While this equation specifies the rate at which environmental air flows into the turbulent mixing region at the edges of an updraft, it does not tell us how this mixing influences the updraft characteristics. For this, one needs to know the details about how mixing occurs.

It is assumed that updraft air is made available for mixing with environmental air at the same rate as environmental air flows into the updraft as defined by (6.74). This mixing of equal portions of environmental and updraft air consists of a large number of subparcels that mix various fractions of environmental and updraft air together, with the resulting subparcels that are positively buoyant being entrained into the updraft and subparcels that are negatively buoyant being detrained into the environment. Some of these subparcels contain mostly updraft air, and so are likely to be positively buoyant, while others contain mostly environmental air, and so are likely to be negatively buoyant. A function $f(x)$ defines how often the various subparcel mixtures are created and is assumed to have a Gaussian form, with

$$f(x) = \frac{1}{0.97\sigma\sqrt{2\pi}} \left[e^{-(x-0.5)^2/2\sigma^2} - e^{-4.5} \right]. \tag{6.75}$$

Here x is the fraction of environmental air mixed in the subparcels and σ is the standard deviation of the distribution and is set to $1/6$ to encompass ± 3 standard deviations between the truncation points of $x = 0$ and $x = 1$. Thus, $f(x)$ is the relative frequency distribution of subparcel mixtures that integrates to a value of 1 over the range $x = 0$, 1 and has a mean value of 0.5. This expression yields the relative rates at which various subparcel mixtures are generated, and is used to determine the total rate at which these subparcels (after mixing) are entrained into the updraft as buoyant parcels, or detrained into the environment as negatively buoyant parcels. Since the number of subparcels that are buoyant after mixing depends upon the environmental and updraft air properties, the entrainment rate is variable and responds to changing environmental conditions. For example, holding everything else constant, moistening the environment yields more subparcels that are buoyant and a larger entrainment rate. Similarly, increasing the temperature difference between the parcel and the environment also leads to more subparcels that are buoyant and a larger entrainment rate.

Sensitivity tests indicate that the convective heating and drying profiles are very sensitive to whether the variable entrainment and detrainment approach is used as compared to the simpler approaches of entrainment only or equal entrainment and detrainment (Fig. 6.27). It appears that the variable entrainment and detrainment approach produces heating and drying profiles that agree well with observations (see Kain and Fritsch 1990). This variable entrainment and detrainment approach is conceptually similar to a stochastic

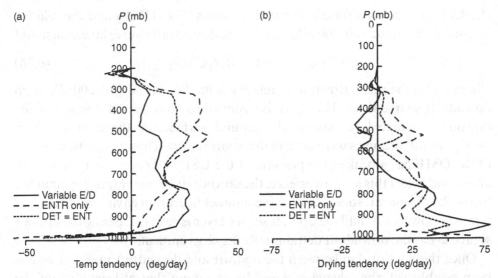

Figure 6.27. Kain–Fritsch scheme (a) heating and (b) drying profiles $(\mathrm{K\,day^{-1}})$ as a function of cloud model entrainment/detrainment type for a GATE composite sounding. From Kain and Fritsch (1990).

mixing model proposed by Raymond and Blyth (1986). Kain (2004) includes a minimum entrainment rate to alleviate problems with widespread light precipitation sometimes seen with low-entrainment clouds as permitted in the original version of the convective scheme.

The inclusion of downdrafts within a convective scheme is important, since they offset the updraft mass flux in the lower troposphere and help to make the convective warming and drying tendencies in this lower portion of the cloud layer more realistic (Kain 2004). Although the original Kain–Fritsch convective scheme uses the level of free sink as the level of downdraft origination (as in Gregory and Rowntree 1990), Kain (2004) notes that the pressure level of the level of free sink can vary significantly from close to 300 hPa to below 850 hPa. This leads to various problems with the predicted heating and drying rates from the convective scheme. Thus, the most recent version of the scheme assumes that the downdraft originates between 150 and 200 hPa above the updraft source layer (Kain 2004). A mass-weighted mean downdraft parcel is defined from the environmental T and q profiles between the updraft source layer and the downdraft origination level. This downdraft parcel is brought to saturation and defines the thermodynamic characteristics of the downdraft.

Downdraft air is saturated above the cloud base, while below the cloud base the relative humidity of the downdraft decreases by 20% per km. The magnitude of the downdraft mass flux (DMF) at the top of the updraft source layer

(USL) is defined as a function of updraft mass flux (UMF) and the relative humidity within the 150–200 hPa layer of the downdraft source layer, such that

$$DMF_{USL} = -2(1 - \overline{RH})UMF_{USL}, \qquad (6.76)$$

where $\overline{RH}$ is the mean (fractional) relative humidity in the 150–200 hPa deep downdraft source layer. The DMF is assumed to go to zero at the top of the downdraft source layer and at the ground surface, or the level at which the downdraft becomes warmer than the environment. Thus, the vertical profile of the DMF has a peak at the pressure of the USL and decreases to zero both above and below this level. In essence, the downdraft begins entraining air at the top of the downdraft source layer and continues to entrain environmental air as it moves downward until the updraft source layer is reached. Below the updraft source layer, the downdraft detrains into the environment.

Once the relationship between the updraft and downdraft mass fluxes has been established, the scheme is closed by assuming that at least 90% of the CAPE in the environment is removed by convection over the time period that the scheme is active. The updraft mass flux, and thus the proportional down-draft mass flux, is increased incrementally until this reduction of CAPE is achieved. In the original scheme version, the environmental CAPE is calculated based upon an undilute ascent of the cloud base parcel, whereas in the latest version the environmental CAPE is calculated for an entraining parcel (Kain 2004). This modification provides for more reasonable rainfall rates and it makes the updraft mass flux field a better predictor of convective intensity (Kain *et al.* 2003).

The remaining significant difference of the Kain–Fritsch convective scheme compared to the other mass flux schemes discussed is the method by which the scheme is activated. This procedure is often called a trigger function in the literature, as it leads to the activation of deep convection. In many ways, the trigger function for this scheme follows the concept of parcel theory in that a parcel must be lifted to its level of free convection for deep convection to develop. This type of trigger function was proposed by Fritsch and Chappell (1980), and remains an important part of the Kain–Fritsch convective scheme today.

Updraft source layers are defined as a 60 hPa deep layer that represents the likely inflow to a convective storm. Beginning at the model surface, vertically adjacent model layers are mixed until the depth of the mixture is at least 60 hPa deep. The mean thermodynamic characteristics of this mixture are calculated, as is the LCL of the mixture and the environmental temperature at this LCL height. Since observations suggest that convective development is favored by background vertical motion, the mixed parcel is given a temperature perturbation δT proportional to the grid-resolved, time running-mean vertical motion (w_g) defined as

$$\delta T = k\left[w_g - c(z)\right]^{1/3}, \tag{6.77}$$

where k is a constant and $c(z) = 0.02\,\text{m s}^{-1}$ if the mixed parcel LCL height (Z_{LCL}) is higher than 2000 m above ground level and $c(z) = 0.02\,(Z_{LCL}/2000)$ if the mixed parcel LCL is lower than 2000 m above ground level. This temperature perturbation is used to eliminate most parcels from further consideration by testing to see if the temperature of the mixed parcel at its LCL (T_{LCL}) plus δT is greater than the environmental temperature (T_{env}) at the mixed parcel LCL height. If

$$T_{LCL} + \delta T < T_{env}, \tag{6.78}$$

then the mixed parcel is no longer considered a possible candidate for starting convection. The base model layer for the updraft source layer is moved up one model level, and a new 60 hPa deep layer is defined and evaluated as a candidate for starting convection. This evaluation of 60 hPa deep layers continues until either deep convection is activated, as described below, or the lowest 300 hPa layer of the atmosphere has been evaluated.

If the mixed parcel $T_{LCL} + \delta T > T_{env}$, then this parcel is a candidate for convection. The mixed parcel is released at its LCL, with its original (unperturbed) temperature and mixing ratio, and given a vertical velocity (w_p) derived from δT and based upon the parcel buoyancy equation. This relationship is specified as

$$w_p = 1 + 1.1[(Z_{LCL} - Z_{USL})\delta T/T_{env}]^{1/2}, \tag{6.79}$$

where Z_{USL} is the height at the base of the updraft source layer. As the parcel rises, the parcel velocity is estimated at each model level above the LCL, after including the effects of entrainment, detrainment, and water loading. If the parcel vertical velocity remains positive over a depth greater than a minimum cloud depth (typically set to 3–4 km), then deep convection is activated. The scheme evaluates the environmental CAPE for this parcel, and determines the appropriate amount of updraft mass flux at the updraft source layer to remove at least 90% of the CAPE.

The process by which an environmental sounding is evaluated by the Kain–Fritsch scheme is illustrated in Figs. 6.28 and 6.29. First, the 60 hPa deep layer closest to the surface is mixed and brought to its LCL. Note that, owing to the well-mixed character of the boundary layer, the parcel does not have much convective inhibition and the trigger function estimates that the

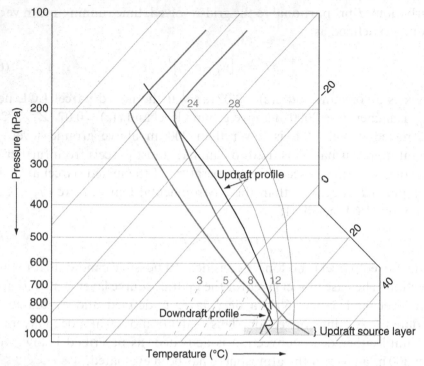

Figure 6.28. Model sounding illustrating the application of the Kain–Fritsch scheme to an initial environmental sounding (thick gray lines), with the updraft source layer shown in the first 60 hPa above ground level. The scheme updraft thermodynamic profile (black line) and the scheme downdraft thermodynamic profile (the black line between the surface and 800 hPa) are indicated. Courtesy of John S. Kain.

parcel can potentially overcome this inhibition. An updraft parcel is defined and its thermodynamic profile specified that includes the effects of entrainment. Starting from its initial upward motion, the parcel moves upward from one layer to the next and in this case the parcel is able to overcome the convective inhibition and reach its LFC. A downdraft initiates approximately 150 hPa above the updraft source layer, with its mass flux being proportional to the updraft mass flux. The updraft mass flux at the cloud base is scaled up, in tandem with the downdraft mass flux, until at least 90% of the CAPE is removed as shown in Fig. 6.29.

The convective tendencies from the Kain–Fritsch scheme are applied over an advective time period (t_c), defined as the time it takes a cloud to advect across a grid cell. The scheme determines this advective time using knowledge of the model grid spacing (Δx) and calculating the mean cloud-layer wind (V_{clayer}) from the cloud base to the cloud top, with

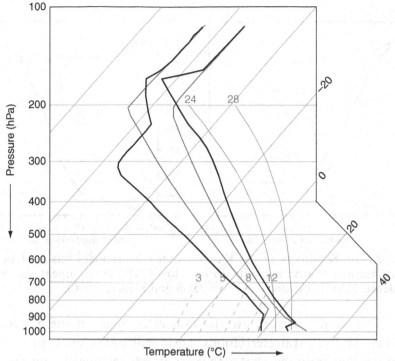

Figure 6.29. Initial environmental sounding (thick gray lines) and final sounding after modification by the Kain–Fritsch scheme (thick black lines). Note that the final sounding is drier, cooler in the planetary boundary layer, and warmer throughout most of the column from the cloud base to 200 hPa. Very little CAPE is present in the final sounding. Courtesy of John S. Kain.

$$t_c = \frac{\Delta x}{V_{clayer}}. \tag{6.80}$$

This is a rare case of a scheme which adjusts its action with the model grid spacing. Assuming convection activates within a model grid cell, the time period over which instability is neutralized via changes to the model profiles of T and q decreases as the grid spacing decreases. However, as the model grid spacing decreases below 10 km, it becomes hard to justify that convection acts to modify the environmental profile as quickly as suggested by the advective timescale, especially for cases with strong winds, so a minimum adjustment timescale of 30 min is assumed. Then the opposite problem arises in which the active convection in a given grid cell should be advected out of the grid cell in less than 30 min. This problem is alleviated somewhat by applying the tendencies only over the advective time period, but with the magnitude of the tendencies calculated using a 30 min adjustment timescale, as done in the

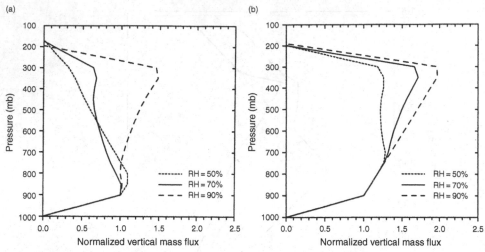

Figure 6.30. Sensitivity of the Kain–Fritsch scheme mass flux profiles to environmental relative humidity for (a) a low CAPE environment and (b) a moderate CAPE environment. From Kain and Fritsch (1990).

scheme. This simple thought exercise illustrates many of the difficulties of applying convective parameterizations at grid spacings below 15 or 10 km. The assumptions that are at the heart of the various schemes may become invalid at small grid spacings, even though the schemes still operate at these resolutions and still produce forecasts. One must ask what is being represented and visualized in model forecasts that use convective parameterization schemes at small grid spacings. The answer is not at all clear.

One of the environmental parameters that the Kain–Fritsch scheme is sensitive to is the environmental relative humidity. Results from Kain and Fritsch (1990) indicate that the normalized vertical mass flux can vary by more than a factor of 2 in upper levels as the relative humidity is varied from 50% to 90% (Fig. 6.30). This is reminiscent of the behavior of the Betts–Miller scheme and its sensitivity to cloud-layer moisture. Thus, although the Kain–Fritsch scheme does not depend explicitly upon cloud-layer moisture to activate, the resulting vertical mass flux is sensitive to this environmental variable. This is because the cloud-layer relative humidity influences whether parcels in some environments can remain buoyant up to the level of the specified minimum cloud depth. This is one example of how one of the variables often used within a deep-layer control scheme clearly influences a low-level control scheme, and why the two types of schemes often share similar sensitivities in their formulations. The atmosphere provides signals regarding convective development, but understanding and reproducing all the complex structures in those signals is quite challenging.

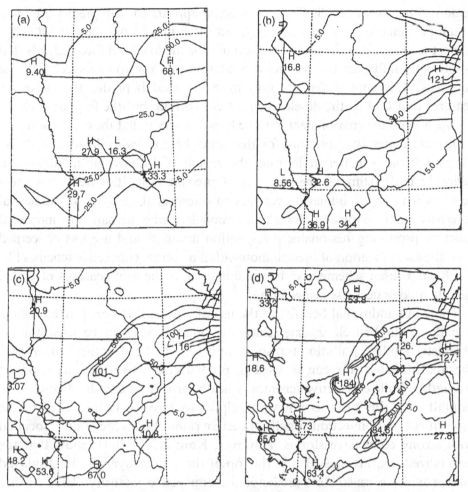

Figure 6.31. Accumulated precipitation (mm) over the 24 h period from 1200 UTC 16 June to 1200 UTC 17 June 1996 from the NCEP Eta model with the Kain–Fritsch scheme at grid spacings of (a) 78, (b) 39, (c) 22, and (d) 12 km. The first isoline is at 5 mm; otherwise the isoline interval is 25 mm. From Gallus (1999).

Simulations with different grid spacing by Gallus (1999) indicate that the precipitation produced by the Kain–Fritsch convective scheme as implemented in the NCEP Eta model is very sensitive to the grid spacing (Fig. 6.31). As the grid spacing decreases, the scheme generally produces larger maximum rainfall amounts and greater structure in the precipitation fields. This is what one would hope, since greater detail is expected as the grid spacing is reduced. However, this is a two-edged sword. Gallus (1999) concludes that while the Kain–Fritsch scheme reproduces the amounts of heaviest precipitation more accurately (undoubtedly with large contributions from grid-scale precipitation), the location is often displaced by several hundred kilometers from the observed region of heaviest rainfall. Thus, the scheme appears to help predict the maximum precipitation amounts correctly, but often fails to put the rainfall in the correct location. The details in the evolution of the low-level forcing fields that so strongly influence the convective initiation are key to the success of the forecasts. This implies that not only must the models predict the mesoscale features that lead to the development of convection, but the feedbacks of the convection on the environment must be handled very well if the convection is to propagate correctly. Note that for the same three cases explored, the Betts–Miller scheme is better at locating the region of heavier precipitation, but seriously underestimates the amounts of precipitation. It is clear that there remains a need for human forecasters to interpret these types of numerical forecasts wisely. However, for climate modelers who perhaps are interested more in producing reasonable precipitation amounts and are less concerned with the exact location of precipitation within a region, convective schemes like the Kain–Fritsch scheme may be quite helpful as the grid spacings of global climate models decrease below 50 km.

There are additional benefits to the use of a mass flux convective scheme, like the Kain–Fritsch scheme, since the mass fluxes can be used for air chemistry studies, pollution transport, and for weather forecast purposes as the updraft mass flux can be used to predict convective intensity (updraft strength). While convective schemes typically provide forecasts of convective rainfall rate, this often is not very helpful to forecasters interested in the intensity of the convection. Small convective rainfall rates are often associated with strong convective storms. However, Kain *et al.* (2003) show that the normalized updraft mass flux at the top of the source layer can be displayed and provides a reasonable predictor for convective intensity. They further suggest that the normalized UMF correlates well with lightning density, severe weather reports, and radar-derived volume-integrated liquid water.

6.4.5 *Emanuel convective scheme*

Emanuel (1991) develops a mass flux convective scheme in which the collective effects of the various subparcels in the cloud are represented by a buoyancy sorting approach instead of a plume model. The cloud base mass flux is determined by the assumption that convective mass fluxes act to keep the subcloud layer neutrally buoyant when displaced upward to just above the top of the subcloud layer (Raymond 1995). Owing to the importance of the subcloud layer, this scheme is categorized as a low-level control scheme. As discussed later, the importance of moisture to the scheme behavior suggests that it also is a moisture-control scheme.

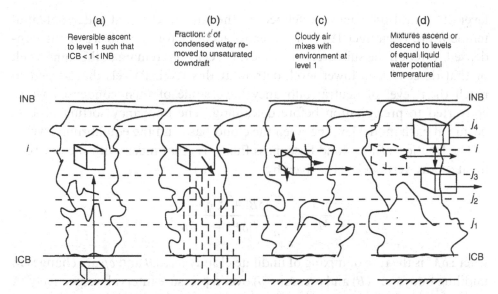

Figure 6.32. Idealized model of the buoyancy sorting that occurs in the Emanuel scheme. Reversible ascent from the subcloud layer to model level *i* between the cloud base (ICB) and the equilibrium level or the parcel level of neutral buoyancy (INB) (a), followed by a fraction of the condensed water converted to precipitation (b). The remaining cloudy air is mixed with the environmental air (c), and then the subparcel mixtures either descend or ascend to their level of neutral buoyancy with the environment (d). From Emanuel (1991).

Mixing in clouds is assumed to be inhomogeneous and to occur episodically, in contrast with entraining plume models in which the mixing is continuous. The buoyancy sorting procedure used is based upon the models of Raymond and Blyth (1986) and Telford (1975). As in the Kain–Fritsch scheme, the mixing of environmental air into a cloud occurs through a finite number of subparcels that contain different fractions of environmental and cloudy air. An equal probability distribution is assumed for the mixing fraction. However, unlike the Kain–Fritsch scheme in which the mixtures either entrain into the updraft or detrain into the environment, the resulting mixtures in the Emanuel scheme either ascend or descend to their level of neutral buoyancy (Fig. 6.32).

The buoyancy sorting process begins as a parcel of air is lifted without mixing from the subcloud layer to an arbitrary level *i* between the cloud base and the parcel equilibrium level. A specified fraction (ε') of the condensed water is converted to precipitation, and the rest retained as cloud water. The cloudy air and an equal amount of environmental air at level *i* are then mixed into various subparcels according to the assumed equal probability distribution of the mixing fraction. Each subparcel mixture then ascends or descends to its level of neutral buoyancy as defined by the subparcel liquid water potential temperature (Emanuel 1994) in order to account for the effects of the condensed water on the subparcel buoyancy. Parcels that remain at the same level, or that descend to a lower level, detrain at this level. Parcels that ascend to reach their level of neutral buoyancy have some of their condensed water converted to precipitation before detraining. The buoyancy sorting process occurs at each model level between the cloud base and the equilibrium level.

The fraction of the cloud base mass flux (M_b) that mixes with the environment at any vertical level is given by

$$\frac{\delta M}{M_b} = \frac{|\delta B| + \Lambda \, \delta p}{\sum_{i=1}^{N} \left(|\delta B| + \Lambda \, \delta p\right)}, \tag{6.81}$$

where δM is the rate of mixing of undilute cloudy air, δB and δp are changes in undilute buoyancy (B) and pressure (p) over a pressure interval, respectively, Λ is a mixing parameter ($0.06 \, \text{hPa}^{-1}$), and N is the number of vertical model levels (Emanuel and Zivkovic-Rothman 1999). The mixing rate can result in either entrainment or detrainment, depending upon the buoyancy of the resulting mixtures.

The evaporation of precipitation both within and below the cloud is allowed to create an unsaturated downdraft of fixed area that is approximated by a plume with no turbulent entrainment. Conservation equations for momentum, heat, and water determine the mass flux, the potential temperature, and the mixing ratio of this unsaturated downdraft at each level. Thus, the downdraft mass fluxes are a unique function of the updraft mass fluxes determined from the buoyancy sorting procedure.

The fraction of cloud water converted to precipitation, or the parcel precipitation efficiency, at each level between the cloud base and the cloud top is a function of temperature. The specification of these efficiencies plus the fractional area of precipitation that falls through unsaturated air are the main closure parameters of the scheme, as they determine the vertical heating profiles. The scheme also rigorously conserves the vertically integrated enthalpy.

The parameters in the Emanuel scheme are optimized by Emanuel and Zivkovic-Rothman (1999) using a single-column model and intensive flux data from 1 month of the Tropical Ocean Global Atmosphere Coupled Ocean–Atmosphere Response Experiment (TOGA/COARE). Tests of the optimized scheme using independent data from the Global Atmospheric Research Program Atlantic Tropical Experiment (GATE) yield results similar to or better than those of two other well-known convective schemes.

6.5 Shallow convection

It is somewhat surprising that, in contrast to deep convection, shallow convective parameterization has received less attention. Shallow cumulus clouds are the most frequently observed of all tropical clouds (Johnson *et al.* 1999) and strongly influence boundary layer depth and boundary layer values of temperature, mixing ratio, and winds. Shallow cumulus clouds also are known to be important to global climate as they strongly influence the radiation budget. Grabowski *et al.* (2006) show that both numerical weather prediction and climate models tend to produce deep convection too early in the daytime portion of the diurnal cycle over land due to these models bypassing the observed shallow convection phase. Yet compared to the fairly sizeable variety in parameterization schemes for deep convection, there exist fewer parameterization schemes designed especially for shallow convection.

We have already seen that the Betts–Miller, Tiedtke, and Gregory–Rowntree schemes have a shallow convection component that is important to consider when trying to understand how the scheme behaves within the numerical weather prediction model. In the Tiedtke and Gregory–Rowntree schemes the shallow convection component is handled similarly to the deep convective component. In the Betts–Miller scheme, the two components are treated very differently. Two other schemes designed only for shallow convection are documented by Deng *et al.* (2003) and Bretherton *et al.* (2004). Both of these schemes are mass flux schemes and use the same buoyancy sorting approach as in the Kain–Fritsch convective scheme. The results of Siebesma and Cuijpers (1995) based upon large-eddy simulations suggest that a single plume mass flux model can provide a very reasonable approximation to the ensemble mean fluxes from a large-eddy simulation. The main differences in the Deng *et al.* (2003) and Bretherton *et al.* (2004) schemes are their closure assumptions and trigger functions. Results from both studies suggest that shallow convective schemes can play an important role in defining boundary layer structure and the amount of cloud.

6.6 Trigger functions

All convective parameterization schemes have what can be called "trigger functions." These functions are the set of criteria in the convective parameterization scheme that determine when, where, and if the scheme is to be activated. Since the processes that produce convection typically occur on very small scales, particularly for thunderstorm formation, one cannot model them explicitly for grid spacings much greater than 1 km. Thus, trigger functions are a very important component of convective parameterization, even if they are not often discussed. Indeed, one can argue that trigger functions are as important as the actual subgrid adjustments in the development and evolution of convection, suggesting a need for more sophisticated trigger functions that account for more of the physical processes that lead to initiation.

A summary of the sensitivity of model simulations to the trigger function definition is provided by Kain and Fritsch (1992). They use the same model and the same convective parameterization scheme – the Kain–Fritsch scheme – but vary the trigger function. They use a trigger function based upon the one outlined in Fritsch and Chappell (1980) (FCT), a trigger based upon the column integrated moisture convergence (AKT), a trigger based upon the amount of convective inhibition (NAT), a trigger based upon the depth between the lifting condensation level of a parcel and its level of free convection (LDT), and a trigger that examines boundary layer forcing (BLT). Results from model simulations of the same mesoscale convective system event illustrate clearly that differences in trigger functions provide for a substantial divergence in the model simulation of convection under strong large-scale forcing for upward motion. In particular, differences in the locations of outflow boundaries, precipitation, and even the warm front are seen (Fig. 6.33). Similar results are found for three variations of the trigger function in Stensrud and Fritsch (1994), in which several mesoscale convective systems develop under weak forcing for upward motion (Fig. 6.34). The amount of convection activated is quite different for the three trigger functions used in this case. These results emphasize the pivotal role that the convective trigger function plays in the development and evolution of convection.

6.7 Discussion

The basic concepts, and more than a few of the important details, of deep-layer and low-level control convective parameterization schemes have been examined. Yet it is clear that there remain a number of important issues involving convective parameterization which are not fully understood. Deep-layer control schemes appear to be based upon relationships that do not always stand up to scrutiny, such as the quasi-equilibrium closure assumption as discussed by Mapes (1997), whereas low-level control schemes require a very accurate mesoscale specification of the environment to yield correct forecasts and include a number of assumptions that are difficult to test with observations. Many facets of deep-layer control schemes appear in low-level control schemes, and vice versa. Emanuel (1991) suggests that many convective parameterization schemes make sweeping assumptions about the distributions of

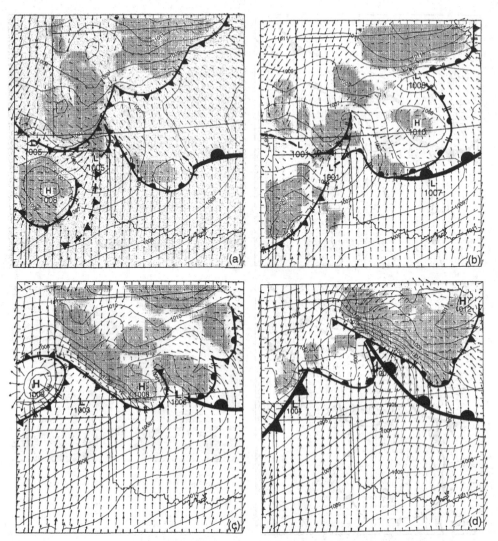

Figure 6.33. Model surface fields at the 12 h forecast time using the FCT (upper left), the AKT (upper right), the LDT (lower left), and a combination of the NAT and BLT (lower right) showing surface wind vectors, sea-level pressure (hPa) every 1 hPa, and where shading indicates areas of convective activity over the past hour. All of the simulations are valid at 0000 UTC 11 June 1985. From Kain and Fritsch (1992).

heating and drying in clouds. He argues that the representation of convection must be built (p. 2314) "solidly on the physics and microphysics of cloud processes as deduced from observations, numerical cloud models, and theory." This certainly highlights the challenges that remain in the realm of convective parameterization.

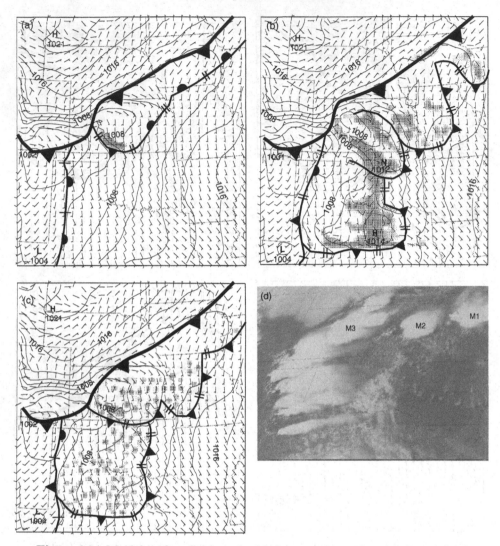

Figure 6.34. Model surface fields at the 9 h forecast time using: (a) the original Fritsch and Chappel (1980) trigger function, (b) the BLT, and (c) a non-organizing boundary layer trigger showing 10 m winds, sea-level pressure every 2 hPa, and hatched regions depicting areas with convective rainfall exceeding 1 mm during the last hour. Positions of the front, outflow boundaries, and a dryline are indicated. Panel (d) is a corresponding visible satellite image showing the observed distribution of deep convection. All times are valid 2100 UTC 11 May 1982. From Stensrud and Fritsch (1994).

One important aspect of convection in many regions of the globe is its diurnal cycle. The observed summertime diurnal cycle of rainfall over the USA east of the Rocky Mountains is compared to the diurnal cycle produced by the NCEP Eta model and the new Weather Research and Forecast (WRF) model by Davis *et al.* (2003). Both models are run at 22 km grid spacing or less, with the Eta model using the BMJ convective scheme and the WRF model using the Kain–Fritsch convective scheme. Results show that the models are unable to reproduce well the west-to-east movement of convective systems during the nighttime hours (Fig. 6.35). Davis *et al.* conclude that the diurnal cycle of rainfall in both models is poorly represented and suggest that these errors are linked to the inadequacies of the convective parameterization schemes.

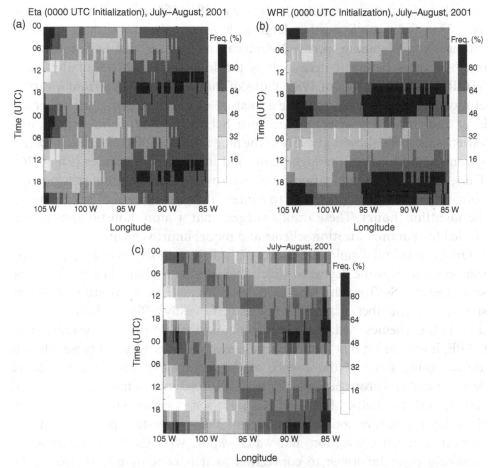

Figure 6.35. Time–longitude rainfall frequency diagram produced from 3 h stage-IV data (bottom), 3 h Eta model forecasts (upper left), and 3 h WRF model forecasts (upper right). All model forecasts start at 0000 UTC. The data are averaged meridionally within strips of 0.1° longitude width between 30° and 48° N during July and August of 2001. The diurnal cycle is repeated to improve the interpretation. Note the darker zone of higher rainfall frequencies seen in the observations (bottom) that slopes downward from left to right, indicating systems that propagate from 105° W to 85° W as time increases. This behavior is missing from both model forecasts. From Davis *et al.* (2003).

While Davis *et al.* (2003) may be right in that convective parameterizations led to the problem in forecasting the diurnal cycle and propagation of deep convection during the summertime, perhaps the better question to consider is how the convective parameterization, boundary layer, and other model schemes interact to produce nocturnally propagating systems. This question is difficult to answer and has been largely ignored, with most studies focusing instead on using the reductionist approach and looking only at the behaviors of a single scheme. However, Stensrud *et al.* (2000a) present results from an ensemble of simulations of the 27–28 May 1985 MCS event. This MCS lasts over 33 h, forms underneath a large-scale ridge axis, and propagates from southeastern Wyoming, across Nebraska and Kansas and Oklahoma, and into Arkansas before it decays. It is the longest-lived MCS of the PRE-STORM field campaign of 1985. Paths of the simulated MCS from a variety of model configurations, that use a mixture of different planetary boundary layer and convective schemes (Arakawa–Schubert, BMJ, Kain–Fritsch, and varieties thereof), show that many of the model simulations capture convective systems that move off the high plains and into the plains states (Fig. 6.36). Thus, some of the combinations of convective parameterization schemes and boundary layer schemes are able to capture the propagation of MCSs during the nighttime hours. These results suggest that a more holistic approach is needed for parameterization scheme and model improvement.

One curious and illuminating aspect of the behavior of convective parameterization schemes is the degree to which their results overlap. The tracks of the simulated MCSs from the long-lived MCS event (Fig. 6.36) are remarkably similar over a number of hours. Considering the vastly different closure assumptions in the schemes used, that range from quasi-equilibrium to the removal of CAPE, it is amazing that the schemes agree at all. Clearly, all of these schemes are capturing some of the essence of the factors that influence convective development and propagation in this weakly forced environment. As mentioned earlier, both the Betts–Miller and Kain–Fritsch schemes have sensitivities to cloud-layer moisture content. Tompkins (2001) suggests a positive feedback between tropical convection and water vapor, with the local environment becoming more favorable to convection as it is concurrently moistened by convection. Similar behaviors are seen in midlatitudes, when numerous convective towers (plumes) ascend into the environment and die before a long-lived cell is able to survive. Thus, there exists some observational and cloud-modeling evidence to suggest that convection is influenced by cloud-layer moisture.

Convective schemes also interact with the grid-scale, explicit microphysical parameterizations in ways that are difficult to predict a priori. Stensrud *et al.* (2000a) show simulations from an MCS that developed under strong large-scale

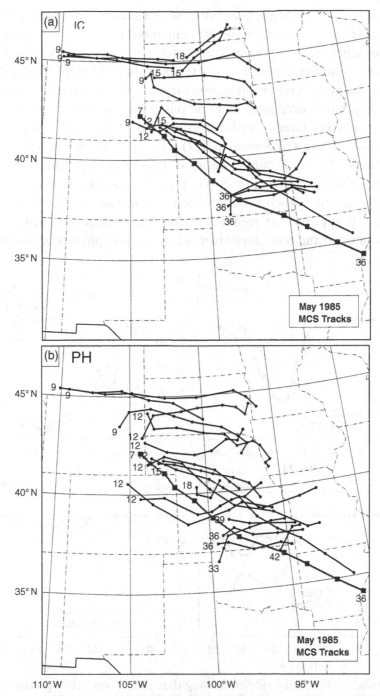

Figure 6.36. Tracks of simulated MCSs from an ensemble of model runs during the 48 h period beginning 1200 UTC 27 May 1985 from the Penn State National Center for Atmospheric Research fifth-generation mesoscale model (MM5) using (a) different initial conditions and (b) a mixture of different boundary layer and convective parameterization schemes. Observed track, derived from radar observations, is shown in gray with solid square boxes indicating the MCS centroid every 3 h. From Stensrud *et al.* (2000a).

forcing for upward motion in association with a shortwave trough. In this case, the MCSs from the ensemble of different mixtures of boundary layer and convective schemes largely agree in producing the initial convective development at the same time and location. However, after 18 h the modeled MCSs diverge, with two different propagation directions being present in the simulations. This solution divergence may be related to the way in which different convective schemes interact with the grid-scale, explicit microphysical parameterizations. Wang and Seaman (1997) show that the percentage of convective rainfall to the total rainfall varies dramatically depending upon which convective scheme is used (Fig. 6.37). This variation leads to differences in both the mesoscale structures of the convection and the depth of the cold pool. Thus, it appears that even in strongly forced large-scale environments, the differences in the boundary layer, explicit microphysics, and convective

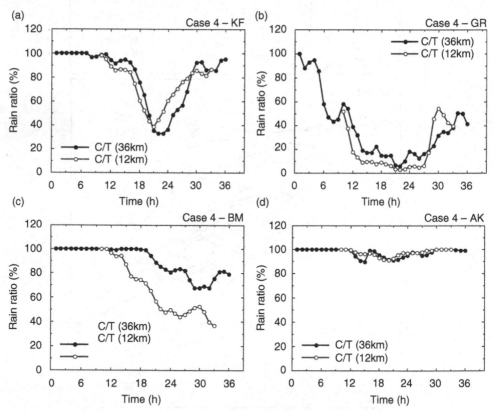

Figure 6.37. Hourly time series showing the ratio of convective precipitation to total precipitation (%) for (a) Kain–Fritsch, (b) Grell (a modified Arakawa–Schubert), (c) Betts–Miller, and (d) Anthes–Kuo (a modified Kuo) schemes at 36 km (dots) and 12 km (open circles) grid sizes for an MCS event from 6–8 May 1985. From Wang and Seaman (1997).

parameterization schemes lead to divergence in the model simulations of convection.

The different evolutions of convection found when using different convective schemes often lead to the belief that convective parameterization is the remaining problem with our numerical forecast models, and that once forecasts with small enough grid spacing are available convective parameterization will no longer be needed and all of these problems will go away. This appears to be a naïve viewpoint. It is a continuation of the historical trend that suggests that smaller grid spacing always results in better forecasts. Yet we already know that this is not necessarily the case, as evidenced by the results of Gallus (1999) and the minute increases in quantitative precipitation forecast skill in the past decades (Fritsch and Carbone 2004). It is not at all clear that moving towards explicit-only forecasts of convection will improve model forecast skill. The models certainly will be more capable of producing realistic atmospheric features and structures, but improvements to forecast skill are a much higher standard of comparison. Never forget that perfect precipitation forecasts result from getting everything else right in the model. Thus, perfect precipitation forecasts require that the radiation, boundary layer, soil, vegetation, and cloud cover schemes also are perfect in order to allow the precipitation processes to respond appropriately. This is a huge challenge, whether the convection is parameterized or explicitly resolved.

While this discussion has focused upon concerns most related to short-range numerical weather forecasts, these same challenges also arise in relation to medium-range, seasonal, and climate predictions. As these models move towards smaller grid spacing, they become very much like the short-range forecast models of today. Since organized deep convection plays such an important role in the hydrological cycle and in modifying the large-scale circulation patterns, it is crucial to represent these organized systems in longer-range predictions. The ability of the numerical models to represent these systems hinges in large part upon the convective parameterization scheme and perhaps its interactions with an explicit microphysics parameterization. Regional climate simulations by Giorgi *et al.* (1993) indicate that summertime precipitation amounts are very sensitive to the convective scheme being used and the closure assumptions.

Emanuel and Zivkovic-Rothman (1999) indicate that water vapor is the strongest feedback factor in all 19 of the global climate models used in externally forced global climate change simulations during the Atmospheric Model Intercomparison Project (AMIP) as of 1990 (Cess *et al.* 1990; Gates *et al.* 1999). They further describe how atmospheric water vapor content depends strongly on the microphysical processes within both convective and stratiform clouds, and advocate more attention to the microphysical aspects of the convective parameterization schemes used in climate models. Indeed, they state that out of the 36 quasi-independent models participating in AMIP almost 75% of them use either a Kuo or Arakawa–Schubert scheme (deep-layer control) or the very simple Manabe *et al.* (1965) scheme. While cloud–radiation feedbacks also play an important role (Cess *et al.* 1990), Emanuel and Zivkovic-Rothman argue that the microphysical processes represented in convective parameterization schemes are important to simulations of global climate change. Results from Albrecht *et al.* (1986) further suggest that the cumulus parameterization problem is intricately linked to parameterization of cloud–radiation interactions, and we are nowhere near resolving shallow cumulus clouds in any of these models. Thus, convective parameterization will be a field of active research for decades to come and it is hoped that the studies done using convective parameterization for short-range forecasts can benefit medium-range, seasonal, and climate predictions and vice versa.

6.8 Questions

1. Plot the sounding data below on a skew-T log-p diagram.

HT	P	TC	TD	SPD	DIR	U	V
791.0	920.5	36.7	16.3	11.0	180	0.0	11.0
965.1	903.0	31.9	12.7	13.0	187	1.6	12.9
1 503.9	850.0	26.1	12.2	15.0	196	4.1	14.4
2 132.6	791.0	20.4	11.9	14.0	191	2.7	13.7
2 819.5	730.0	13.8	10.4	7.0	214	3.9	5.8
2 889.4	724.0	16.0	1.7	5.0	218	3.1	3.9
3 175.4	700.0	15.2	−8.2	1.0	204	0.4	0.9
5 098.8	554.0	−0.9	−15.9	1.0	290	0.9	−0.3
5 906.3	500.0	−8.4	−19.0	5.0	304	4.1	−2.8
6 713.2	450.0	−15.4	−24.4	7.0	269	7.0	0.1
7 091.2	428.0	−16.3	−34.5	12.0	267	12.0	0.6
7 597.0	400.0	−19.7	−37.3	12.0	264	11.9	1.3
8 409.8	358.0	−26.4	−43.0	13.0	270	13.0	0.0
8 960.1	329.8	−30.0	−34.0	14.2	284	13.8	−3.5
9 387.1	310.6	−30.4	−32.9	17.5	273	17.5	−1.0
9 630.6	300.1	−32.7	−35.5	17.4	270	17.4	0.1
10 501.1	264.7	−40.2	−40.0	18.0	259	17.6	3.3
10 888.0	250.0	−43.8	−50.0	18.5	255	17.8	4.8
12 355.9	200.0	−53.3	−60.0	33.2	251	31.4	10.8
13 723.4	161.0	−62.5	−70.0	24.9	282	24.3	−5.1
14 160.5	150.0	−62.1	−75.0	20.8	294	18.9	−8.6
15 157.9	127.4	−67.1	−80.0	25.8	300	22.4	−12.8

2. Calculate the saturation points (SPs) of the sounding (use different colored pencils).

3. Label any distinct mixing lines that occur over layers at least 100 hPa deep (if any) and calculate the value of β for those mixing lines. Do at most three of these layers, if there are that many.

4. Calculate the cloud base (LCL) and the cloud top (equilibrium temperature level) for a representative parcel. How is this parcel defined? Discuss why the way in which this parcel is defined is important.

5. Draw the first-guess reference temperature and humidity profiles for the Betts–Miller convective scheme. Please use a second skew-T log-p diagram for this plot. Using knowledge of this scheme, adjust the BMJ profiles to conserve enthalpy and plot the resulting profiles. Will the scheme activate? Discuss how this decision was made. If the scheme will not activate, then what would it take to make the scheme activate? (Note that a more general, approximate answer is desired. There is no need to code up the entire scheme!)

6. Assume that the Kain–Fritsch scheme activates at this grid point. Draw an undilute updraft profile and an undilute downdraft profile. Knowing that the final profile is a mixture of the updraft, downdraft, and environmental profiles, estimate a final reference profile for the scheme. Draw these estimated reference temperature and moisture profiles for the KF scheme, and explain how the three contributing profiles are mixed. Please use a third skew-T log-p diagram for this plot. (Note the details on how to do this calculation were not provided formally, and since the KF scheme uses a buoyancy sorting procedure, one cannot reproduce the resultant profiles exactly without writing a lengthy computer program. However, several examples were provided and the general working of the KF scheme was summarized, so it is possible to provide a general sketch of what the reference profiles look like.)

7. How do the adjusted profiles from the Betts–Miller and Kain–Fritsch schemes compare? Under what circumstances would one be preferred over the other? Why?

8. Rework Questions 1–7 using a sounding from a more tropical environment.

9. If possible, examine hourly soundings from an operational model forecast in a region of convection. From this examination, find soundings just prior to, during, and after convection has developed in the model. Outline the convective scheme used in this model forecast, and use these soundings to illustrate the behavior of this scheme.

10. Present an argument for why one would choose a deep-layer over a low-level control convective scheme, or vice versa. Why are both types of convective schemes in use today?

7

Microphysics parameterizations

7.1 Introduction

As discussed more fully in the previous chapter, moist convection plays a very important role in the atmosphere. Moist convection is a key link in the El Niño–Southern Oscillation that influences global circulation patterns, while organized mesoscale regions of convection are also known to modify the local and large-scale environments across the globe. Moist convection produces clouds, some of which can persist for days, influencing the absorption and scattering of solar radiation and the absorption of terrestrial radiation. Clouds also affect the concentration of aerosol particles through scavenging, precipitation, and chemical interactions.

Cloud formation is accomplished primarily by upward vertical air movement in cloud-free regions leading to patches of air that have relative humidities in excess of 100%. Once the relative humidity is above 100%, cloud droplets can form producing clouds. The microphysical processes that govern cloud particle formation, growth, and dissipation on very small scales play an important role in how moist convection develops and evolves. Cloud microphysical processes are very important to predictions of the atmosphere at temporal scales ranging from minutes to centuries, owing to the effects of latent heat release due to the phase changes of water and the interactions between clouds and radiation (GEWEX 1993).

Cloud microphysical processes represent an important uncertainty in climate modeling. Increases in aerosols due to either anthropogenic or natural causes produce an increase in cloud droplet concentration. For a fixed liquid water content, this leads to a decrease in droplet size and to an increase in cloud optical thickness and hence cloud albedo (Twomey 1977). This is particularly important over the oceans where there are fewer aerosols. However, the decreased cloud droplet size also reduces the precipitation efficiency, thereby increasing the cloud liquid water content, cloud thickness, and cloud lifetime (Albrecht 1989). Some aerosol particles also act to decrease cloud reflectance (Kaufman and Nakajima 1993). These indirect aerosol effects strongly influence fractional cloudiness and albedo, producing a net cooling effect, yet they represent one of the largest uncertainties in studies of global climate change. An added complication arises due to the aerosol particle chemical composition being important to its activation as a cloud droplet (Shulman *et al.* 1997; Raymond and Pandis 2002), although activation appears to be mainly determined by aerosol particle size (Dusek *et al.* 2006).

Cloud microphysical processes also are important to very short-lived atmospheric phenomena, such as thunderstorms. Brooks *et al.* (1994a, b) suggest that the strength and lifetime of low-level mesocyclones (a rotating vortex of 2–10 km diameter within a convective storm) are a function of the balance between outflow development and low-level baroclinic generation of vorticity. Outflow strength is determined by the evaporation of precipitation falling from the convective storm and thus is a key factor in low-level mesocyclones that can produce tornadoes. While microphysical processes are involved in all aspects of the storm lifecycle, this example highlights how the processes that occur at the molecular scale alone can influence much larger phenomena.

These studies highlight the need to include microphysical processes within numerical weather prediction models. As model grid spacing decreases, it becomes possible to model cloud development and evolution explicitly by incorporating additional equations into a model that represent the various water substance phases and cloud particle types. This is referred to as an explicit microphysics parameterization, in which clouds and their associated processes are represented directly on the model grid. This parameterization approach contrasts with convective parameterization, where only the cumulative effects of clouds are represented on the grid (an implicit parameterization). It is generally believed that moving away from convective parameterization toward the explicit representation of cloud processes is beneficial, although many uncertainties have yet to be explored fully.

The distinction between implicit and explicit parameterization of cloud processes is often blurred in actual models. As discussed in Chapter 6, many models that are run at 10–40 km grid spacing use some combination of convective parameterization and explicit cloud microphysics parameterization. The convective parameterization scheme activates deep convection prior to the occurrence of water vapor saturation at a grid point, while the resolvable-scale model equations are assumed to develop the appropriate larger than convective-scale precipitation processes. Thus, the implicit and explicit representations of convection act in combination to produce the convective activity.

Figure 7.1. Stratocumulus clouds observed over Massachusetts photographed from a jet airplane. Notice the apparently repeatable patterns within features that are highly irregular.

Zhang *et al.* (1988) show that using both implicit and explicit schemes is a good method for handling mixed convective and stratiform precipitation systems and does not lead to double counting the heating or moistening produced by convection.

Since clouds are highly irregular, fractal entities (see Fig. 7.1), one wonders at what model grid spacing clouds can be resolved accurately. Weisman *et al.* (1997) suggest that 4 km grid spacing is just sufficient to resolve the circulations associated with squall lines. However, for predictions that include isolated thunderstorms, results from experimental model runs at 4 km grid spacing produced for the NOAA Storm Prediction Center (SPC) indicate that a 4 km grid spacing is too large. The convection produced at 4 km tends to develop upscale quickly, thereby producing too many organized convective systems when compared with observations. These results suggest that the largest grid spacing at which clouds may be accurately represented is unknown. While it is certainly less than 4 km, the behavior of thunderstorms is known to change as the grid spacing is decreased from 2 km to 500 m

(Adlerman and Droegemeier 2002) and the behavior of squall lines is known to change as the grid spacing is decreased from 1 km to 125 m (Bryan *et al.* 2003). While these behavioral changes may not greatly modify the cloud properties of the thunderstorm or squall line, they certainly modify the evolution of features that are important to severe weather warning operations, such as mesocyclones and gust fronts. Thus, it is likely that the required grid spacing is strongly tied to the atmospheric phenomena one wishes to predict. For example, very small grid spacing may be needed to predict the detailed evolution of rainfall from a precipitating mesoscale convective system. Only time and experience will yield the guidance needed to make good decisions regarding grid spacing choices.

Numerical modelers are moving quickly to incorporate microphysics parameterization schemes in models as the grid spacing goes below about 30 km. Whether these schemes are part of a hybrid approach that includes convective parameterization or a completely explicit approach, there is a need to include microphysics parameterizations in most operational forecast models in use today. Model developers face considerable challenges in their efforts to implement microphysics parameterization schemes that are both realistic and computationally affordable.

Two challenges to microphysics parameterizations are the number of phase changes of water, and the number of different interactions between cloud and precipitation particles that must be considered. The phase changes of water that can occur in the atmosphere are

    vapor → liquid (condensation)
    liquid → vapor (evaporation)
    liquid → solid (freezing)
    solid → liquid (melting)
    vapor → solid (deposition)
    solid → vapor (sublimation)

and do not occur at thermodynamic equilibrium. Instead, the forces of surface tension for water drops and the surface free energy for solid particles must be taken into account. To illustrate the second challenge, imagine being in an elevator with glass walls moving upward from the ground surface underneath a severe thunderstorm. Large raindrops and hail are hitting the elevator as it sits on the ground. As the elevator begins to move upward, the hail and rain persist, but raindrops of somewhat smaller sizes are also seen. At the cloud base, a dramatic transition occurs in that tiny cloud droplets are everywhere and act to obscure visibility. It is still raining and hailing, but the variety of sizes of these cloud particles is overwhelming. As the elevator moves higher

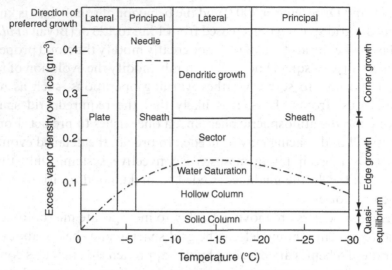

Figure 7.2. Diagram of ice crystal habit (form) as a function of temperature (°C) and excess vapor density above ice saturation over a flat ice surface as determined by measurements of T. Kobayashi. From Fletcher (1962). Bailey and Hallett (2004) provide an improved habit diagram for temperatures below − 20 °C and extend the diagram to a temperature of −70 °C.

and higher, the hail stops as the hail shaft is exited. The temperature outside soon dips below freezing, and large snowflakes with the occasional raindrop are now seen. As the elevator continues upward, a wide variety of frozen particles (known as habits) are observed that look like needles, plates, snow-flakes, and large clumped snowflakes (Fig. 7.2). This mixture of different particle shapes and sizes also contributes to the challenge of parameterizing microphysics, since these shapes and sizes influence how the particles interact with each other. Different vertical profiles of particles are seen in different parts of the thunderstorm, further complicating the situation. Yet parameter-izations must be relatively simple if they are to be used in numerical models with predictive capabilities.

In this chapter a number of microphysics parameterizations are explored that typically are used for numerical model simulations with relatively small grid spacings. It is assumed in developing these schemes that there is no subgrid-scale variability in the microphysics variables within the model grid cells. However, this assumption cannot be used in climate models that use larger grid spacings and is an arguable assumption for many situations even for grid spacings below 30 km. Another approach is to allow for subgrid-scale variability in the microphysics parameterization by also calculating the frac-tion of the grid cell filled with cloud. In this way the grid cell is divided into clear and cloudy portions. This type of approach typically uses a simpler microphysics parameterization combined with a diagnostic or predictive equation for cloud cover. As one might imagine, the specification of fractional cloud cover within a grid cell can lead to large changes in cloud–radiation interactions. Thus, an overview of microphysics parameterizations is presented in this chapter, and the additional complication of predicting cloud cover is saved for Chapter 9 in order to also include the effects of radiation in the discussion.

7.2 Particle types

Before examining the various microphysics parameterization schemes, it is important to overview briefly the physical processes that are important in the formation of the main microphysical particle types observed. These different hydrometeor types include cloud droplets, raindrops, ice crystals, aggregates, sleet, and rimed ice particles.

7.2.1 Cloud droplets

Liquid cloud droplets form when water saturation is exceeded over a range of temperatures from above freezing to about $-40\,°C$. The Clausius–Clapeyron equation describes the equilibrium state for a system of water vapor over a flat liquid surface of pure water in which condensation and evaporation occur at identical rates. Since cloud droplets have a curved surface, an adjustment to the Clausius–Clapeyron equation is needed to account for the force of surface tension due to droplet curvature. This equation for the equilibrium vapor pressure (e_s) over the surface of a cloud droplet of radius r was derived by William Thomson, later Lord Kelvin, in 1870. It states that

$$e_s(r) = e_s(\infty)\, e^{2\sigma/rR_v\rho_w T}, \tag{7.1}$$

where σ is the surface tension (approximately $0.075\,\mathrm{kg\,s^{-2}}$ over the range of meteorologically relevant temperatures), ρ_w is the density of water, R_v is the gas constant for water vapor ($461\,\mathrm{J\,kg^{-1}\,K^{-1}}$), T is the temperature of the air, and $e_s(\infty)$ is the saturation vapor pressure over a flat liquid surface given by the Clausius–Clapeyron equation. This equation is often rearranged to express a saturation ratio S, such that

$$S = \frac{e_s(r)}{e_s(\infty)} = e^{2\sigma/rR_v\rho_w T}. \tag{7.2}$$

A saturation ratio with a value of 1 implies that the atmosphere is just saturated (i.e., the relative humidity equals 100%). Observed maximum saturation ratios in the atmosphere are typically less than 1.01, or 1% supersaturation. A numerical simulation by Clark (1973) suggests that supersaturations may reach 10% in localized regions of strong updrafts with precipitation, although these simulated high supersaturation values have yet to be verified or refuted by observations.

One would expect, and observations also indicate, that the first cloud droplets to form within a given air parcel are very small droplets. Yet the equilibrium vapor pressure $e_s(r)$ over a cloud droplet surface calculated using (7.1) increases as the droplet radius gets smaller. For typically initial cloud droplet radii, the required saturation ratio is above 2, or above 200% relative humidity! Since the maximum observed values of supersaturation in the atmosphere are much smaller, the formation of cloud droplets consisting of pure water is not common. The formation of cloud droplets instead depends most often on the presence of small particles in the atmosphere of micron size that have an affinity for water. Thus, it is not pure water that forms cloud droplets, but water in solution. The dissolved solute lowers the equilibrium vapor pressure. Thus, these particles, or aerosols, are centers for condensation and so are called cloud condensation nuclei (CCN). While the size and concentration of CCN vary across a large range, from 10^{-3} to 100 µm in size and orders of magnitude in concentration, they are very common in the atmosphere.

The presence of CCN necessitates a further modification to the equation for the saturation ratio. This leads to the expression (Rogers 1976)

$$S = \frac{e_s(r)}{e_s(\infty)} = \left(1 - \frac{b}{r^3}\right)e^{2\sigma/rR_v\rho_w T}, \tag{7.3}$$

where $b = 3im_S M_W/4\pi\rho_L M_S$, m_S is the solute mass (kg), M_S is the molecular weight of the solute, M_W is the molecular weight of a water molecule (18.016), and i is the degree of ionic dissociation (e.g., $i = 2$ for a dilute NaCl solution). For a given temperature T and solute type and mass, the resultant curve of saturation ratio S versus droplet radius r (called a Köhler curve) shows several interesting features (Fig. 7.3). First, for small radii, the solution effect dominates and for very tiny droplets the drop can be in equilibrium with the environment at values of S less than 1.0 (or relative humidities of less than 100%). This behavior (called deliquescence) accounts for haze in the atmosphere that restricts visibility and can turn the sky white instead of blue. For large radii, the surface tension effect dominates. In between these two extremes is where cloud droplet life is very interesting. When the slope of the curve is

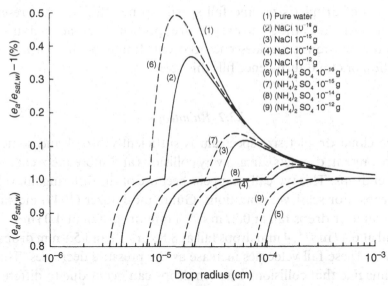

Figure 7.3. Equilibrium vapor pressure over an aqueous solution drop as a function of drop radius (cm) for various amounts of NaCl (solid lines) and $(NH_4)_2SO_4$ (dashed lines) in solution at $20\,^{\circ}C$. From Pruppacher and Klett (2000).

positive, an increase in S is required for the drop to grow in size. However, once the slope equals zero, a critical point occurs. If the drop grows in size beyond where the slope is zero (the peak in the curve), it continues to grow without an increase in S and reaches cloud droplet sizes of at least $10\,\mu m$.

The radius at which the peak of the curve is reached is called the critical radius r^*. Drops with radii smaller than r^* only grow in response to increases in environmental relative humidity (or S), and therefore often are called haze particles. Also note that the greater the mass of the solute, the larger the value of r^* becomes. When a drop exceeds r^*, then the simple relationships discussed so far suggest that it grows indefinitely. However, in reality many water drops compete for the available water vapor and the environmental relative humidity decreases and not all drops continue to grow.

The initial growth of a cloud droplet is due to condensation and is proportional to $(S-1)/r$. Thus, as the droplet gets bigger its rate of growth slows. Because of this effect, the timescale required for condensation to produce raindrops is much longer than the timescale over which actual clouds produce precipitation. Another, more efficient mechanism for droplet growth must be acting, namely collision and coalescence, to greatly accelerate droplet growth. However, the collision and coalescence mechanism requires the presence of a distribution of droplet sizes and fall speeds to be effective. At present it is uncertain exactly how the cloud droplet size spectrum is broadened sufficiently to allow for collision and coalescence to occur. It may be that the aerosol size distribution or cloud turbulence fills this role.

7.2.2 Raindrops

Once the cloud droplet size spectrum is sufficiently broad and some larger drops are present, droplets can grow by collisions and subsequent coalescences with other drops. This typically happens because of the differing fall velocities of the drops. For sea-level conditions, Gunn and Kinzer (1949) indicate that 0.1 mm diameter drops fall at $0.27\,\mathrm{m\,s^{-1}}$, 1 mm drops fall at $4.03\,\mathrm{m\,s^{-1}}$, 2 mm drops fall at $6.49\,\mathrm{m\,s^{-1}}$, 4 mm drops fall at $8.83\,\mathrm{m\,s^{-1}}$, and 5.8 mm drops fall at $9.17\,\mathrm{m\,s^{-1}}$. These fall velocities increase as the pressure decreases. Thus, it is easy to imagine that collisions between drops can occur due to differences in fall velocities. Unfortunately, a droplet collision does not imply necessarily that the two droplets simply coalesce to form a single drop. Instead, a number of outcomes are possible. Some droplets collide and coalesce into a larger droplet. Other droplets collide and then bounce apart. Droplets also may collide and coalesce temporarily, and then separate into droplets similar to their initial pre-collision sizes or break into a number of smaller droplets. Large drops, in particular, tend to separate after collision to form two large drops and a number of smaller satellite drops.

The concept of efficiencies is used to define the likelihood that two drops actually collide and coalesce. The collision efficiency is the fraction of drops with radius r that collide with a larger drop of radius R as the larger drop overtakes the smaller drops (Figs. 7.4 and 7.5). The coalescence efficiency is then the fraction of drops that remain together after the collision. Finally, the collection efficiency is the collision efficiency multiplied by the coalescence efficiency. To obtain the total change in the size of the collector drop with initial radius R, it is necessary to integrate over all drop sizes less than or equal to R. In general, it is found that collection efficiencies for small drops are small and the efficiency increases with drop size up to drops with radius of around 1 mm. Above this point the collection efficiency depends upon the relative sizes of r and R. Small drops tend to move with the air flow around the larger drops and this leads to lower collection efficiencies when the difference in droplet size becomes large. This decrease in efficiency also happens when the droplets are similar in size.

Raindrops are limited in size owing to both collisions with other drops and the drops becoming unstable. As raindrops increase in size, they are less and

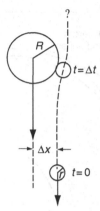

Figure 7.4. Schematic diagram depicting a large raindrop with radius R overtaking a smaller raindrop with radius r. The drops are separated at time $t = 0$, but collide at time $t = \Delta t$ owing to the larger drop having a faster fall speed. The drops may or may not coalesce after collision. The centers of the drops are initially separated by distance Δx, but this changes as the drops get closer owing to the flow around the larger raindrop. After Rogers (1976).

less likely to permanently coalesce when they collide and instead may break upon collision into several smaller drops. In addition, aerodynamically induced circulations occur within the raindrop as it falls. For drop diameters above 3 mm the effects of surface tension are no longer certain to hold the drop together in the face of the drop rotational energy. Grazing collisions can produce a spinning drop that elongates and quickly breaks apart. The tendency for raindrops to break up as their diameters increase explains in part the negative exponential distribution of raindrops (Rogers 1976).

Rain can develop in a cloud within 15 min after cloud formation, although it may take longer before rain is observed at the ground. Since condensation alone is unable to produce large enough drops in such a short time period, collision and coalescence are the dominant mechanisms for producing raindrops so quickly. However, when the cloud becomes sufficiently deep, cold cloud precipitation processes occur and melting snow and graupel may provide most of the rain production.

When raindrops freeze, or when large melted snowflakes refreeze as they fall through a layer with temperatures below freezing near the ground surface, sleet or ice pellets may be produced. Sleet consists of small, generally transparent, solid grains of ice. Sleet that reaches the ground and freezing rain (rain that falls as liquid but freezes upon contact with the ground or other objects) are significant winter forecast concerns as they lead to very slick and dangerous surface conditions.

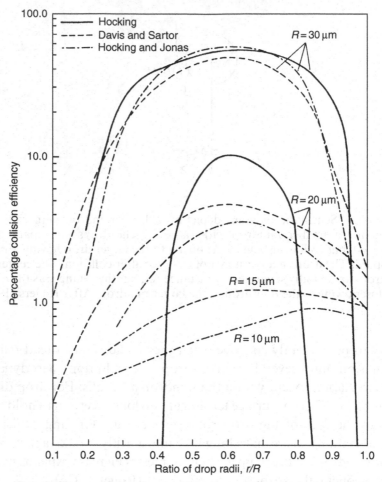

Figure 7.5. Collision efficiencies as a function of the raindrop size ratio (r/R) calculated by Hocking (1959), Davis and Sartor (1967), and Hocking and Jonas (1970). From Mason (1971).

7.2.3 *Ice crystals and aggregates*

When the temperatures within the cloud are below freezing, ice crystals may form. However, the freezing of cloud droplets may not occur immediately when the temperature drops below freezing and liquid drops have been observed down to temperatures near $-40\,°\mathrm{C}$. This unusual situation occurs because, unlike pools of water where a single ice nucleation event anywhere in the pool causes the entire pool to freeze, each individual cloud droplet must have an ice nucleation event. Evidence suggests that the homogeneous freezing of liquid drops happens when temperatures are below $-40\,°\mathrm{C}$.

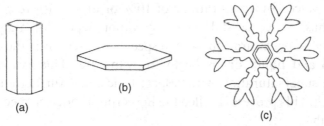

Figure 7.6. Sketches of the three habits of ice crystals: (a) column, (b) plate, and (c) dendrite. From Rogers (1976).

For temperatures warmer than $-40\,°C$, the formation of ice requires the presence of an ice nucleus, just like CCN are needed for the formation of cloud drops. Four processes are believed to lead to ice nucleation: vapor-deposition nucleation, condensation-freezing nucleation, immersion-freezing nucleation, and contact-freezing nucleation. Vapor-deposition nucleation is the direct transfer of water vapor to an ice nucleus, resulting in the formation of an ice crystal. Condensation-freezing nucleation is the condensation of water onto an ice nucleus to create an embryonic drop, which is followed by the freezing of the embryonic drop. Immersion-freezing is the freezing of a supercooled liquid drop on an ice nucleus that is immersed within the liquid of the drop. Finally, contact-freezing nucleation is the freezing of a supercooled drop by an ice nucleus that comes into contact with the drop surface.

Ice crystals typically are observed starting at temperatures of $-15\,°C$. There are three main forms, or habits, of ice crystals: columns, plates, and dendrites (Fig. 7.6). The habit of ice crystal growth is temperature, ice supersaturation, and initial nucleation process dependent (Bailey and Hallett 2004) and changes as the ice crystal moves through the cloud and experiences different environmental conditions. Note that larger drops tend to freeze at warmer temperatures.

The availability of ice nuclei (IN) also is temperature dependent. As the temperature decreases below $0\,°C$, more IN are available to help produce ice crystals. However, there remains a large discrepancy between the number of IN and the number of ice crystals observed, with more ice crystals being observed than would be expected based upon the observed number of IN. This discrepancy decreases with decreasing temperature, but can be as large as several orders of magnitude (10^4–10^5) in some cases (e.g., Hobbs and Rangno 1985).

Once ice crystals are produced, they can grow by vapor-deposition if the environment is supersaturated with respect to ice. Since saturation vapor pressure with respect to water is higher than saturation vapor pressure with respect to ice, any cloud that is saturated with respect to water is supersaturated with respect to ice. Supersaturations of 10% or more with respect to ice are common. Thus, as ice crystals grow by vapor deposition they reduce the environmental supersaturation with respect to water until the air becomes subsaturated and liquid drops begin to evaporate. This evaporation then increases the supersaturation with respect to ice, and further encourages ice crystal growth. This process is called the Bergeron–Findeisen mechanism of ice crystal growth.

The processes of collision and coalescence can also produce clusters of ice crystals, called aggregates. Snowflakes are formed via this aggregation process. The collision and coalescence process for ice crystals is more complicated than for liquid drops owing to the various habits of the individual ice crystals and the possibility of mechanical interlocking or sticking after collision. In addition, whereas the coalescence efficiency for cloud drops is near unity, this is not necessarily the case for ice crystals. Laboratory experiments suggest that the coalescence efficiencies are higher at warmer temperatures ($> -5\,°C$) and are a function of crystal habit. As aggregates fall through a cloud, typically at fallspeeds of less than $1\,m\,s^{-1}$, one can imagine them turning and changing velocity as different faces of the aggregates interact with the airflow. All this variability leads to uncertainty in how to describe the aggregation process in mathematical terms.

7.2.4 Rimed ice particles, graupel, and hail

Riming occurs as ice crystals collide and coalesce with supercooled cloud droplets at environmental temperatures below freezing (Fig. 7.7). When these collisions occur, the supercooled droplets freeze rapidly. As long as the features of the original ice particle can be distinguished, the ice particle is called a rimed particle. However, when the initial particle shape can no longer be distinguished, then the ice particle is typically considered a graupel particle. Graupel particles typically fall at speeds of $1-3\,m\,s^{-1}$ (Nakaya and Terada 1935), depending in part upon the density of the graupel particle. Graupel particle density varies across a large range owing to variations in the denseness of the frozen drops on the ice crystal. Graupel particles also serve as embryos for hailstones, which fall an order of magnitude faster at $10-50\,m\,s^{-1}$. A large fraction of convective rainfall is meltwater graupel in strong thunderstorms and thus is produced via cold cloud processes.

Graupel-sized particles are produced initially both by ice crystal riming and drop freezing. Of these two processes, ice crystal riming is by far the slowest since it takes time for small ice crystals to grow by vapor deposition and aggregation until they are large enough to have an appreciable fallspeed.

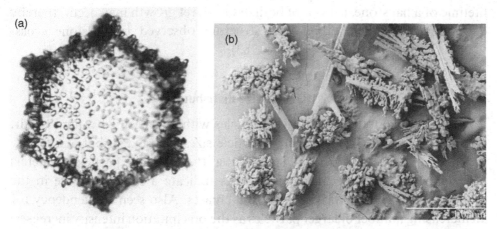

Figure 7.7. Picture of a rimed snowflake (left) and rimed column (right). From Hobbs *et al.* (1971) and the Electron Microscopy Unit of the Beltsville Agricultural Research Center of the United States Department of Agriculture, respectively.

Once they begin to fall, then riming can be very effective in growing the ice crystal into a graupel particle as supercooled cloud droplets are collected. In contrast, drop freezing by contact-freezing nucleation is a more rapid process of graupel generation. Small ice crystals collide with supercooled raindrops, leading to drop freezing and the instant formation of graupel-sized particles (Lamb 2001). A feedback mechanism may develop in which riming contributes both to the growth of graupel particles and to the production of secondary ice particles from splintering. These secondary ice particles grow by vapor deposition and add to the population of small ice particles that can lead to contact-freezing nucleation (Hallett and Mossop 1974). However, this process is not active in all clouds.

As graupel particles fall through the liquid cloud they continue to grow via riming. In cases of extreme riming, hailstones are formed. Hailstones are typically several centimeters in diameter, but observations indicate that hailstones as large as 10–15 cm can occur (Knight and Knight 2005). The latent heat of fusion from freezing accreted droplets heats the surface of the stone, thus slowing the rate of droplet freezing and influencing the growth of the hailstone. Owing to this latent heat release, a growing hailstone is often several degrees warmer than its environment. When the hailstone temperature remains below freezing, "dry" growth occurs as all collected droplets freeze upon contact. When the hailstone temperature rises to 0 °C, "wet" growth occurs as the collected droplets no longer freeze on contact. Instead, some of the collected water from the droplets is frozen, while the remaining unfrozen water is either lost by shedding or incorporated into the hailstone to form spongy ice. The liquid fraction of large hailstones may exceed 20%. During the lifetime of a hailstone, periods of both dry and wet growth may occur, thereby developing the layered onion-like structure observed in hailstone cross-sections (Rogers 1976; Ziegler *et al.* 1983).

7.3 Particle size distributions

When sampling all the liquid or ice particles within a specified volume of air, observations show a distribution of particle sizes. These distributions specify the number of particles per unit size interval (typically the particle diameter) per unit volume of air. All measurements indicate a rapid decrease in the number of particles as the particle size increases. Also seen is a tendency for an increasing number of larger particles as the precipitation intensity increases.

A function that approximates the distribution of particles well is an inverse exponential function, first suggested by Marshall and Palmer (1948). Thus, particle distributions can be approximated by

$$n(D) = n_0 e^{-\lambda D}, \tag{7.4}$$

where D is the particle diameter (m), n is the number of particles per unit volume (m^{-4}), λ is the slope parameter that defines the fall off of particles as the diameter increases (m^{-1}), and n_0 is the intercept parameter that defines the maximum number of particles per unit volume at $D = 0$ size.

While observations of particle distributions typically depart from the pure negative exponential, this simple expression tends to be the limiting form when samples of the distributions are averaged (Fig. 7.8). This is true for raindrops, snow crystals, graupel, and hail. Thus, the Marshall–Palmer distribution is a common assumption in many parameterizations of cloud microphysical processes. The gamma distribution also is used to describe particle distributions, varing from the inverse exponential distribution mainly for very small droplet sizes. Smith (2003) argues that the observational limitations at very small droplet size produce larger uncertainties in the particle distributions than the differences in the bulk properties between the gamma and inverse exponential distributions.

For those interested in pursuing the details of cloud physics, the books by Rogers (1976), Cotton and Anthes (1989), Rogers and Yau (1989), Houze (1993), Pruppacher and Klett (2000), are wonderful resources. For present interests, suffice it to say that the microphysical processes are complex, require approximations for numerous interactions, and are founded upon a less than perfect observational and theoretical base. Yet this understanding is sufficient to begin the parameterization process with some hope of success.

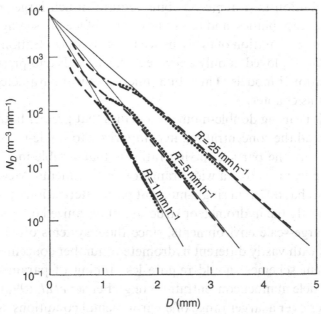

Figure 7.8. Observed drop-size distributions (dots) for different rain rates (R in mm h^{-1}) compared with best-fit inverse exponential curves (solid lines) and drop distributions reported by others (dashed lines). From Marshall and Palmer (1948).

7.4 Bulk microphysical parameterizations

Microphysical parameterizations typically are grouped into "bulk" and "bin" approaches. Bulk approaches use a specified functional form for the particle size distributions and generally predict the particle mixing ratio (total mass per unit volume of air), although some bulk approaches can, in addition, predict the total particle concentration (Ziegler 1985; Ferrier 1994; Meyers *et al.* 1997). Schemes that predict only the particle mixing ratio are called single-moment schemes, while those that predict the particle mixing ratio and concentration are called double-moment schemes. Triple-moment schemes are available, but are used in only a very few research models (e.g., Clark 1974). Often the particle size distribution is approximated by the inverse exponential distribution in bulk approaches, although gamma functions and log-normal functions also have been used to describe some hydrometeor distributions (e.g., Clark 1976; Ziegler 1985; Ferrier 1994). In contrast, a bin approach does not use a specified function for the particle distribution, and instead divides the particle distribution into a number of finite size or mass categories (Berry 1967; Kogan 1991; Ovtchinnikov and Kogan 2000; Lynn *et al.* 2005). This division of the particle distribution into numerous bins requires much larger memory and computational capabilities, and poor knowledge of ice phase physics hampers the accurate representation of evolving ice particle concentrations. Therefore, bin models are employed in only a few research models and presently are not used in operational models. Thus, bulk microphysical parameterizations are the focus of this chapter.

One benefit to using double-moment schemes that predict both the particle mixing ratio and the concentration, in comparison to single-moment schemes that predict only the particle mixing ratio, is that double-moment schemes should be applicable across a wider range of environments. McCumber *et al.* (1991) suggest that different single-moment parameterizations must be used to simulate correctly the hydrometeor structure of organized convective systems in different large-scale environments, since these systems often have two or more regions with vastly different hydrometeor number concentrations. Since double-moment schemes should require less tuning of parameters that are related to particle number concentrations (e.g., Ferrier *et al.* 1995), they should perform better over a larger range of environmental conditions. Similarly, the benefits of using a bin approach in comparison to a bulk approach are that the scheme includes more specific parameterizations of various microphysical processes and the interactions between particles. However, these schemes also are more computationally expensive as they require a larger number of calculations.

One common and important assumption in many of the schemes examined is that cloud water and the smallest cloud ice particles are monodisperse and do not move relative to the flow, so they are simply advected with the flow both horizontally and vertically (see Fig. 7.9). Precipitating particles (raindrops, snow, graupel, and hail), on the other hand, have significant fall speeds and move relative to the flow. This clean separation in drop size between cloud water and rainwater, for example, is seen in numerical simulations of cloud droplet growth and precipitation development after collision and coalescence acts over time to broaden the drop size distribution (Fig. 7.10; Berry 1965; Cotton 1972; Berry and Reinhardt 1974a).

Another important assumption is the type of function used to approximate the distribution of particles within a volume. Many schemes assume a Marshall–Palmer-type inverse exponential distribution in which the intercept parameter n_{0x} is specified and assumed to be constant throughout the simulation. Thus, the slope parameter λ_x varies as the mixing ratios change, where

$$\lambda_x = \left(\frac{\pi \rho_x n_{0x}}{\rho q_x}\right)^{1/4}. \tag{7.5}$$

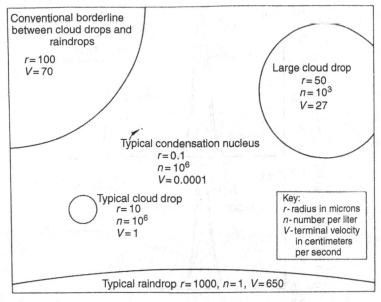

Figure 7.9. Schematic diagram of the relative sizes, concentrations, and fall velocities of some of the particles involved in warm cloud processes. From McDonald (1958), reprinted with permission from Elsevier.

Here x refers to the particle type ($x = r$, s, g for rain, snow, or graupel, respectively), ρ_x is the density of the particle, ρ is the atmospheric density, and q_x is the mixing ratio. Other choices for the function used to define the distributions are gamma functions (e.g., Ziegler 1985; Ferrier 1994). It is often helpful to note that the value of λ_x^{-1} yields the mean particle diameter. In double-moment schemes, the constant n_{0x} is replaced by the variable n_x.

The various bulk microphysical parameterization schemes in the literature can vary in both the approximations used to describe the interplay between the different particles and the number of interactions (both phase and habit changes) included in the parameterization scheme. For example, the Dudhia (1989) simple ice scheme incorporates 12 different interactions between water vapor, cloud water, rainwater, ice, and snow. The Lin *et al.* (1983) and Reisner *et al.* (1998) schemes both include 32 different interactions between water vapor, cloud water, rainwater, ice, snow, and graupel. While most of the bulk microphysical schemes in use today include both water and ice processes, there remain some schemes or models that only include warm phase microphysics (i.e., cloud water and rainwater only). Johnson *et al.* (1993) and Gilmore *et al.* (2004a) indicate that differences in both the amount of rainfall and the cold pool strength occur when ice processes are included in model simulations of

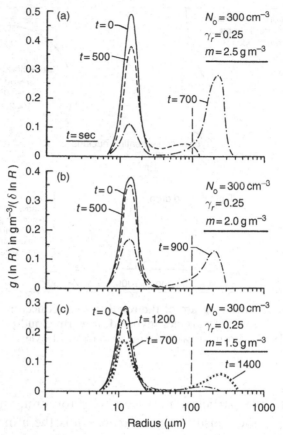

Figure 7.10. Computed distributions of the water mass density at various times between initial time ($t=0$) and 1400 s (indicated on each panel) for an initial cloud droplet concentration of 100 cm^{-3}, a radius of dispersion of 0.25, and liquid water contents of 2.5 (top), 2.0 (middle), and 1.5 (bottom) g m^{-3}. The Kessler (1969) threshold radius separating cloud droplets from raindrops is shown by the vertical dashed line. Note the fairly distinct separation of sizes between cloud drops and raindrops, with few drops being in between the two distributions. From Cotton (1972).

thunderstorms as compared to simulations with only warm-phase microphysics. Thus, it remains important to understand the details in the schemes that one uses to forecast or simulate atmospheric processes.

The equations that govern the evolution of the microphysical variables all follow a similar structure, namely

$$\frac{\partial q_x}{\partial t} = -ADV(q_x) + TURB(q_x) + (P_1 + P_2 + P_3 + P_4 + P_5 + \cdots), \quad (7.6)$$

where q_x is any microphysical variable (e.g., mixing ratios of water vapor, cloud water, rainwater, ice, snow, and graupel), ADV represents the advective processes, $TURB$ represents the turbulent processes, and P_i represents the various tendencies from the microphysics parameterization. These tendencies can be either a positive or a negative contribution to the total tendency for q_x. For example, condensation produces a positive cloud water tendency, whereas evaporation produces a negative cloud water tendency. In addition, the tendencies in the microphysical variables often involve latent heat release, and so their effects also must be incorporated into the temperature tendency equation. Thus, while this chapter examines how the various microphysical interaction terms are parameterized, the connections to the other model variables (temperature, water vapor mixing ratio) also need to be made even if they are not discussed specifically.

Perhaps the best approach to understanding how these bulk microphysical schemes operate is to explore in detail how some of the particle interaction approximations are derived. There are a large number of microphysical parameterizations in the literature, and it is impossible to discuss them all. Instead, the approach taken here is to take an overview of the general ideas and some similarities between a handful of commonly used schemes to provide the reader with a basic understanding of how the broader class of schemes operate. We begin with the process of condensation.

7.4.1 Condensation

Most single-moment bulk microphysical schemes parameterize condensation following Asai (1965), and the following largely follows this discussion. The tendency for producing supersaturation in the atmosphere is nearly completely offset by the condensation of many tiny cloud drops. Thus, when water vapor condenses and cloud droplets are formed, we have the supersaturation condition that

$$q_v - q_{vs} = \delta M > 0, \tag{7.7}$$

where q_v is the water vapor mixing ratio ($\mathrm{kg\,kg^{-1}}$), q_{vs} is the saturation water vapor mixing ratio ($\mathrm{kg\,kg^{-1}}$), and δM represents the total possible condensed water ($\mathrm{kg\,kg^{-1}}$). However, in actuality, δM is divided into two parts ($\delta M = \delta M_1 + \delta M_2$): condensed water δM_1 (cloud water) and an increase in the water vapor mixing ratio to be stored in the air δM_2, owing to the latent heat release from condensation that produces a warmer air temperature and a greater saturation water vapor mixing ratio.

The warming due to condensation can be expressed as

$$\delta\theta_1 = \frac{L_v}{c_p}\left(\frac{p_0}{p}\right)^{\kappa}\delta M_1,$$ (7.8)

where L_v is the latent heat of vaporization, c_p is the specific heat at constant pressure, p_0 is the surface pressure (Pa), p is the pressure (Pa), and $\kappa = R_d/c_p = 0.286$. Now, the Clausius–Clapeyron equation states that

$$\frac{de_s}{dT} = \frac{L_v e_s}{R_v T^2},$$ (7.9)

where e_s is the saturation vapor pressure (Pa), R_v is the individual gas constant for water vapor ($461\,\mathrm{J\,kg^{-1}\,K^{-1}}$), and T is temperature (K). One can replace the vapor pressure by a mixing ratio using the approximation

$$e = \frac{q_v p}{\varepsilon},$$ (7.10)

which leads to a modified version of the Clausius–Clapeyron equation

$$d\left(\frac{q_{vs}p}{\varepsilon}\right) = \frac{L_v(q_{vs}p/\varepsilon)}{R_v T^2}\,dT.$$ (7.11)

If the pressure is assumed to be constant, then this equation simplifies to

$$dq_{vs} = \frac{L_v q_{vs}}{R_v T^2}\,dT.$$ (7.12)

From the definition of potential temperature, we have

$$T = \theta\left(\frac{p}{p_0}\right)^{\kappa}.$$ (7.13)

Thus, the dT in the modified Clausius–Clapeyron equation can be replaced by $d\theta$, such that

$$dq_{vs} = \frac{L_v q_{vs}}{R_v \theta^2}\left(\frac{p_0}{p}\right)^{\kappa}\,d\theta.$$ (7.14)

Finally, replace $d\theta$ with $\delta\theta_1$ to represent the warming effect of condensation, and dq_{vs} with δM_2 to represent the increased saturation mixing ratio due to warming. This leads to

$$\delta M_2 = \frac{L_v^2}{c_p R_v}\left(\frac{p_0}{p}\right)^{2\kappa}\frac{q_{vs}}{\theta^2}\,\delta M_1,$$ (7.15)

and so the ratio of $\delta M_1/\delta M$ is

$$\frac{\delta M_1}{\delta M} = r_1 = \frac{1}{\left[1 + (L_v^2/c_p R_v)(p_0/p)^{2\kappa}(q_{vs}/\theta^2)\right]}. \tag{7.16}$$

Thus, the increase in cloud water over the model time step (Δt) due to condensation (P_{COND}) is

$$P_{COND} = (r_1\,\delta M)/\Delta t. \tag{7.17}$$

Over the same time interval, the value of q_v must then decrease by the same amount, so that the air is just saturated when the latent heating from condensation is included in the temperature tendency equation. This approach is used within the schemes of Rutledge and Hobbs (1983), Dudhia (1989), and Reisner *et al.* (1998). A similar adjustment approach for the condensation tendency term is used by Tripoli and Cotton (1980) and Schultz (1995), although the adjustment factor used by Schultz is constant ($r_1 = 0.75$) and does not vary with the environmental conditions. Asai (1965) indicates that r_1 varies from 0.25 to 0.9 for a lapse rate of $6.5\,^{\circ}\mathrm{C}\,\mathrm{km}^{-1}$ and temperatures between freezing and $30\,^{\circ}\mathrm{C}$. If the supersaturation of the environment is less than 1, then evaporation occurs using the same parameterization but with the sign reversed.

Now that cloud droplets can form in the model, we look to see how rain water is created. As discussed earlier, rain is first formed from the cloud droplets through a binary collision–coalescence process called "autoconversion."

7.4.2 Autoconversion

Autoconversion is the process where cloud droplets collide and coalesce with each other and eventually form raindrops. Many parameterizations follow Kessler (1969) in specifying a relationship that resembles

$$P_{AUTO} = \max[k_1(q_c - q_{c\_threshold}), 0], \tag{7.18}$$

where q_c is the cloud water mixing ratio ($\mathrm{kg\,kg}^{-1}$), k_1 is a conversion rate, and $q_{c\_threshold}$ is a threshold value for q_c ($\mathrm{kg\,kg}^{-1}$) below which autoconversion does not occur. Dudhia (1989) and Reisner *et al.* (1998) both use $k_1 = 0.001\,\mathrm{s}^{-1}$ and $q_{c\_threshold} = 0.0005\,\mathrm{kg\,kg}^{-1}$.

Lin *et al.* (1983) use a slightly different expression for autoconversion that is based upon the relation suggested by Berry (1968). Gilmore *et al.* (2004b) rederive the Lin *et al.* (1983) microphysics scheme, providing additional detail. Instead of a linear relationship between the autoconversion rate and q_c, Lin *et al.* (1983) suggest a quadratic relationship such that

$$P_{AUTO} = \frac{1 \times 10^{-3}[\rho(q_c - q_{c\_threshold})^2]}{\{1.2 \times 10^{-4} + [1.569 \times 10^{-18} N_1/D_0\rho(q_c - q_{c\_threshold})]\}}, \quad (7.19)$$

where N_1 is the number concentration of cloud droplets $(1 \times 10^9 \, m^{-3})$ and D_0 is the dispersion of the cloud droplet distribution and is assumed to be 0.15. The threshold value for the cloud water mixing ratio is set to $0.002 \, kg \, kg^{-1}$ in Lin *et al.* (1983), a value four times larger than that of Dudhia (1989) or Reisner *et al.* (1998). A similar approach is used by Ferrier (1994), except that the auto-conversion threshold is a function of the droplet number concentration.

Another parameterization for autoconversion is proposed by Tripoli and Cotton (1980). This approach again uses a threshold value for q_c below which there is no conversion. However, they also include a factor that represents the mean collision frequency for cloud drops that become raindrops after collid-ing. This leads to the expression

$$P_{AUTO} = \frac{0.104 g E \rho_0^{4/3}}{\mu(n_c\rho_w)^{1/3}} q_c^{7/3} h(q_c - q_{c\_threshold}), \quad (7.20)$$

where the collection efficiency $E = 0.55$, the mean cloud droplet concentra-tion $n_c = 3 \times 10^8 \, m^{-3}$, the cloud water mixing ratio threshold $q_{c\_threshold} = 0.0005 \, kg \, kg^{-1}$, ρ_0 is the density of the hydrostatic reference state, and μ is the dynamic viscosity

$$\mu = 1.72 \times 10^{-5}[393/(T+120)](T/273)^{3/2}, \quad (7.21)$$

where T is in kelvin and μ has units of $kg \, m^{-1} \, s^{-1}$. The Heaviside stepfunction $h(x)$ equals 0 when $x < 0$, equals 0.5 for $x = 0$, and equals 1 for $x > 0$. This initially produces a more gradual conversion rate than compared to the Kessler approach, but produces a larger conversion rate at high cloud water mixing ratios. The autoconversion parameterization of Berry and Reinhardt (1974b) as adapted by Walko *et al.* (1995) also tends to produce lower con-version rates than the Kessler approach (Thompson *et al.* 2004).

Cotton and Anthes (1989) indicate that the conversion rates from several parameterizations of autoconversion differ by several orders of magnitude. Since the rain formation process is dominated by accretion, the collection of cloud droplets by raindrops, the magnitude of the autocollection rate may not be important. However, Cotton (1972) and Cotton and Anthes (1989) also mention that it takes time for cloud droplets to transform into raindrops. This "aging" period is not included in any of these parameterization schemes, and may lead to the early production of rain water too low in the cloud (Simpson and Wiggert 1969; Cotton 1972). Straka and Rasmussen (1997) suggest incorporating a separate tendency equation for the age of a process or the condition of parcels (in a Lagrangian sense) that could be used to delay the autoconversion process in Eulerian models and may yield better results. Using a double-moment scheme, Ziegler (1985) explicitly accounts for non-linear coalescence rates and finds the timing of rain formation to be reasonable.

In a similar manner, Schultz (1995) defines

$$P_{AUTO} = (l_c - l_{c\_threshold}), \qquad (7.22)$$

where l_c is the specific content (kg m^{-3}) of the cloud water and $l_{c\_threshold}$ ~ 0.0007 kg m^{-3}. The specific content is related to the mixing ratio by a density scaling, where $l_c = \rho q_c$. Schultz uses specific contents instead of mixing ratios for processes that are independent of the air density, such as collection and diffusion. In the autoconversion approximation, the conversion only occurs if the rain water specific content l_r is zero – otherwise it is not allowed. Thus, Schultz is assuming that autoconversion may be important for the initial creation of raindrops, but not afterwards. Lin *et al.* (1983) suggest that their high plains thunderstorm simulation is more realistic when the autoconversion term is turned off, perhaps owing to the increased importance of cold cloud processes in these high-based thunderstorms.

7.4.3 Accretion

Accretion is the process by which a liquid drop collides and coalesces with smaller liquid drops as it falls. Kessler (1969) assumes that a raindrop falling through a layer of cloud droplets sweeps out a cylindrical volume as it falls, thereby having a chance to collect all the cloud droplets in its path. Thus, he defines the rate of rain mass accumulation due to the accretion of cloud droplets from a single raindrop as

$$\frac{d(m(D))}{dt} = \frac{\pi D^2}{4} E V_r \rho q_c, \qquad (7.23)$$

where $m(D)$ is the mass of a raindrop of diameter D, E is the collection efficiency, V_r is the raindrop fall speed (m s^{-1} and assumed to be positive), ρ is the atmospheric density, and q_c is the cloud water mixing ratio. For $E = 1$, the raindrop accumulates all the cloud droplets in its path.

Equation (7.23) is a continuous growth equation, since for every time step there is a change in the mass of the droplet. However, as discussed by Telford (1955), the increase in droplet mass actually is a discrete process related to the specific number of cloud droplets collected. Telford (1955) takes into account

the discrete nature of the accretion process and develops a stochastic coalescence equation. The incorporation of the discrete nature of coalescence can
lead to the development of a complete raindrop spectrum much faster than using
the continuous growth assumption (Telford 1955). Ziegler (1985) develops a
bulk parameterization based upon the stochastic coalescence equation.

To convert (7.23) to a bulk parameterization, integrate (7.23) over all raindrop sizes assuming a distribution for the raindrops. In many parameterizations, cloud drops are assumed to be monodispersed (all drops in the
distribution are set to the same size, although the size can vary based upon
the mixing ratio). This is one of the parameterizations for which the function
used to define the distribution of particles is important. Kessler (1969), and
many others, assume a Marshall–Palmer-type inverse exponential distribution, which then leads to the bulk expression for accretion. Thus, the rate at
which the inverse exponential raindrop distribution accretes the monodispersed cloud droplet distribution is given by

$$P_{ACCR} \equiv \int_0^\infty \frac{d(m(D))}{dt} n(D) \, dD = \int_0^\infty \frac{\pi D^2}{4} E V_r \rho q_c n_{0r} e^{-\lambda D} \, dD, \qquad (7.24)$$

where n_{0r} is the intercept parameter for the raindrop distribution and λ is the
slope of the raindrop distribution. If one assumes $V_r = aD^b$, then with knowledge of the gamma function Γ we obtain

$$P_{ACCR} = \frac{\pi E}{4} \rho q_c n_{0r} a \frac{\Gamma(3+b)}{\lambda_r^{3+b}}. \qquad (7.25)$$

The parameterizations of Lin *et al.* (1983), Dudhia (1989), and Reisner *et al.*
(1998) are all based upon this derivation, and differ only in how they include
corrections to the accretion rate as a function of height or pressure. Liu and
Orville (1969) conduct a least-squares analysis of the fall speed data of Gunn
and Kinzer (1949) and find that the constants a and b are $841.996 \, \text{m}^{1-b}\text{s}^{-1}$ and
0.8, respectively (Gilmore *et al.* 2004b). However, raindrop fall speeds also
vary with pressure, becoming larger as pressure decreases.

In contrast, Tripoli and Cotton (1980) re-express the inverse exponential
raindrop distribution (7.5) as a function of the characteristic raindrop radius,
yielding a constant value for λ and an intercept parameter n_{0r} that varies with
rainwater content (see Fig. 7.11). Under this assumption, the terminal velocity
of rain is independent of the rain density. They conclude that the accretion rate is

$$P_{ACCR} = 0.884 E \left(\frac{g \rho_0}{\rho_w R_m} \right)^{1/2} q_c q_r, \qquad (7.26)$$

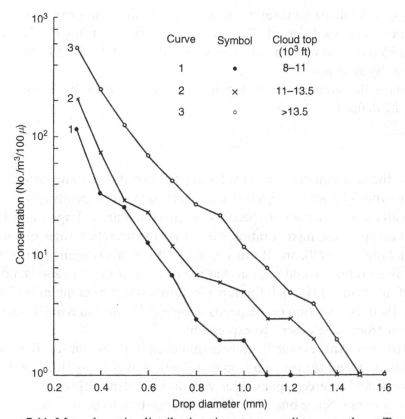

Figure 7.11. Mean drop size distributions in warm cumuli over southeast Texas where drop diameters must be >250 μm. Data are categorized according to estimated cloud tops (shown in 10^3 ft). Note how the distribution slope appears nearly constant, while the intercept parameter changes for the data associated with different cloud tops. From Klazura (1971).

where R_m is a characteristic drop radius taken as 2.7×10^{-4} m in Tripoli and Cotton (1980). This yields a lower accretion rate than Kessler with $E = 1.0$ when the water mass density is less than 2.28×10^{-3} kg m^{-3}. But the differences are small for higher water mass densities. They also define a variable E based upon the work of Langmuir (1948) in which

$$E = \begin{cases} 0 & \text{for } STK < 0.0833 \\ \dfrac{STK^2}{(STK + 0.5)^2} & \text{for } STK \geq 0.0833 \end{cases} \tag{7.27}$$

where STK is the Stokes number and

$$STK = \frac{0.22 \rho_w V_r(R_m) q_c^{2/3}}{R_m \mu} \left(\frac{0.239}{n_c/3}\right)^{2/3}. \tag{7.28}$$

Here ρ_w is the density of water and n_c is the mean cloud drop concentration (assumed to be $3 \times 10^8 \, \mathrm{m}^{-3}$). The value of E calculated using (7.27) is greater than 0.95 for cloud water mixing ratios greater than $5.3 \times 10^{-7} \, \mathrm{kg \, kg}^{-1}$ and so is generally at or near 1.

Perhaps the simplest expression for accretion is given by Schultz (1995), where he defines

$$P_{ACCR} = 17 l_r l_c. \tag{7.29}$$

Again, the rate equations in Schultz are for specific content instead of the mixing ratio. The general idea is that the accretion rate depends upon both the rain water and cloud water specific contents, similar to Tripoli and Cotton (1980) except using mixing ratios. Note that the accretion rates of Lin *et al.* (1983), Dudhia (1989), and Reisner *et al.* (1998) are also dependent upon these two mixing ratios (q_r and q_c), but that the rain water mixing ratio is hidden in the definition of λ_r in (7.5). If this is made more explicit by combining (7.5) with (7.25), then the accretion rate depends upon $q_c q_r^{0.95}$, and so is not dramatically different from these other two expressions.

Cotton and Anthes (1989) indicate that even if one is comfortable with the function assumed for the raindrop distribution, it is likely that the distribution slope and intercept parameter vary independently throughout the lifetime of a cloud. Since one or the other is specified to be a constant in these bulk schemes, errors are likely. They argue that this is especially true for mesoscale convective systems that have both a convective line and a distinct trailing or forward stratiform precipitation region. The stratiform portion of the convective system is likely to have a different microphysical structure and history, and so any raindrop distributions assumed for the convective line may not be realistic for the stratiform region and vice versa. Differences in raindrop distributions between stratiform and convective regions of tropical mesoscale convective systems are seen from observations (Tokay and Short 1996). This observed horizontal and temporal variability in the microphysical parameters is certainly one challenge to these types of schemes.

7.4.4 Evaporation

Byers (1965) develops an equation governing the evaporation/condensation of a single raindrop. Both the effects of the vapor pressure difference between the drop and the ambient environment and the latent heat flux are taken into account. It is shown that

$$\frac{dm(D)}{dt} = \frac{2\pi D(S-1)}{A+B},\tag{7.30}$$

where S is the supersaturation, and

$$A = \frac{L_v^2}{K_a R_v T^2}, \qquad B = \frac{R_v T}{e_S(T) D_{diff}}.\tag{7.31}$$

Here D_{diff} is the diffusivity of water vapor in air and K_a is the thermal conductivity. However, an additional effect is due to the ventilation of the drop as it falls and this effect also should be incorporated. Byers uses the results of Kinzer and Gunn (1951) to account for the ventilation, while others may use the results of Beard and Pruppacher (1971). Thus, the resulting equation for the evaporation of a raindrop that includes a ventilation factor F becomes

$$\frac{dm(D)}{dt} = \frac{2\pi D(S-1)F}{A+B}.\tag{7.32}$$

The ventilation factor F is often defined based upon the Schmidt number (S_c) and the Reynolds number (Re), such that

$$F = 0.78 + 0.31 S_c^{1/3} Re^{1/2}.\tag{7.33}$$

The Schmidt number is the ratio of kinetic viscosity to the diffusivity of water vapor and represents the relative ease of molecular momentum and mass transfer, while the Reynolds number is the ratio of inertial to viscous forces and indicates whether the flow is turbulent or laminar. Again, the total evaporation from all drops is then found by multiplying the equation for the evaporation of a single raindrop by the function that describes the raindrop distribution, and then integrating over all drop sizes. This approach is the basis for most of the evaporation expressions in various bulk microphysics schemes. The resulting equation typically resembles the following one taken from Lin *et al.* (1983), with

$$P_{EVAP} = \frac{2\pi(S-1)n_{0r}}{\left[\rho\left(\frac{L_v^2}{K_a R_v T^2} + \frac{1}{\rho q_{vs}\psi}\right)\right]}$$
$$\times \left[0.78\lambda^{-2} + 0.31 S_c^{1/3}\Gamma\left(\frac{b+5}{2}\right)a^{1/2}\nu^{-1/2}\left(\frac{\rho_0}{\rho}\right)^{1/4}\lambda_r^{-(b+5)/2}\right],\tag{7.34}$$

where most of the variables are as previously defined, a and b are from the expression of fall velocity ($V_r = aD^b$), ν is the kinematic viscosity of air, and ψ is the diffusivity of water vapor in air.

Owing to their different assumptions, Tripoli and Cotton (1980) and Schultz (1995) have different expressions for evaporation. The different behavior of the raindrop distribution assumed in Tripoli and Cotton leads to a slightly different equation, but the evaporation rate is still a function of $(S - 1)q_r$ and is still based upon the results of Byers (1965). In contrast, Schultz (1995) again simplifies the process to a rate equation directly related to $(S - 1)$, but with the evaporation allowed in a given time step limited to a specified amount.

For warm rain processes only, the parameterizations for condensation, autocollection, accretion, and evaporation are the foundation of what is needed in a numerical model. Added to this is the assumed raindrop distribution functional, and the only remaining item of importance is the mass-weighted fall speed. All precipitating fields are assumed to fall at their mass-weighted fall speed, defined as

$$\overline{V}_r = \frac{\int_0^\infty n_r(D)m(D)V_r(D)\,dD}{\int_0^\infty n_r(D)m(D)\,dD}. \tag{7.35}$$

If we assume that raindrops are spheres, then the mass of a raindrop is $\pi D^3 \rho_w / 6$. We again assume that the fall speed of an individual raindrop can be approximated as $V_r = aD^b$, and that the raindrop distribution is represented by an inverse exponential, leading to

$$\overline{V}_r = \frac{a\Gamma(4 + b)}{6\lambda_r^b}. \tag{7.36}$$

Expressions for the mass-weighted fall speed of rain differ based upon the approximations used for defining the constants a and b for the fall speed of an individual raindrop, and whether or not they add a density correction factor.

7.4.5 Ice initiation

The basic idea behind many parameterizations of ice initiation is that ice nuclei are activated to form ice crystals in ice supersaturated conditions, since the observed concentrations of ice nuclei appear to be sufficient to explain ice crystal concentrations in some atmospheric clouds. Thus, from predicted or known ice nucleus concentrations it is possible to calculate the concentration of vapor-activated ice crystals. Given the ice crystal concentration, knowledge of the mass of a typical ice crystal is sufficient to calculate a value for the cloud ice mixing ratio. So, when the air is supersaturated with respect to ice and the air temperature is below freezing, then most parameterization schemes assume that cloud ice forms when in the presence of ice nuclei. This assumption allows observations of ice nuclei to be used as the basis for these schemes. Homogeneous nucleation that does not require ice nuclei is also included in some schemes (DeMott *et al.* 1994; Walko *et al.* 1995).

Fletcher (1962) derives a formula for the number of ice nuclei per unit mass of air in which

$$n_c = 0.01 \exp[0.6(273.16 - T)], \tag{7.37}$$

where n_c (m^{-3}) is the number of ice nuclei present at temperature T below freezing. Over the high plains of the USA, Bowdle *et al.* (1985) suggest a similar formula for the number of ice nuclei, but with a leading constant of 0.02 and an exponential factor of 0.3 instead of 0.6. Rauber (2003) shows that (7.37) is a reasonable fit to the available observations of ice nuclei concentrations for temperatures between -10 and $-25\,°C$, although there is a fair amount of variability in the concentrations measured.

Dudhia (1989) uses the Fletcher expression for ice initiation, whereas Reisner *et al.* (1998) indicate that Fletcher's expression overestimates ice nucleation at very cold temperatures and so the value of T is not allowed to be below 246 K. The initiation rate of cloud ice is then described as

$$P_{ICE} = \left(\frac{m_i n_c}{\rho} - q_i\right)\frac{1}{\Delta t}, \tag{7.38}$$

where q_i is the cloud ice mixing ratio (kg kg^{-1}), m_i is the mass of a typical ice particle (1×10^{-12} kg), and Δt is the model time step (seconds).

Meyers *et al.* (1992) propose a different equation for ice crystal concentration, suggesting that

$$n_c = 1000 \exp\left[-0.639 + 12.96\left(\frac{q_v}{q_{vsi}} - 1\right)\right], \tag{7.39}$$

where q_{vsi} is the saturation water vapor mixing ratio over ice. This expression is used by Schultz (1995) to parameterize ice nucleation, but is not allowed if ice is already present. This is done because ice nucleation is a much slower process than deposition growth of ice crystals. Meyers *et al.* (1992) demonstrate the importance of the ice crystal concentration by comparing their formula for ice crystal concentration to that of the Fletcher (1962) formulation in a model simulation (Fig. 7.12). They also suggest that the Fletcher relationship probably overpredicts ice crystal concentrations for temperatures below $-25\,°C$.

A third equation for ice crystal concentration is given by Cooper (1986) which provides for ice crystal concentrations that often are in between the values determined by Fletcher (1962) and Meyers *et al.* (1992). Thompson

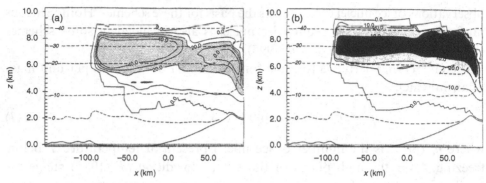

Figure 7.12. Simulated ice crystal concentrations at 4 h with (a) the Meyers formulation for ice crystal concentration and (b) a formula based upon the Fletcher (1962) equation. Maximum value in (a) is $76 \, l^{-1}$ and in (b) is $900 \, l^{-1}$. Isotherms are indicated every $10\,°C$. All distances are in km. From Meyers *et al.* (1992).

et al. (2004) use the Cooper (1986) relationship in their bulk microphysics parameterization.

7.4.6 Ice and snow aggregation

Ice crystals aggregate together to form snow, and snow particles may aggregate together to form graupel or hail (although graupel and hail formation and growth are strongly dominated by riming). These processes are often parameterized in a very similar fashion. The basic idea behind this parameterization is based upon the collision–coalescence process for cloud droplets to yield raindrops as proposed by Kessler (1969). Lin *et al.* (1983) suggest that the conversion rate from cloud ice to snow from aggregation of the cloud ice can be represented by

$$P_{AGGS} = 0.001 \exp[0.025(T - T_0)](q_i - q_{i\_threshold}), \qquad (7.40)$$

where q_i is the cloud ice mixing ratio ($\mathrm{kg\,kg^{-1}}$), T_0 is the freezing temperature, and the threshold value for the cloud ice mixing ratio used is $1 \times 10^{-3}\,\mathrm{kg\,kg^{-1}}$. Dudhia (1989) follows Rutledge and Hobbs (1983), who develop a similar expression, namely

$$P_{AGGS} = \left[q_i - \frac{m(500\,\mu m)n_c}{\rho}\right] \frac{1}{\Delta t}, \qquad (7.41)$$

where $m(500\,\mu m)$ is the mass of a 500 μm ice crystal ($9.4 \times 10^{-10}\,\mathrm{kg}$), and n_c is the ice nuclei concentration (7.37) from Fletcher (1962).

Graupel or hail formation from the aggregation of snow crystals can be defined in a very similar fashion. Lin *et al.* (1983) have

$$P_{AGGG} = 0.001 \exp[0.09(T - T_0)](q_s - q_{s\_threshold}), \qquad (7.42)$$

where q_s is the snow mixing ratio, T_0 is the freezing temperature, and $q_{s\_threshold} = 6 \times 10^{-4}\,\text{kg}\,\text{kg}^{-1}$. The temperature dependence of the rate coefficient has the same form as that used for the aggregation of ice crystals to form snow. The initiation of graupel through collisions of snow crystals is considered unimportant by Rutledge and Hobbs (1984), who only allow graupel to form from collisions between liquid water and ice.

Reisner *et al.* (1998) base their aggregation approximation on the study of Murikami (1990), and define the aggregation of ice crystals to form snow as

$$P_{AGGS} = \frac{q_i}{\Delta \tau_1}, \qquad (7.43)$$

where

$$\Delta \tau_1 = -\frac{2\rho_i}{\rho q_i a_i EX} \log \left(\frac{(3\rho q_i / 4\pi \rho_i n_i)^{1/3}}{r_{S0}} \right)^3. \qquad (7.44)$$

Here ρ_i is the density of ice, $E = 0.1$ is the collection efficiency, $X = 0.25$ is the dispersion of the fall velocity of cloud ice, $r_{S0} = 0.75 \times 10^{-4}\,\text{m}$ is the radius of the smallest snowflake, n_i is the predicted number concentration of cloud ice, and $a_i = 700$ is the constant in the snow fall speed equation. The size of the smallest snowflake in this parameterization is slightly more than three times smaller than the assumed snowflake size used in Rutledge and Hobbs (1983) and Dudhia (1989).

Finally, snow is produced in the Schultz (1995) scheme when cloud ice exceeds a specified content. Thus, he sets

$$l_s = (l_i - l_{i\_threshold}), \qquad (7.45)$$

where $l_{i\_threshold} = 0.1 \times 10^{-3}\,\text{kg}\,\text{m}^{-3}$.

7.4.7 Accretion by frozen particles

Accretion by frozen particles is the process by which precipitating frozen particles (snow, graupel, or hail) collect other ice or liquid particles as they fall. There are a number of different water forms that can be accreted by frozen particles. For example, graupel can accrete cloud ice, snow, rain water, and cloud water. The accretion by graupel or snow of rain water and cloud water is commonly called riming. Riming dominates the formation and growth of graupel and hail. As with the accretion of cloud water by raindrops, the general derivation follows Kessler (1969) and it is assumed that a precipitating particle falling through a given layer sweeps out a cylindrical volume as it falls, thereby having a chance to collect all the droplets in its path. The basic derivation follows the same process as accretion of cloud droplets by raindrops discussed above and so is not repeated.

The rate of accretion by graupel/hail of cloud ice is represented in Lin *et al.* (1983) as

$$P_{ACGI} = \frac{\pi E n_{0g} q_i \Gamma(3.5)}{4\lambda_g^{3.5}} \left(\frac{4g\rho_g}{3C_D\rho}\right)^{1/2},$$ (7.46)

where $E = 0.1$ is the collection efficiency of graupel for cloud ice, q_i is the cloud ice mixing ratio ($kg\,kg^{-1}$), λ_g is the graupel slope parameter, ρ_g is the graupel density, n_{0g} is the graupel intercept parameter, and $C_D = 0.6$ is a drag coefficient for hail. A nearly identical expression is used for the accretion of cloud water by graupel (riming), but with a collection efficiency of 1.0 and q_i replaced by q_c in the equation. Houze (1993) indicates that the collection efficiency for riming is not well-known either from a theoretical or an observational perspective, but is thought to be very high and so an efficiency of unity is often assumed.

Some differences are found in the various expressions for accretion, largely owing to how the fall speed is defined. For example, Rutledge and Hobbs (1983) derive the accretion of cloud ice by snow as

$$P_{ACSI} = \frac{\pi \rho a_s q_i E n_{0s} \Gamma(b_s + 3)}{4\lambda_s^{b_s+3}} \left(\frac{p_0}{p}\right)^{0.4},$$ (7.47)

where $E = 0.1$ is the collection efficiency for cloud ice by snow, and a_s and b_s are the constants in the fall speed equation for snow. Owing to the similar form of the fall speed equation, this expression is nearly identical to the accretion equation for cloud water by rain water as expected.

Additional terms are added when both species are falling, such as the accretion of snow by graupel or the accretion of rain by graupel (riming). For example, for the accretion of rain water by graupel, Lin *et al.* (1983) use

$$P_{ACGR} = \pi^2 E n_{0g} n_{or} |\overline{V}_g - \overline{V}_r| \left(\frac{\rho_w}{\rho}\right) \left(\frac{5}{\lambda_r^6 \lambda_g} + \frac{2}{\lambda_r^5 \lambda_g^2} + \frac{0.5}{\lambda_r^4 \lambda_g^3}\right).$$ (7.48)

Thus, not only is the accretion rate a factor of the intercept and slope parameters for the two precipitation types, but also is a factor of the difference in fall speeds. Some schemes use a polynomial fit to the fall speed relationship for raindrops, in which the fall speed is a function of the drop diameter, and this then alters the accretion rate equation. Murikami (1990) modifies this expression slightly to account for continued accretion when the mass-weighted fall speeds of the colliding hydrometeor species are similar, although Ferrier *et al.* (1995) find that this modification has little influence.

Unlike the accretion of cloud water by rain water, which can be complete and leave no cloud water behind, cloud ice typically remains within clouds that extend above the freezing level (Schultz 1995). Thus, Schultz checks to make certain that the cloud ice specific content of $0.001 \, kg \, m^{-3}$ remains once cloud ice is formed. His accretion equation is

$$P_{ACSI} = \frac{5}{\rho}\left(1 - \frac{273.1 - T}{50}\right)l_i l_s, \tag{7.49}$$

where the temperature dependence is related to the collection efficiency of the aggregation process.

Reisner *et al.* (1998) incorporate an equation for the rate of conversion from snow to graupel via accretion (riming) of cloud water. A so-called "three-body process" is parameterized in which the interaction of two habits produces a third, yielding

$$P_{AGGG} = 2\alpha\Delta t \frac{3\rho_0 \pi n_{0s}(\rho q_c)^2 E^2 \Gamma(2b_s + 2)}{8\rho(\rho_g - \rho_s)\lambda_s^{2b_s+2}}, \tag{7.50}$$

where ρ_g and ρ_s are the densities of graupel and snow, respectively, E is the collection efficiency for snow collecting cloud water, and n_{0s} is the intercept parameter for snow. A more standard equation for the accretion of cloud water by snow is determined, as in Rutledge and Hobbs (1983), and any increase in the accretion of cloud water by snow that is not converted directly into graupel is used to increase the snow mixing ratio.

Finally, Schultz (1995) parameterizes an accretion rate for graupel as simply the specific content $(kg \, m^{-3})$ of the colliding particles multiplied together with a constant conversion parameter. Thus, graupel is allowed to accrete cloud water and snow.

7.4.8 Deposition

The growth of ice crystals, and also their sublimation to water vapor, are parameterized following Byers (1965). Ice crystal, snow, and graupel growth via deposition occur when the environment is supersaturated with respect to ice. The treatment of deposition follows very closely the growth and evaporation of rain water and cloud water. However, the variety of shapes of ice crystals necessitates the diffusion process being handled in a different manner, since a simple spherical shape cannot be assigned. Thus, the diffusion of water vapor to a given crystal is approximated as if it is a current flowing to an object in an electric field. Following Rutledge and Hobbs (1983), the growth rate by vapor deposition of an ice crystal is defined as

$$\frac{dm}{dt} = \frac{C(S_i - 1)/\varepsilon_0}{A'' + B''},$$ (7.51)

where

$$A'' = \frac{L_v}{K_a T}\left(\frac{L_s}{R_v T} - 1\right)$$ (7.52)

and

$$B'' = \frac{R_v T}{\chi e_{si}}.$$ (7.53)

Here ε_0 is the permittivity of free space ($8.854 \times 10^{-12}\,\mathrm{C^2\,N^{-1}\,m^{-2}}$), K_a is the thermal conductivity of air ($2.43 \times 10^{-2}\,\mathrm{J\,m^{-1}\,s^{-1}\,K^{-1}}$), χ is the diffusivity of water vapor in air ($2.26 \times 10^{-5}\,\mathrm{m^2\,s^{-1}}$), and e_{si} is the saturation vapor pressure for ice. Rutledge and Hobbs (1983) assume a hexagonal plate-like ice crystal shape, so that $C = 4D_I\varepsilon_0$, where D_I is the average diameter of the ice crystal. They then relate D_I to the mass of the plate-like ice crystal, which is computed from the cloud ice mixing ratio and the number of ice crystals specified from the formula of Fletcher (1962). Assuming a single size for all ice crystals, this yields an equation for the depositional growth of cloud ice,

$$P_{DEPI} = \frac{4\overline{D}_I(S_i - 1)n_c}{A'' + B''}.$$ (7.54)

A similar approach is taken by Koenig (1971), the results of which are used in the microphysics scheme of Reisner *et al.* (1998). This derivation yields a slightly different expression, in which the depositional growth of cloud ice is given as

$$P_{DEPI} = \frac{q_v - q_{vsi}}{q_{vsw} - q_{vsi}} a_1(m_i)^{a_2} n_i \frac{1}{\rho},$$ (7.55)

where q_{vsi} is the saturation vapor pressure for ice, q_{vsw} is the saturation vapor pressure for water, m_i is the mass of a pristine ice crystal ($m_i = q_i/n_i$), n_i is the ice crystal number concentration, and a_1 and a_2 are temperature-dependent constants from Koenig (1971) that are used over a range from 0 to $-35\,^\circ$C.

A related approach is taken by Schultz (1995), who defines the depositional growth of ice crystals as a function of the supersaturation and the ice crystal mass already present in the atmosphere. Thus, he defines

$$P_{DEPI} = 25(l_v - l_{vsi})l_i, \tag{7.56}$$

where the rate is in terms of the specific content. These values are then converted back to mixing ratios after all the calculations are complete.

When snow is present in air that is supersaturated with respect to ice, then snow particles can grow by deposition as well. However, snow particles are not monodisperse and so one must again assume a distribution function. If an inverse exponential function is assumed for the snow size distribution, then the deposition growth equation can be integrated over all sizes to obtain (Rutledge and Hobbs 1983)

$$P_{DEPS} = \frac{4(S_i - 1)n_{0s}}{A'' + B''} \left[\frac{0.65}{\lambda_s^2} + 0.44 \left(\frac{a_s \rho}{\mu} \right)^{1/2} \left(\frac{p_0}{p} \right)^{0.2} \frac{\Gamma\left(\frac{b_s+5}{2}\right)}{\lambda_s^{(b_s+5)/2}} \right]. \tag{7.57}$$

This expression for deposition is used by Reisner *et al.* (1998), except for neglecting the pressure correction term, and a very similar expression is used by Lin *et al.* (1983). Some of the differences in the expressions are due to different assumed particle shapes, with Lin *et al.* (1983) assuming spherical particles and Rutledge and Hobbs (1983) assuming hexagonal particles.

The deposition growth of graupel or hail yields expressions that are very similar to those for the deposition growth of snow, but with different intercept and slope parameters, slightly different constants related to the expressions for particle fall speed, and different coefficients related to particle ventilation factors.

A subtle but important factor to consider in the construction of microphysics parameterizations is the order in which the various physical processes are evaluated. For example, if cloud ice deposition is evaluated before snow deposition, and the saturation ratio is adjusted after each evaluation, then there is less vapor available for snow deposition in comparison to cloud ice deposition. Some schemes address this problem by evaluating all the terms at once, while others neglect it.

7.4.9 Melting

Calculations of the melting rates of hailstones and graupel are derived first by Mason (1956). A hailstone or graupel particle falling through air gains heat from the environment via conduction and convection. If its surface temperature, assumed to be 0 °C during melting, is below the dewpoint of the environment, then additional heat may be produced by the condensation of water vapor on the surface of the particle. This heat is then used to melt the particle. In contrast, if the environment is very dry the particle may lose heat due to evaporation. Similar processes occur for snow. Assuming an inverse exponential distribution for the snow size distribution and integrating over all sizes yields an expression for the rate of the melting of snow (Lin *et al.* 1983)

$$
\begin{aligned}
P_{MLTS} = {} & -\frac{2\pi}{\rho L_f}[K_a(T - T_0) - L_v\varphi\rho(q_{vsi}(T_0) - q_v)]n_{0s} \\
& \times \left[\frac{0.78}{\lambda_s^2} + 0.31 S_c^{1/3}\Gamma\left(\frac{b_s + 5}{2}\right)c^{1/2}\left(\frac{\rho_0}{\rho}\right)^{1/4}\frac{1}{\nu^{1/2}\lambda_s^{(bs+5)/2}}\right] \\
& -\frac{c_w(T - T_0)}{L_f}(P_{SACW} + P_{SACR}),
\end{aligned}
\tag{7.58}
$$

where L_f is the latent heat of fusion, T is the environmental temperature (K), T_0 is the temperature of freezing, K_a is the thermal conductivity of air, φ is the diffusivity of water vapor in air, $q_{vsi}(T_0)$ is the water saturation mixing ratio at T_0, ν is the kinematic viscosity of air, c_w is the specific heat of water, and P_{SACW} and P_{SACR} are the accretion of cloud water and raindrops, respectively, by the falling snow. The first term on the right-hand side with $K_a(T - T_0)$ represents the conduction of heat from the air, while the second term represents the heat added from condensation on the surface of a melting particle (Musil 1970). Rutledge and Hobbs (1983) and Reisner *et al.* (1998) use very similar expressions, but neglect the condensation effects. Mason (1956) shows that the distance hail falls before melting completely is reduced by several hundred meters when condensation effects are not included.

One of the differences between melting hail and graupel is that hail particles, being of solid ice density, shed all meltwater immediately. However, as graupel is composed of low-density rime (i.e., it contains air spaces between frozen droplets), graupel particles initially soak in the meltwater and subsequently shed the excess meltwater after the rime has saturated. Such effects present an additional challenge to the explicit prediction of microphysical variables.

Dudhia (1989) assumes that all frozen particles melt as they pass through the freezing level, and so the melting rate is directly proportional to the precipitation mixing ratios. Cloud ice only melts if the grid-scale vertical motion is subsiding, since cloud ice follows the model flow field and has no relative fall speed. Schultz (1995) assumes that the melting rate is a function of temperature only, with increasing rates as the temperature increases above freezing. He also

checks to be certain that melting does not cause the environmental temperature to go below freezing. Rutledge and Hobbs (1984) and Ferrier (1994) both
account for the heat transfer that enhances melting as cloud water and rain are
collected by the falling graupel particles. Model simulations indicate that
melting is an important process that influences the development and structure
of midlatitude convective systems (Tao *et al.* 1995).

7.5 Discussion

While this discussion of the various microphysical interaction terms is not
complete, and has left out a number of terms, the general procedures for
calculating the rates of conversion are similar across all schemes and processes.
Thus, the preceding discussion gives a flavor of the typical parameterization
methods and the included physical processes. It is important to realize that the
various microphysical schemes tend to differ significantly in the number of
interactions between the microphysical particles that are represented
(Fig. 7.13). Some schemes are fairly simple and only account for a handful
of interactions, while others are much more sophisticated. The scheme available or selected for use in a given model likely depends upon the phenomena

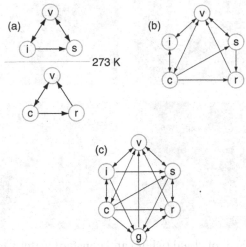

Figure 7.13. Illustration of the microphysical processes available in (a)
Dudhia (1989), (b) Reisner *et al.* (1998), and (c) Lin *et al.* (1983). Arrows
denote the direction in which the particles are allowed to interact. The letter
v denotes water vapor, i denotes ice, s denotes snow, c denotes cloud water,
r denotes rain water, and g denotes a combination of graupel and hail. The
line in (a) marked 273 K denotes the freezing level, with frozen precipitation
processes only above this level and warm precipitation processes only below
this level.

one would like to forecast or simulate, and the availability of computational resources to run the scheme in a timely manner.

Numerical simulations using bulk microphysics schemes have been quite successful in reproducing many observed features of individual clouds and cloud systems. For example, Ziegler *et al.* (1997) show that a model with 1 km minimum grid spacing can at times reproduce the storm-scale features of convective storms near the dryline (Fig. 7.14), beginning with swelling and towering cumulus along the dryline that develop over time into a thunderstorm with

INITIATION STAGE   ACTIVE STAGE

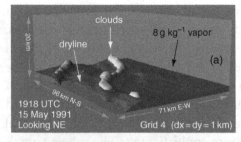

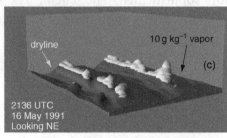
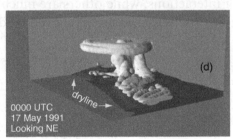
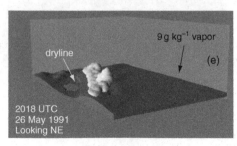
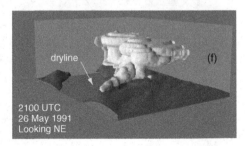

Figure 7.14. Surfaces of cloud (white) and constant water vapor mixing ratio (gray) in perspective view at the initiation and active stage of simulated dryline convection from three days during 1991 using a multigrid cloud-scale model. The depictions are valid at (a) and (b) 1918 and 2100 UTC 15 May, (c) and (d) 2136 UTC 16 May and 0000 UTC 17 May, and (e) and (f) 2018 and 2100 UTC 26 May, respectively. The values of water vapor mixing ratio isosurfaces are indicated in each initiation stage figure. Simulations shown here employ a bulk single-moment warm rain parameterization. From Ziegler *et al.* (1997).

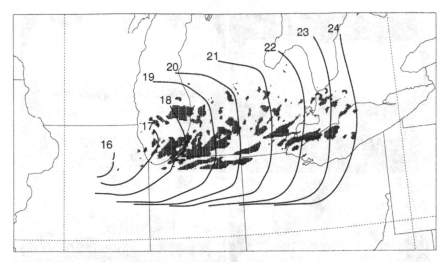

Figure 7.15. Approximate hourly location of a surface gust front (solid lines) associated with a simulated derecho-producing convective system and grid points from the 2.22 km grid that have winds $\geq 26\,\mathrm{m\,s^{-1}}$ at the surface through the 24 h simulation time. Simulation shown here uses a bulk single-moment ice microphysics parameterization. From Coniglio and Stensrud (2001).

overshooting tops and a spreading anvil. Coniglio and Stensrud (2001) simulate a derecho-producing convective system and find that the swaths of severe surface winds ($\geq 25\,\mathrm{m\,s^{-1}}$) produced by the model are very realistic in their distribution (Fig. 7.15). Lee and Wilhelmson (1997) use a bulk scheme to reproduce Florida thunderstorms with land spouts (Fig. 7.16). Finally, Bryan *et al.* (2003) use a warm-rain-only bulk scheme at very small grid spacing (125 m) to reproduce very detailed features in a squall line that contrast with the much smoother fields seen when using 1 km grid spacing (Fig. 7.17). Thus, it is clear that models with bulk microphysical schemes can reproduce many of the observed features of deep convection.

However, we are learning that there are numerous limitations to bulk microphysics schemes that only predict the mixing ratio. Gilmore *et al.* (2004a) nicely describe several of these limitations. While the constants empirically derived for most of the rate equations involving microphysical processes are fairly well specified through a history of observational studies (e.g. Pruppacher and Klett 2000), it is widely regarded that there exists a range of possible values for two graupel parameters that are defined a priori and held constant during the model simulations. These two parameters are the intercept parameter for the graupel inverse exponential size distribution and the graupel density. Reasonable selections for these two parameters can yield substantial

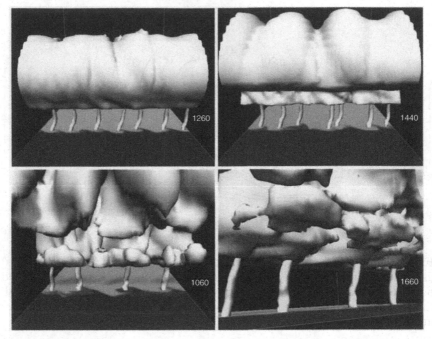

Figure 7.16. Three-dimensional renderings of model simulation output showing the evolving ensemble of leading-edge vortices, storms, and outflow boundaries at 21 (upper left), 24 (upper right), and 31 min (bottom panels) into the simulation. White vertical tubes indicate a vertical vorticity in excess of $0.1 \, \text{s}^{-1}$. The shades along the surface depict the temperature of surface outflow (darker shades denote colder temperatures). Cloud isosurfaces shown for $0.2 \, \text{g} \, \text{kg}^{-1}$ values of cloud water. Simulations shown employ only bulk single-moment warm rain processes following Kessler (1969). From Lee and Wilhelmson (1997).

and operationally important differences in numerical simulations of thunderstorms in terms of storm structure, severity, and intensity (Fig. 7.18).

In actual thunderstorms, the value of the intercept parameters n_{0g} can vary widely within a single storm and among storms (Fig. 7.19) in the same background environment (Dennis *et al.* 1971; Federer and Waldvogel 1975; Spahn 1976; Knight *et al.* 1982). Values of n_{0g} suggested by observations range from 10^3 to 10^5 for hail and as high as 10^{10} for graupel (Knight *et al.* 1982). The density of graupel particles also varies significantly, and can range from 50 to $890 \, \text{kg} \, \text{m}^{-3}$ while hail density varies from 700 to $900 \, \text{kg} \, \text{m}^{-3}$ in observed storms (Pruppacher and Klett 2000).

The values of n_{0g} and ρ_g are important because they influence many of the microphysical process parameterization schemes. For example, as n_{0g} and ρ_g increase, the mass-weighted terminal velocity decreases. This is because the

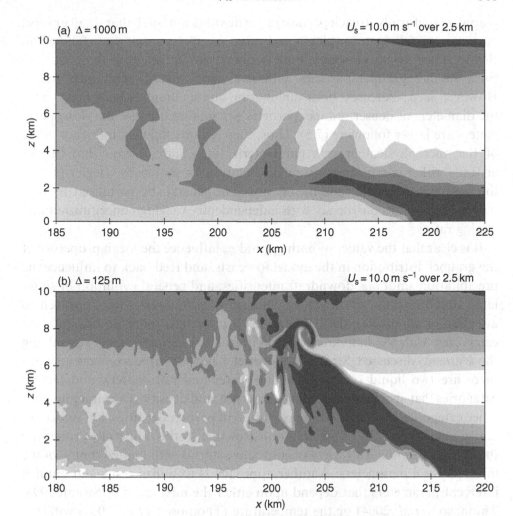

Figure 7.17. Squall line-normal cross-sections of equivalent potential temperature (K) from weak shear simulations at 180 min using (a) 1000 m grid spacing and (b) 125 m grid spacing. Cross-sections taken along $y = 49$ km. Warm rain processes only following the bulk single-moment approach of Kessler (1969). Note the dramatic change in squall line structure as the grid spacing is reduced. From Bryan *et al.* (2003).

particle distribution becomes more heavily weighted towards the smaller particles, which fall more slowly. Changes in fall speed influence the vertical distribution of the graupel particles over time. As ρ_g increases, the graupel particle distribution has fewer larger particles and becomes more heavily weighted towards smaller particles. In addition, the mass-weighted fall speed is applied uniformly to each particle in the distribution, such that small graupel particles are falling too quickly and large graupel particles are falling too slowly. Then as n_{0g} increases, both sublimation and melting rates increase since smaller graupel particles sublimate more readily and melt faster. However, as ρ_g increases, both sublimation and melting rates decrease, since the diameters of denser graupel particles are smaller because the slope parameters are larger following (7.5). Thus, these two parameters influence many of the microphysical process parameterizations and in ways that can be at times either complementary or destructive. As shown by Ziegler (1988), the intercept and slope parameters must, in general, be permitted to vary independently for consistency with independently varying concentration and mixing ratios.

It is clear that the values of both n_{0g} and ρ_g influence the mean properties of the graupel distribution in the model forecasts, and feed back to influence the precipitation amounts, downdraft intensities, and general evolution of simulated thunderstorms as shown by Gilmore *et al.* (2004a). One approach to avoiding or minimizing these limitations is to include more ice precipitation categories, with the individual microphysical rate terms broadly resembling those already discussed. Straka and Mansell (2005) describe a scheme in which there are two liquid categories (cloud water and rain water), and ten ice categories that are characterized by habit, size, and density. The larger number of ice categories allows for a range of different particle densities, fall velocities, and greater complexity of precipitation growth histories, hopefully promoting the ability to simulate a variety of convective storms with limited tuning of the microphysical parameters. Another approach is to utilize snow and graupel intercept parameters that depend upon either the mixing ratio (Swann 1998; Thompson *et al.* 2004) or the temperature (Thompson *et al.* 2004) within a single-moment bulk microphysics scheme.

In a wintertime environment, results from a simulation of an orographic precipitation event indicates that the bulk microphysics parameterization used produces too much supercooled cloud water aloft and too little snow compared to observations (Colle *et al.* 2005a). While the total rainfall is reasonably well predicted, observations suggest that snow deposition and aggregation dominated the generation of surface precipitation in contrast to the model results that indicate riming and cloud water accretion being dominant. Colle *et al.* (2005a) further show model sensitivity to the slope intercept value for the snow size distribution, further emphasizing the challenges to accurate bulk microphysics parameterizations. Other comparisons of model simulations and observations of microphysics parameters by Garvert *et al.* (2005) and Colle *et al.* (2005b) indicate that while the overall simulation of the cloud field is

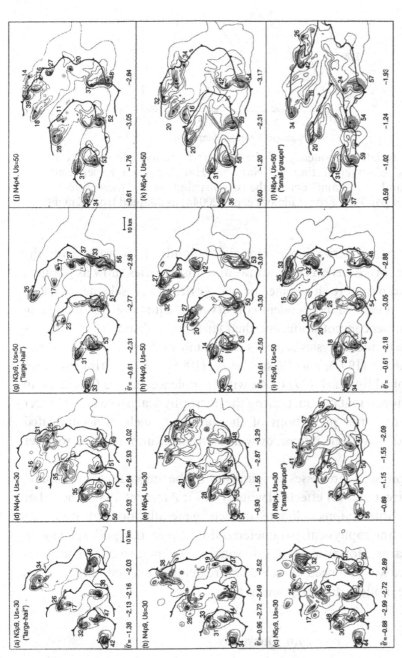

Figure 7.18. Evolution of mid-level thunderstorm structure shown at 30, 60, 90, and 120 min for 30 m s^{-1} of half-circle hodograph shear in (a)–(f) and 50 m s^{-1} of half-circle hodograph shear for (g)–(l). Results from six variations in the bulk three-class single-moment microphysics scheme are illustrated. The total precipitation mixing ratio at $z = 4.75$ km is shown using 2 g kg^{-1} intervals starting at 1 g kg^{-1}. Regions of updraft greater than 5 m s^{-1} at this level are shaded. Maximum updraft values indicated in m s^{-1}. The barbed line shows the position of the surface gust front. The minimum area-averaged potential temperature for each time is shown at the bottom of each panel. Cases are labeled $Na\rho b$, where a indicates the exponent in the slope intercept ($n_{ox} = 4 \times 10^a$ m^{-4}), and b is the first digit in the graupel density, $b00$ kg m^{-3}. Thus, $N3\rho9$ has a slope intercept of 4×10^3 m^{-4} and a graupel density of 900 kg m^{-3}. From Gilmore et al. (2004a).

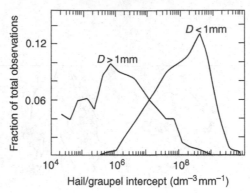

Figure 7.19. Relative frequency of the hail and graupel slope intercept parameter n_{0g} observed during the National Hail Research Experiment. "$D > 1$ mm" and "$D < 1$ mm" denote the two distributions for these different sized groups of particles. From Gilmore *et al.* (2004a) as adapted from Knight *et al.* (1982).

reasonable, the parameterization overpredicts cloud liquid water over the windward slopes, produces excessive snow, and misclassifies snow as graupel.

Beyond including more hydrometeor categories in a scheme, another improvement is to include independent conservation equations for the mixing ratio and the number concentration, and their tendencies. Examples of this class of double-moment microphysics schemes are found in Ziegler (1985), Ferrier (1994), Meyers *et al.* (1997), and Reisner *et al.* (1998), while a triple-moment scheme is developed by Clark (1974). As with turbulence closure (discussed in Chapter 5), it is hoped that parameterizing the microphysical processes at higher moments leads to better predictions of the variables, such as accumulated precipitation and downdraft intensities which are most important to weather forecast users.

While microphysical processes are known to be important to climate, through the indirect aerosol effects (Twomey 1977; Albrecht 1989) and other cloud–radiation interactions, climate models generally do not use such detailed explicit microphysical parameterizations since the grid spacing in these models is large. However, as the grid spacing of climate models continues to decrease, there will be a need to include more detailed explicit microphysics schemes in these models at some point in time.

7.6 Questions

1. Plot the saturation ratio S for pure water as a function of the cloud droplet radius.
2. In the calculation of the saturation ratio S, include the effects of a solute and replot the curve for various amounts of solute mass. Assume the solute is sodium chloride

(NaCl) and that the solute has mass values of 1×10^{-16}, 1×10^{-15}, 1×10^{-14}, 1×10^{-13}, 1×10^{-12}, 1×10^{-11}, and 1×10^{-10} g.

3. Plot the Marshall–Palmer raindrop size distributions for $n_{or} = 1 \times 10^{-4}\,\mathrm{m}^{-4}$, $\rho = 1.0\,\mathrm{kg\,m}^{-3}$, and rain water mixing ratios between 1×10^{-8} and $1 \times 10^{-3}\,\mathrm{kg\,kg}^{-1}$. Plot the curves at regular intervals of the rain water mixing ratio. At what value of the rain water mixing ratio do more than 10 drops with diameters greater than 5 mm occur for a bin size of 1 mm?

4. Calculate the autoconversion rate using the Kessler, Lin *et al.*, and Tripoli and Cotton expressions over a likely range of cloud water mixing ratios. How large are the differences in autoconversion rates?

5. Derive the Kessler form of the accretion equation (7.23)–(7.25). Show all steps.

6. Compare the Kessler, Tripoli and Cotton, and Schultz accretion rates. Assume that $a = 841.996\,\mathrm{m}^{1-b}\,\mathrm{s}^{-1}$ and $b = 0.8$ in the fall speed equation. At what value of the rain water mixing ratio are the schemes most different?

7. Calculate the distance a raindrop can fall in an environment with a constant relative humidity of 80% and a constant temperature of 278 K for drop sizes of 0.5, 2, and 5 mm. Assume the ventilation factor $F = 1$ to simply the calculations. Assume that the thermal conductivity $K_a = 2.5 \times 10^{-2}\,\mathrm{J\,m}^{-1}\,\mathrm{K}^{-1}\,\mathrm{s}^{-1}$ and that the diffusivity of water vapor in the air $D_{diff} = 2.4 \times 10^{-5}\,\mathrm{m}^2\,\mathrm{s}^{-1}$.

8. Compare ice aggregation to form snow rate equations from Lin *et al.*, Dudhia, and Murikami for a reasonable range of values for the ice mixing ratio. State any assumptions made. How large are the differences in the snow production rate from these three approaches?

8

Radiation parameterizations

8.1 Introduction

Radiation is the ultimate driver of atmospheric circulations, since radiation passes through the atmosphere and reaches the Earth's surface in amounts that are unequally distributed in space and time. This unequal energy distribution, due in part to the Earth's spherical shape, produces horizontal gradients in temperature, which produce atmospheric motions. Radiation not only determines the Earth's climate, but also plays a significant role in local energy budgets by providing the largest energy source terms. Radiation is unique among atmospheric processes since it can transport energy without a medium, yet it interacts with gases, liquids, and solids in very different ways.

Changes in the mean annual net radiation of a fraction of 1% can lead to significant changes in global climate when this change persists over a number of years. This highlights the importance of accurate radiation parameterizations to global climate models that are being used both to understand how increasing greenhouse gas concentrations affect future climate and to provide guidance to policy makers across the world. However, radiation also is important in the day-to-day weather events that influence our lives. Just think of a chilly fog-filled morning that breaks into a sunny and warm afternoon and the effects of radiation on the weather we experience become clear. Radiation is a key player in the atmosphere, both on very short and very long timescales. Thus, radiation needs to be parameterized accurately under a wide variety of atmospheric conditions.

Radiation parameterizations are intended to provide a fast and accurate method of determining the total radiative flux at any given location. These calculations provide both the total radiative flux at the ground surface, which is needed for the surface energy budget, and the vertical radiative flux divergence, which is used to calculate the radiative heating and cooling rates of a

306

given atmospheric volume. We know from Chapters 2 and 3 that the magnitude of the terms in the surface energy budget can set the stage for moist deep convection and are crucial to the formation of low-level clouds. In addition, the vertical radiative flux divergence can produce substantial cooling, particularly at the tops of clouds, which can have a strong dynamic effect on cloud evolution. While one can also use the total radiative flux to calculate the horizontal radiative flux divergence, this term is often neglected in numerical models owing to scaling arguments. Once the upward (F_U) and downward (F_D) radiative flux densities ($W\,m^{-2}$) are determined, the heating rate for a given layer of the atmosphere is defined as

$$\frac{\partial T}{\partial t} = \frac{1}{\rho c_p} \frac{\partial}{\partial z} (F_D - F_U). \tag{8.1}$$

The challenge of radiation parameterizations is to find ways to calculate F_U and F_D efficiently and accurately. This is because global climate models (GCMs) have found that the radiation parameterization calculations can easily consume most of the computer resources needed for the model simulations. As is seen throughout this and the following chapter, we are still far from handling radiation well when clouds are present and some difficulties remain in clear skies as well.

The spectral distributions of solar and terrestrial irradiance received at sea level through a cloud and haze-free atmosphere (Fig. 8.1) indicate that several simplifications are possible for radiation parameterizations. First, the shortwave (solar) and longwave (terrestrial) portions of the spectra are distinct and, therefore, can be treated separately. Second, many of the gases that absorb either solar or longwave radiation (H_2O, CO_2, O_2, O_3, etc.) are associated with specific wavelength bands (Fig. 8.2). While there are some wavelength bands for which multiple gases absorb energy, the number of bands for which only a single gas is important is a fairly large fraction of the total.

Most parameterization schemes are either highly empirically driven approaches that use bulk expressions for gaseous absorption and clouds, or approaches that use two-stream or related approximations. Two-stream approximations attempt to represent the total radiative flux in two streams: one for the downward component and one for the upward component. Thus, many of the schemes available today either use bulk column properties to parameterize the shortwave and longwave contributions separately, or use a two-stream approach to calculate the shortwave and longwave contributions for both the upward and downward components. Before we discuss the specifics of various approaches to simplifying the radiation calculations, a few concepts are reviewed.

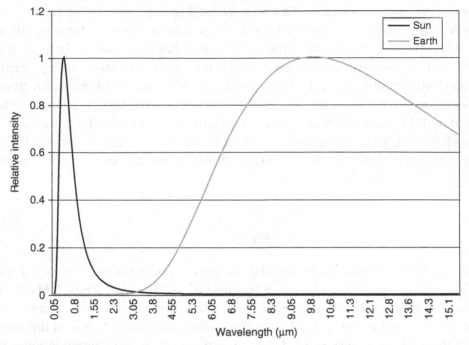

Figure 8.1. Radiative flux calculated using the Planck function and normalized to 1 as a function of wavelength for the sun ($T = 6000$ K, black line) and the Earth ($T = 290$ K, gray line). Note the lack of overlap between the two curves, which allows for the shortwave (sun) and longwave (Earth) components to be treated separately.

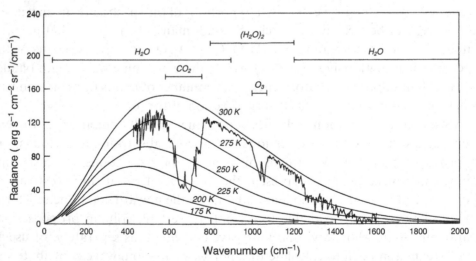

Figure 8.2. Terrestrial longwave spectra calculated for selected temperatures, with the various absorption bands indicated. Also shown is the actual emission spectrum taken by the Nimbus IV IRIS instrument near Guam on 27 April 1970. From Liou (1980), reprinted with permission from Elsevier.

8.2 Basic concepts

Electromagnetic radiation can be described as an ensemble of waves of varying wavelength that all travel through a vacuum at the same speed – the speed of light. The totality of all of these waves with different wavelengths represents the electromagnetic spectrum (see Wallace and Hobbs 1977; Liou 1980; Goody and Yung 1995). The human eye is sensitive to only a small wavelength band within this spectrum called visible light. But waves with both smaller (X-ray and ultraviolet) and longer (near-infrared, infrared, and microwave) wavelengths also are important to understanding the atmospheric energy budget. All materials with temperatures above absolute zero emit electromagnetic radiation continually.

The electromagnetic spectrum produced by a given source depends upon its composition and physical state. Solids and liquids produce a continuous electromagnetic spectrum, as illustrated by the smoothly varying curves in Fig. 8.2, as do incandescent gases under extremely high pressure such as the sun. However, luminous gases at low pressure, such as the polar aurora, produce spectra that consist of distinct lines. These lines occur because an isolated molecule can only emit and absorb energy in discrete units called photons, in contrast to the strongly interacting molecules found in liquids, solids, and gases at extreme pressure that together emit and absorb nearly all incident electromagnetic radiation. An isolated molecule can transition to a higher energy level by absorbing electromagnetic radiation, or can transition to a lower energy level by emitting electromagnetic radiation. But only discrete changes in energy are allowed. Since the energy of a photon of radiation depends upon its wavelength, the discrete nature of the various energy transitions in an isolated molecule leads to a spectrum of distinct absorption or emission lines that are very narrow and defined by the allowed changes in energy level, separated by gaps in the spectrum for which no absorption or emission is possible.

When a continuous electromagnetic spectrum, such as that produced by the sun, passes through cool gases, such as found in the atmosphere, the observed spectrum that reaches the ground is influenced by the selective absorption of radiation by the gas molecules encountered. These molecules absorb the electromagnetic radiation at distinct wavelengths, producing absorption lines in the spectrum. The resulting spectra show the continuous electromagnetic spectrum of the source interrupted by bands or lines that are a result of the selective absorption. The width and shape of these absorption lines are influenced by the atmospheric pressure and temperature through the effects of Doppler and pressure broadening (Liou 1980). Thus, the absorption lines associated with the gases that constitute the atmosphere, with liquid water and ice from clouds, and with aerosols from natural and anthropogenic sources, are important to determining the amount of radiation that is transferred through the atmosphere and absorbed by it.

8.2.1 Blackbody radiation

As discussed thoroughly by Liou (1980), Planck made two assumptions for atomic oscillators in 1901 that led to the development of what is now called Planck's law. He first assumed that an oscillator can only have specific energies, i.e. that the energy is quantized. Second, he assumed that oscillators only radiate energy in discrete jumps or quanta. These assumptions led to the development of the Planck function B_ν

$$B_\nu(T) = \frac{2h\nu^3}{c^2(e^{h\nu/k_B T} - 1)},\tag{8.2}$$

where ν is the frequency, T is the absolute temperature, k_B is Boltzmann's constant ($1.3806 \times 10^{-23}\,\mathrm{J\,K^{-1}}$), c is the velocity of light ($3.0 \times 10^8\,\mathrm{m\,s^{-1}}$), and h is the Planck constant ($6.6262 \times 10^{-34}\,\mathrm{J\,s}$). This can also be written in terms of wavelength $\lambda\ (= c/\nu)$ such that

$$B_\lambda(T) = \frac{c_1}{\lambda^5(e^{c_2/k_B \lambda T} - 1)},\tag{8.3}$$

where $c_1 = 1.191 \times 10^{-16}\,\mathrm{W\,m^{-2}\,sr^{-1}}$ and $c_2 = 1.4388 \times 10^{-2}\,\mathrm{m\,K}$. The Planck function defines the emitted monochromatic intensity for a given frequency (or wavelength) and temperature of the emitting substance. Recall that intensity, or radiance, is the radiant power per unit solid angle (steradian) and implies directionality in the radiation stream. Blackbody radiant energy increases with temperature, whereas the wavelength of maximum intensity λ_m decreases with increasing temperature ($\lambda_m = a/T$, where T is temperature in K and a is a constant). This relationship between the wavelength of maximum intensity and temperature is called Wien's displacement law. Note that the two curves of emitted monochromatic intensity in Fig. 8.1 are determined directly from (8.3), while the decrease in the wavelength of maximum intensity with temperature can be seen from Wien's displacement law.

A solid angle, $d\Omega$, as used in the definition of radiance, or intensity, is a surface area on a unit sphere (defined to have radius of 1). The equation for a solid angle is

$$d\Omega = \frac{dA_s}{r^2},\tag{8.4}$$

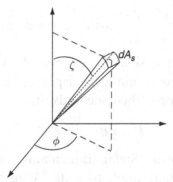

Figure 8.3. An illustration of a solid angle as defined using polar coordinates. A pencil of radiation extends from the center of a sphere through area dA_s.

where dA_s is the surface area on a sphere of radius r (Fig. 8.3). Solid angles have units of steradians, which are analogous to the use of radians for a circle. The surface area dA_s is found to be

$$dA_s = (r\, d\zeta)[r\sin(\zeta)\, d\phi] = r^2 \sin(\zeta)\, d\zeta\, d\phi, \qquad (8.5)$$

where ζ and ϕ are the zenith and azimuth angles, respectively. Thus, we have

$$d\Omega = \frac{dA_s}{r^2} = \sin(\zeta)\, d\zeta\, d\phi. \qquad (8.6)$$

If this equation is integrated over an entire hemisphere, then

$$\Omega_h = \int_0^{2\pi} \int_0^{\pi/2} \sin(\zeta)\, d\zeta\, d\phi = 2\pi \text{ steradians}. \qquad (8.7)$$

Now, if the amount of radiance passing through a given horizontal plane parallel to the Earth's surface is desired, then the component of radiation normal to this surface is needed. Knowledge of the zenith angle definition allows us to define this normal component as

$$F_\lambda = I_\lambda \cos(\zeta), \qquad (8.8)$$

where I_λ is the intensity, or radiance, for a given wavelength. If this relationship is integrated over an entire hemisphere, then

$$F = \int_0^{2\pi} \int_0^{\pi/2} I_\lambda \cos(\zeta) \sin(\zeta)\, d\zeta\, d\phi, \qquad (8.9)$$

and if the emitted radiation is isotropic, then this expression further simplifies to

$$F = I_\lambda \int_0^{2\pi} \int_0^{\pi/2} \cos(\zeta) \sin(\zeta) \, d\zeta \, d\phi = \pi I_\lambda. \tag{8.10}$$

If the Planck function is similarly integrated over all frequencies/wavelengths, and all directions within a hemisphere, and it is assumed that the emitted radiation is isotropic, then one finds that

$$F = \pi B(T) = \sigma T^4. \tag{8.11}$$

This is the well-known Stefan–Boltzmann law, where σ is the Stefan–Boltzmann constant ($\sigma = 5.67 \times 10^{-8}\,\mathrm{W\,m^{-2}\,K^{-4}}$). This expression represents the maximum amount of radiative energy that an object can emit at a given temperature. Since this quantity is integrated over an entire hemisphere of directions, it represents the irradiance. In general, the irradiance depends upon the orientation of the surface.

8.2.2 Radiative transfer

As a pencil of radiation traverses a layer in the atmosphere, it will be weakened by its interaction with various atmospheric constituents (Fig. 8.4). The decrease in the intensity I_λ of the radiation at wavelength λ is observed to follow

$$dI_\lambda = -k_\lambda \rho I_\lambda \, ds, \tag{8.12}$$

where ρ is the density of the gas, k_λ is the absorption coefficient for radiation of wavelength λ (which is a measure of the fraction of gas molecules that are absorbing radiation at λ), and ds is the thickness of the layer.

We can integrate this equation to obtain the intensity I_λ after traversing a distance s in the absorbing material, yielding

$$I_\lambda(s) = I_\lambda(0)e^{-\int_0^s k_\lambda \rho \, ds}, \tag{8.13}$$

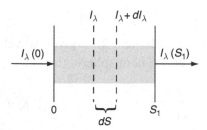

Figure 8.4. Depletion of the radiant intensity I_λ as it traverses an absorbing medium. After Liou (1980) reprinted with permission from Elsevier.

where $I_\lambda(0)$ is the intensity of the radiation at wavelength λ upon entering the atmospheric layer in question, and $I_\lambda(s)$ is the intensity upon exiting this layer. This relation is known as Beer's law, or the Beer–Bouguer–Lambert law in Liou (1980). This expression is often rewritten as

$$I_\lambda(s) = I_\lambda(0)e^{-\tau_\lambda},\tag{8.14}$$

where

$$\tau_\lambda = \int_0^s k_\lambda \rho \, ds,\tag{8.15}$$

is called the optical depth, or optical thickness, depending upon the context (Liou 1980). The optical depth is a measure of the cumulative depletion that a beam of radiation experiences as it passes through a given layer.

Geometry also plays a role in calculating the optical depth used in models. It is convenient to measure the distance normal to the surface of the Earth, yet the direction of radiation is often at some angle to this upwardly directed normal. This angle is again our good friend the zenith angle. As the zenith angle increases from zero, a beam of radiation passes through more and more gas molecules as it traverses across a given layer of thickness dz. Thus, to change the integral defining the optical depth from ds to dz, we need to take into account the zenith angle. This leads to

$$\tau_\lambda = \frac{1}{\cos \zeta} \int_{z_1}^{z_2} k_\lambda \rho \, dz.\tag{8.16}$$

For any electromagnetic radiation I_λ that passes through a medium, such as the atmosphere, some of the energy is absorbed, some is scattered, and some is transmitted through the medium. Thus, the conservation of energy requires that

$$\frac{I_\lambda(\text{absorbed})}{I_\lambda} + \frac{I_\lambda(\text{reflected})}{I_\lambda} + \frac{I_\lambda(\text{transmitted})}{I_\lambda} = a_\lambda + r_\lambda + t_\lambda = 1,\tag{8.17}$$

where a_λ is the monochromatic absorptivity, r_λ is the monochromatic reflectivity, and t_λ is the monochromatic transmissivity. When a_λ is non-zero, then one also has to incorporate the emission of radiation from the medium through which the radiation is passing. This yields, following Liou (1980), the relationship

$$\frac{dI_\lambda}{k_\lambda \rho \, ds} = -I_\lambda + B_\lambda(T) + J_\lambda,\tag{8.18}$$

where T is the absolute temperature of the medium, $-I_\lambda$ represents the loss of radiance from attenuation (absorption plus scattering), $B_\lambda(T)$ represents the emission of radiance from the medium, and J_λ represents a second source of radiation from scattering into the line segment ds.

For longwave radiation, the scattering of radiation is negligible (Liou 1980), yielding the simplified equation of radiative transfer

$$\frac{dI_\lambda}{k_\lambda \rho \, ds} = -I_\lambda + B_\lambda(T). \tag{8.19}$$

If we assume a plane-parallel atmosphere (i.e., the atmosphere varies much more in the vertical than the horizontal direction and that the Earth is large enough to be considered flat) and account for the orientation of the radiation beam to the upwardly pointing normal, and use our definition of the optical depth from[1] τ, where $d\tau_\lambda = -k_\lambda \rho ds$, then we find that the equation of radiative transfer becomes

$$-\frac{dI_\lambda}{d\tau_\lambda \cos \zeta^{-1}} = -I_\lambda + B_\lambda(T). \tag{8.20}$$

Solutions commonly are found separately for upward and downward radiances. The general procedure to obtain a solution is to multiply the equation by the integrating factor $e^{-\tau_\lambda / \cos \zeta}$ and to integrate over the layer. After combining the two I_λ terms on the left-hand side, the equation becomes

$$-\int_{z_1}^{z_2} d\left[I_\lambda(z)e^{-\tau_\lambda(z)/\cos \zeta}\right] = \int_{z_1}^{z_2} B_\lambda(T)e^{-\tau_\lambda(z)/\cos \zeta} \, d\tau_\lambda. \tag{8.21}$$

If we then integrate the left-hand side term from z_1 to z_2 and divide both sides by $e^{-\tau_\lambda(z_2)/\cos \zeta}$ we find that after rearrangement we are left with

$$I_\lambda(z_2) = I_\lambda(z_1)e^{-[\tau_\lambda(z_1)-\tau_\lambda(z_2)]/\cos \zeta} + \int_{z_1}^{z_2} B_\lambda(T)e^{-[\tau_\lambda(z_1)-\tau_\lambda(z_2)]/\cos \zeta} \, d\tau_\lambda. \tag{8.22}$$

The first term on the right-hand side again represents the attenuation of the outgoing radiation, while the second term is the internal atmospheric contribution over the layer (Liou 1980).

Since the shortwave and longwave portions of the spectral distributions of irradiance are distinct, we can develop separate parameterizations for each component. We begin with discussions of the longwave component, since this is the one that is most expensive computationally.

---

[1] Optical depth increases in the opposite sense to z, and hence the negative sign.

8.3 Longwave radiative flux

8.3.1 Empirical methods

As discussed earlier, there are two basic approaches to the parameterization of radiation that have been used in models. The simplest and least computationally demanding, but likely to be the least accurate, is an empirical approach that relates bulk properties to the radiative flux. For example, numerous methods have been developed to estimate the downwelling longwave radiation at the ground from surface observations (Monteith and Unsworth 1990). One of the simplest approaches is to assume

$$Q_{Ld} = c + d\sigma T_a^4, \tag{8.23}$$

where c and d are constants calculated from observations and T_a is the 2 m air temperature. Unsworth and Monteith (1975) find that $c = -119 \pm 16\,\mathrm{W\,m^{-2}}$ and $d = 1.06 \pm 0.04$, using observations from the English Midlands. Similar values are determined over Australia by Swinbank (1968). Extensions to these types of statistical correlations of radiative fluxes with weather variables are possible for cloud conditions as well (Unsworth and Monteith 1975). They are most accurate under average weather conditions and are not appropriate for use at all times and places.

A similar approach is used by Anthes *et al.* (1987) to calculate the net longwave radiation at the surface, where they define

$$Q_{L_{net}} = \varepsilon_g \varepsilon_a \sigma T_a^4 - \varepsilon_g \sigma T_g^4. \tag{8.24}$$

Here T_g is the ground temperature (K), T_a is the air temperature approximately 40 hPa above the ground surface, ε_g is the ground or soil emissivity (typically 0.9–1.0), σ is the Stefan–Boltzmann constant, and ε_a is the atmospheric longwave emissivity. The downward longwave component is based upon the work of Monteith (1961) who shows that

$$\varepsilon_a = 0.725 + 0.17 \log_{10} w_p, \tag{8.25}$$

in which w_p is the total column precipitable water in centimeters. While computationally very efficient, these empirical approaches provide no information on the radiative flux divergence above the ground that can be dynamically important to cloud formation and breakup. In general, these approaches also neglect gaseous emissions from sources other than water vapor.

When clouds are present, the approach of Anthes *et al.* (1987) simply increases the downwelling longwave component using an enhancement factor, such that

$$Q_{Ld} = Q_{Ld_{clear}} \left(1 + \sum_{i=1}^{3} c_i n_i \right), \tag{8.26}$$

where n_i is the cloud fraction for a given atmospheric layer (low, middle, high), and the c_i values are given by

Cloud level	c_i
Low ($i = 1$)	0.26
Middle ($i = 2$)	0.22
High ($i = 3$)	0.06

The cloud layers are typically defined based upon the maximum relative humidity between specified pressure levels. The influences of clouds on radiation are discussed more fully in the following chapter on cloud cover.

8.3.2 Two-stream methods for clear skies

The other approach to parameterizing the longwave radiative flux is based upon solving the radiative transfer equation as discussed by Liou (1980). The discussion in this section is largely based upon Stephens (1984) and Ellingston *et al.* (1991). The equations appropriate for longwave flux are

$$F_U(z) = \int_0^\infty \pi B_\nu(0) \tau_\nu^f(z, 0) \, d\nu + \int_0^\infty \int_0^z \pi B_\nu(z') \frac{d\tau_\nu^f}{dz'}(z, z') \, dz' \, d\nu, \tag{8.27}$$

$$F_D(z) = \int_0^\infty \int_z^\infty \pi B_\nu(z') \frac{d\tau_\nu^f}{dz'}(z, z') \, dz' \, d\nu, \tag{8.28}$$

where F_U and F_D are the upward and downward fluxes through height z, B_ν is the Planck function, and τ_ν^f is the diffuse transmission function defined over a hemisphere. The first term in F_U represents the attenuation of the longwave radiation emitted from the Earth's surface, while the second term represents the emittance of longwave radiation by the atmosphere. The single term in F_D represents the atmospheric contributions only. The diffuse transmission function is written as

$$\tau_\nu^f(z, z') = 2 \int_0^1 \tau_\nu(z, z', \mu) \, \mu \, d\mu, \tag{8.29}$$

where[2] $\mu = \cos(\zeta)$ and

$$\tau_\nu(z, z', \mu) = \exp\left[-\frac{1}{\mu}\int_{u(z)}^{u(z')} k_\nu(p, T)\, du\right], \tag{8.30}$$

in which $k_\nu(p, T)$ is the absorption coefficient (a function of pressure p and temperature T) and u is the concentration of the attenuating gas along the path from z to z'. The different parameterizations for longwave radiation differ in how these four integrals are calculated, owing to the need to minimize the computational costs. However, several common assumptions are found within these schemes.

The first common assumption is that one can replace the integration over all zenith angles in the $\tau_\nu^f(z, z')$ equation with the simplification

$$\tau_\nu^f(z, z') \propto \tau_\nu(z, z', 1/\beta), \tag{8.31}$$

where β is the diffusivity factor and is equal to 1.66. This states that the transmission of flux through the slab from z to z' is equivalent to the transmission of a beam along the zenith angle $\zeta = \cos^{-1}(1/\beta)$. This has been found to be a very reasonable and useful approximation (Rodgers and Walshaw 1966; Liu and Schmetz 1988).

The second group of common assumptions involves the integration of the absorption coefficient over the optical path. While it is known that k_ν is a function of p and T, laboratory data used to calculate k_ν are determined using constant values of p and T instead of the rapid changes observed in the atmosphere. Two common approximations used to determine $\tau_\nu(z, z', \mu)$, the scaling and the two-parameter approximation, both assume that absorption along a non-homogeneous path can be approximated by absorption along a homogeneous path using adjusted values of p and T.

The scaling approximation assumes that the absorption coefficient depends only upon the value of k_ν at reference p and T and the concentration of the attenuating gas u. Thus, the scaling approximation assumes

$$k_\nu(p, T) = k_\nu(p_0, T_0)\tilde{u} = k_\nu(p_0, T_0)\int_{u(z)}^{u(z')} \left(\frac{p}{p_0}\right)^n \left(\frac{T_0}{T}\right)^m du, \tag{8.32}$$

where m and n are constants that are specified for various absorbing species and range from 0 to 1.75 for n and from 0 to 11 for m. The variable $\tilde{u}$ is the adjusted concentration of the attenuating gas u. The variables p_0 and T_0 are the

---

[2] In radiative transfer discussions, this polar angle is measured relative to a beam pointing upwards, so $\zeta = 0$ and $\mu = 1$ for a beam pointing upwards, and $\zeta = \pi$ and $\mu = -1$ for a beam pointing downwards.

reference pressure and temperature. Goody (1964) provides further information on this approximation.

The two-parameter approximation was proposed by Curtis (1952), Godson
(1954), and Van de Hulst (1945) and so is often called the VCG approximation.
It allows for more accurate approximations of the path integral by using two
parameters to relate the absorption along a non-homogeneous path to that of
a corresponding homogeneous path. It assumes that the mean transmission
between two levels is the same as if all the absorbing gas along the path were at
a constant pressure (Rodgers and Walshaw 1966). Thus, it adjusts both the
path u and the pressure p according to

$$\tilde{p}\tilde{u} = \int p \, du$$

$$\tilde{u} = \int du. \tag{8.33}$$

This adjustment is found by matching the absorption in the strong and weak
limits. Thus, instead of the path integral only depending upon the absorber
concentration as in the scaling approximation, it now depends upon both the
absorber concentration and the mean pressure.

A third method also has come into use recently. Instead of scaling a reference set of k_ν values as a function of u and p, a third method linearly
interpolates between stored sets of k_ν values that have been previously calculated over the full range of atmospheric conditions. The generation of these
stored sets of k_ν values is a significant overhead cost, and requires the use of
a very computationally expensive line-by-line radiative transfer model.
Presently, this approach is used only in the rapid radiative transfer model
(RRTM) as discussed in Mlawer *et al.* (1997).

Now that we have approximations that allow for easy calculation of the
transmissivity function, the integrations over frequency remain. This is where
the differences in the parameterizations are more clearly seen. The problem of
integrating over frequency is more complex than simply averaging k_ν over
some interval $\Delta\nu$. As illustrated by Stephens (1984), there are four distinguish-
able frequency scales that must be taken into account when making this
calculation. These scales range from the smoothly varying Planck function
to the rapidly varying absorption lines for the different absorbers (Fig. 8.5). As
discussed by Liou (1980), the examination of high-resolution spectroscopy
shows that the emissions of certain gases are composed of a large number of
characteristic spectral lines. These lines occur because each quantum jump
between fixed energy levels within an electron results in emission or absorption

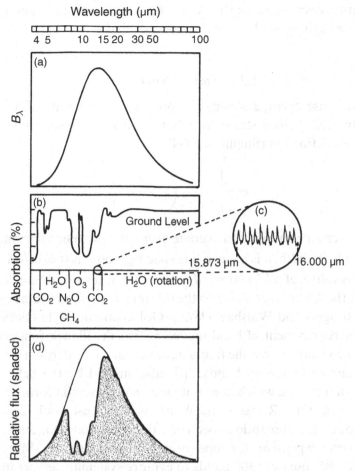

Figure 8.5. Schematic diagram illustrating the different frequency scales involved in the calculation of atmospheric longwave flux. These frequency scales are (a) the Planck curve, (b) atmospheric gaseous absorption spectrum for longwave radiation reaching the ground surface, (c) the individual absorption lines and line separations found when looking at very small frequency intervals, and (d) the convolution of the Planck function and the atmospheric absorption spectrum to give the atmospheric longwave flux (shaded area). From Stephens (1984).

of a characteristic frequency, which then appear as absorption lines. Two general parameterization approaches are used to surmount the challenges provided by these widely varying functions included in the frequency integration. The first approach is to divide the Planck function into a number of discrete intervals, define the absorption characteristics of each interval separately, and sum the resulting values (narrow-band models). The second method is to convolve the absorption and Planck functions and integrate this quantity over large portions of the longwave spectrum (wide-band models). We begin by looking at the narrow-band models.

8.3.3 Narrow-band models

Band models use average absorption properties for bands of lines that are specified by well-defined statistical relationships. The resulting transmission functions are defined as (Stephens 1984)

$$\tau_{\bar{\nu}} = \exp\left[\frac{-\bar{S}u\beta}{\bar{d}}\left(1 + \frac{\bar{S}u\beta}{\pi\bar{\alpha}}\right)^{-1/2}\right], \qquad (8.34)$$

where the overbar represents an average over $\Delta\nu$, $\bar{S}$ is a mean line intensity, $\bar{\alpha}$ is a mean line half-width, $\bar{d}$ is a mean line spacing, and β is the diffusivity factor. The various values of $\bar{S}/\bar{d}$, and $\pi\bar{\alpha}/\bar{d}$ for the two water vapor bands, the 15 μm CO_2 band, the 9.6 μm O_3 band, and the 6.3 μm water vapor vibration band are found in Rodgers and Walshaw (1966), Goldman and Kyle (1968), and Wu (1980). One requirement of band models is that the Planck function must be treated as a constant across the frequency interval selected, often leading to the need for narrow frequency bands. It is also limited by the behavior of the transmission function, which does not take an exponential form across broad spectral bands. While Rodgers and Walshaw (1966) used only 21 intervals to span the longwave absorption spectrum, the computational efforts required are still very expensive for operational considerations. Morcrette and Fouquart (1985) use over 300 bands to explore systematic errors in longwave radiation calculations.

An alternative approach to the narrow-band method that is demonstrably faster and more accurate is the correlated-k, or k-distribution, method. This method uses the fact that the transmission within a relatively wide spectral interval is independent of the ordering of the values of the absorption coefficient k_ν with respect to ν. Thus, for an assumed homogeneous atmosphere, the transmission depends on the fraction of the selected interval $f(k)$ that is associated with a particular value of k. This approach groups frequency intervals according to line strengths, and the transmission function is rewritten as

$$\tau_{\bar{\nu}}(z, z') = \frac{1}{\Delta\nu}\int_{\Delta\nu} e^{-k_\nu u}\, d\nu = \int_0^\infty f(k)e^{-ku}\, dk. \qquad (8.35)$$

Thus, instead of integrating over frequency, one integrates over the absorption coefficient values (Fig. 8.6). Errors associated with the correlated-k technique

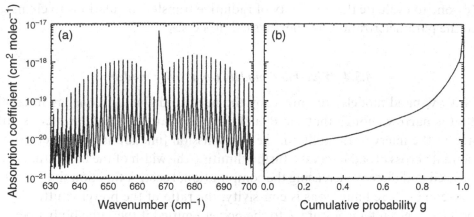

Figure 8.6. Absorption coefficients due to carbon dioxide for a layer at 507 hPa in the middle of summer (a) as a function of wavenumber, and (b) after being rearranged in ascending order for the spectra range 630–700 cm^{-1}. From Mlawer *et al.* (1997).

are at least a factor of two smaller than most other band models, and the technique is computationally faster as well. Lacis and Oinas (1991) suggest that the correlated-k technique produces results to within 1% of a line-by-line radiative transfer model (LBLRTM), while Fu and Liou (1992) explore the mathematical and physical conditions under which the method is valid and find that it works well for atmospheric radiative transfer. Differences in cooling rates between the rapid radiative transfer model (RRTM), which uses this approach, and a LBLRTM are typically less than 0.15 K day^{-1} for clear skies (Mlawer *et al.* 1997).

The main differences between the various narrow-band models are the number of frequency bands used in the calculations, whether the scaling or two-parameter approximation is used for the optical path integration, and the data sets used to generate the transmission functions for the spectral intervals. However, other details can also be important. As pointed out by Stephens (1984), while the absorption in the "atmospheric window" between 8 and 14 μm is weak, it is important to reproduce it accurately as significant amounts of radiant energy are exchanged between the ground surface, clouds, and space. Another concern is the method by which the transmission of two different gases that absorb radiation in the same spectral interval (e.g., H_2O and CO_2) is handled. Please refer to Stephens (1984) and Ellingston *et al.* (1991) for further discussions and details on these narrow-band methods. In addition, Edwards and Slingo (1996) develop a two-stream radiation parameterization in which the spectral resolution of the code is variable, allowing for one to evaluate the sensitivity of radiative transfer calculations to changes in the parameterizations of the physical processes.

8.3.4 Wide-band models (emissivity models)

Narrow-band models are limited by the need to define a frequency interval that is narrow enough that the Planck function can be treated as a constant across the interval. In addition, the transmission function is no longer exponential across broad intervals, further limiting the width of the bands that can be used and thereby increasing the computational cost. A function that largely overcomes these limitations is emissivity, the ratio of the power emitted by a body at a given temperature T to the power emitted if the same body obeyed Planck's law.

Emissivity ε is thus defined as

$$\varepsilon(z, z') = \frac{1}{\sigma T^4} \int_0^\infty A_\nu(z, z') \pi B_\nu(T) \, d\nu, \tag{8.36}$$

where $A_\nu (= 1 - \tau_0)$ is the absorptivity of the gas. This allows us to rewrite the radiative flux equation following Stephens (1984) as

$$F_U(z) = \sigma T_g^4 (1 - \varepsilon(z, 0)) + \int_0^z \sigma T^4(z') \frac{d\varepsilon}{dz'}(z, z') \, dz', \tag{8.37}$$

$$F_D(z) = \int_z^\infty \sigma T^4(z') \frac{d\varepsilon}{dz'}(z, z') \, dz'. \tag{8.38}$$

Note that the integration over frequency has now vanished from the radiative flux equations! Emissivity has physical significance and can be measured, although there is a fair amount of ambiguity in these measurements when gas emissions overlap. However, the solution of these flux equations is easy given the emissivity as a function of the absorber concentration u (Charlock and Herman 1976; Sasamori 1968; Staley and Jurica 1970).

As an example, Rodgers (1967) develops an upward and downward emissivity as

$$\varepsilon(u) = \sum_{n=0}^N b_n (\ln(u))^n, \tag{8.39}$$

where u is the water vapor path and b_n are constants that depend upon temperature. For u less than $10 \, \mathrm{g \, m^{-2}}$, an alternative expression is used such that

$$\varepsilon(u) - \sum_{n=1}^{N} a_n u^{n/2}, \tag{8.40}$$

where a_n are constants that depend upon temperature. The variable N is the number of terms in the polynomial expansion and equals 4 in Rodgers (1967). This approach is used by Dudhia (1989) in a mesoscale weather prediction model.

One challenge to the broad-band method is that it is not always simple to obtain expressions of emissivity as a function of absorber concentration u that are sufficiently accurate to obtain precise values of $d\varepsilon/du$ or $d\varepsilon/dz$. If instead a modified emissivity is defined as

$$\varepsilon'(z, z') = \int_0^\infty A_\nu(z, z') \frac{dB_\nu(T)}{d\sigma T^4(z')} \, d\nu = \frac{R(z, z')}{4\sigma T^3}, \tag{8.41}$$

where $R(z, z')$ is the mean absorptivity parameter (Elsasser and Culbertson 1960), then an integration by parts of the original emissivity form of the flux equation yields

$$F_U(z) = \sigma T_g^4 + \int_0^z \varepsilon'(z, z') \frac{d\sigma \, T^4(z')}{dz'} \, dz', \tag{8.42}$$

$$F_D(z) = \int_z^\infty \varepsilon'(z, z') \frac{d\sigma \, T^4(z')}{dz'} \, dz'. \tag{8.43}$$

While ε' is not directly related to ε, it also doesn't greatly differ from ε. Indeed, Ramanathan *et al.* (1983) define the relationship

$$\varepsilon = \frac{\varepsilon'}{0.847 u^{0.222}}, \tag{8.44}$$

where u is the absorber concentration, which allows us to convert from one emissivity to the other. Rodgers (1967) shows that the flux calculations are most accurate when ε' is used to calculate F_U and ε is used to calculate F_D.

Another widely used emissivity-based approach is the simplified exchange method (SEM) of Fels and Schwarzkopf (1975) and Schwarzkopf and Fels (1991). This method recognizes that, in many situations, the dominant contribution to atmospheric cooling rates at any given height is from the "cool-to-space" (CTS) term (Fig. 8.7) as discussed by Rodgers and Walshaw (1966). The total cooling rate Q can then be divided into two parts

$$Q = Q_{ex} + Q_{CTS}, \tag{8.45}$$

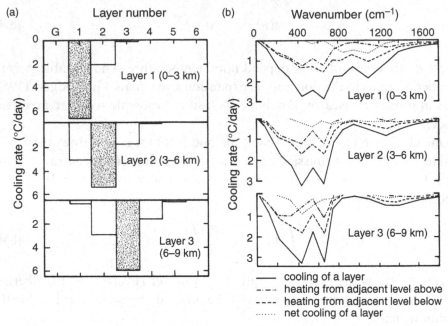

Figure 8.7. The contribution to the specified 3 km deep layer longwave radiative cooling from the layer itself (shaded) and the absorption of longwave radiation from adjacent 3 km deep layers (left). Shaded areas denote cooling, whereas the open areas denote warming from the adjacent layers. Also shown is the spectral distribution of the cooling/heating for the three 3 km deep layers (right). Cooling to space is the largest contribution in each of the three layers shown. From Stephens (1984) as modified from Wu (1980).

where Q_{ex} is the exchange term and Q_{CTS} is the cool-to-space term. Since Q_{CTS} is the dominant term, it needs to be calculated very accurately. In contrast, Q_{ex} can be calculated using approximate techniques with little loss of accuracy. This division of the calculation into accurate and approximate approaches is the key characteristic of this approach (Fig. 8.8). Thus, one can calculate Q_{ex} using

$$Q_{ex} = Q^\varepsilon - Q^\varepsilon_{CTS}, \tag{8.46}$$

where both Q^ε and Q^ε_{CTS} are determined using broad-band emissivity methods. The Q_{CTS} term is calculated using

$$\frac{\partial T}{\partial t} = \frac{g}{c_p} \sum_n B_n(T) \frac{\partial \tau_n}{\partial p}(0, p), \tag{8.47}$$

for an isothermal atmosphere at temperature T, where n is the number of frequency bands. This equation is used to determine the cooling rates at every

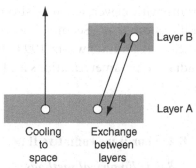

Figure 8.8. Schematic diagram of the two different contributions to the radiative cooling of a layer in the atmosphere. One contribution is the cooling to space, which occurs mainly in the transparent regions of the absorption spectrum. The other contribution is by mutual exchange between layers, say layers A and B, as illustrated. After Stephens (1984).

level in the model, even when T is not constant. The computational efficiency of this CTS term compared to other emissivity methods is that the cooling rates are a function of T and only depend upon the absorber concentration above the level in question. Thus, the CTS term is calculated using multiple bands, while the broad-band emissivity method is used for the exchange term. This yields computational efficiency and accurate calculations.

The above discussion outlines various approaches to longwave radiative flux calculations that apply to any given layer of the atmosphere. To incorporate these equations into a numerical weather prediction model, the values of F_U and F_D are calculated at each vertical model level. Then the heating rate is calculated from the flux divergence between every two vertical levels using a finite-difference form. Stephens (1984) suggests that around a dozen or so vertical levels are needed for relatively accurate calculations if there are no significant discontinuities in moisture or temperature in the vertical direction. But computational efficiency is always a concern, leading to efforts to develop very fast radiation parameterizations (e.g., Harshvardhan *et al.* 1987).

Ellingston *et al.* (1991) discuss results from an international project to compare radiation codes used in climate models under clear-sky conditions. Fifty-five separate cases are used to compare the various methods, representing a range of conditions. Results indicate that the line-by-line models and the narrow-band models agree to about ±2% for fluxes at the atmospheric boundaries and to about ±5% for the flux divergence in the troposphere. However, only five wide-band models are found to match the performance of the narrow-band models. Many of the less detailed parameterizations display a spread of 10–20% in their calculations. However, many changes have occurred since 1991. First, the parameterizations have constantly evolved over time, such that results with newer schemes should be an improvement over those shown by Ellingston *et al.* Second, the Atmospheric Radiation Measurement Program (Stokes and Schwartz 1994) has assisted in funding further refinements to radiation parameterizations and more detailed comparisons with observations.

8.4 Shortwave radiative flux

8.4.1 Empirical methods

As with longwave methods, the simplest of the shortwave radiation parameterization schemes only predict the total shortwave radiation that reaches the Earth's surface. These schemes do not incorporate the effects of shortwave absorption on atmospheric diabatic heating. They can be based upon detailed comparisons with more complex shortwave radiative transfer models, or upon comparisons with observations. Regardless of the specific approach, their main advantage is their computational efficiency.

Anthes *et al.* (1987) present a fairly typical approach to calculating the amount of shortwave radiation that reaches the surface. They approximate

$$Q_S = S_0(1 - a)\tau \cos(\zeta), \tag{8.48}$$

where S_0 is the solar constant, a is the albedo, ζ is the solar zenith angle, and τ is the shortwave transmissivity. This transmissivity term is based upon the work of Benjamin (1983) for multiple reflections and is defined as

$$\tau = \tau_a[\tau_s + (1 - \tau_s)(1 - b)]/(1 - X_R a), \tag{8.49}$$

where τ_a is the absorption transmissivity, τ_s is the scattering transmissivity, b is the backscatter coefficient, and X_R is the multiple reflection factor where

$$X_R = \tau_{ad}(1 - \tau_{sd})b_d. \tag{8.50}$$

The τ_{ad}, τ_{sd}, and b_d terms refer to absorption, scattering, and backscatter parameters for diffuse radiation. These clear-air transmissivities (τ_a, τ_s, τ_{ad}, τ_{sd}) are functions of path length and precipitable water as found from the Carlson and Boland (1978) radiative transfer model.

Benjamin (1983) also presents a scheme to alter the shortwave radiation based upon the amount of cloud cover. Three layers of cloud are allowed, depending upon the atmospheric pressure: low clouds for pressures greater than 800 hPa, middle clouds for pressures between 800 and 450 hPa, and high clouds for pressures below 450 hPa. He then defines

Table 8.1. *Values of absorption and scattering transmissivity parameters for determining the total transmissivity when clouds are present from Benjamin (1983).*

Cloud level	τ_{ai}	τ_{si}
Low ($i = 1$)	0.80	0.48
Middle ($i = 2$)	0.85	0.60
High ($i = 3$)	0.98	0.80

$$\tau_{ac} = \prod_{i=1}^{3}[1 - (1 - \tau_{ai})]n_i, \tag{8.51}$$

$$\tau_{sc} = \prod_{i=1}^{3}[1 - (1 - \tau_{si})]n_i, \tag{8.52}$$

where n_i is the cloud fraction (0 to 1) for each of the cloud layers and is based upon the maximum relative humidity value in each layer, and the absorption transmissivities for each layer are predetermined and shown in Table 8.1. The total transmissivity for a cloudy atmosphere is then defined by

$$\tau = \frac{\tau_{ac}\tau_{sc}\tau_a[\tau_s + (1 - \tau_s)(1 - b)]}{1 - X_C a}, \tag{8.53}$$

where

$$X_C = \tau_{ad}\tau_{ac}(1 - \tau_{sd}\tau_{sc})\overline{b_d}, \tag{8.54}$$

and

$$\overline{b_d} = \frac{b_d(1 - \tau_{sd}) + (1 - \tau_{sc})}{(1 - \tau_{sd}) + (1 - \tau_{sc})}. \tag{8.55}$$

It should be noted that applying fixed, predetermined shortwave characteristics to clouds is problematic. The radiative properties of clouds change with the changing cloud character and with zenith angle. Thus, determining just what values for transmissivity should be used is not clear and these approaches without doubt do not work well in all circumstances.

Savijärvi (1990) presents a slightly more detailed approach for clear skies, defining

$$Q_S = S_0 \sin(h) \Big[1 - 0.024(\sin(h))^{-0.5} - 0.11aa(u/\sin(h))^{0.25}$$
$$- as(0.28/(1 + 6.43\sin(h)) - 0.07a) \Big], \tag{8.56}$$

where

$$u = \int_{p_{sfc}}^{0} q\Big(\frac{p}{1013}\Big)^{0.85} \Big(\frac{273}{T}\Big)^{0.5} \frac{dp}{g} \tag{8.57}$$

is a scaling approximation that represents a linearly pressure-scaled vertical water vapor amount (cm), and h is the local hour angle of the sun. This equation accounts for O_3, H_2O, and CO_2 depletion as well as Rayleigh scattering. The parameters aa and as (both ≥ 1) represent a crude inclusion of aerosol absorption and scattering, respectively. Savijärvi suggests that the best results are found with $aa = 1.2$ and $as = 1.25$ for continental industrialized areas.

Unlike the Anthes *et al.* approach, Savijärvi also includes a simple parameterization for solar heating in the atmosphere, using

$$\frac{\partial T}{\partial t} = S_0 \Big(\frac{q}{c_p}\Big) \Big(\frac{p}{p_0}\Big) \Big[y\frac{u}{\sin(h)} + 1.67ay(u_*)\sin(h) \Big] + 1.7 \times 10^{-6}[\sin(h)]^{0.3}, \tag{8.58}$$

where u_* is u calculated from the surface to the top of the atmosphere, and

$$y = \begin{cases} 0.029u^{-0.81} & u \geq 0.05 \text{ cm} \\ 0.050u^{-0.63} & u < 0.05 \text{ cm}. \end{cases} \tag{8.59}$$

The Atwater and Ball (1981) constant cloud transmission functions are recommended when clouds are present.

Finally, the last example for the empirical approaches is from Dudhia (1989), who provides a slightly more complex one-stream approach. He defines

$$F_D(z) = \mu S_0 - \int_z^\infty (dS_{cs} + dS_{ca} + dS_s + dS_a), \tag{8.60}$$

where S_{cs} and S_{ca} are the decreases in irradiance from cloud scattering and absorption, and S_s and S_a are the decreases in irradiance from clear-air scattering and absorption. The parameterization assumes that the cloud fraction is either 0 (no cloud) or 1 (overcast) for each grid point.

The scattering and absorption coefficient values are bilinearly interpolated from Stephens' (1978b) tabulated functions of μ and the natural logarithm of the liquid water path for cloudy conditions and from the Lacis and Hansen (1974) absorption function for clear-air conditions (which requires μ and the water vapor path). Dudhia indicates that the total effect of a cloud layer above a height z is obtained from the above function as a percentage of the downward solar flux that is absorbed or reflected by clouds. Then at height $z - \Delta z$, a new total percentage is determined that incorporates the absorption and scattering effects of the layer Δz. Thus, the clear-air effect above z is removed since clouds are present. Clear-air scattering is assumed to be uniform and proportional to the atmospheric mass path length, allowing for zenith angle effects, with 10% scattering occurring in one atmosphere. The atmospheric heating rate is computed from the vertical change in the absorption terms.

8.4.2 Two-stream methods in clear skies

The parameterization of shortwave radiative flux has many similarities to the parameterizations for longwave radiative flux discussed previously. However, instead of being grouped into general types (narrow-band, broad-band), the schemes are named (Eddington, delta-Eddington, quadrature, hemispheric, two-stream) based upon the choices made in approximating the effective zenith angle for the stream directions and how and when the single-scattering phase function is approximated. This definitely lends a different flavor to the literature.

Unlike longwave radiation, both absorption and scattering are important processes for shortwave radiation and need to be included. Stephens (1984) indicates that Rayleigh scatter is dominant only for the shorter wavelengths, while liquid water absorption in clouds occurs only for the longer wavelengths, leading to a natural division on either side of the 0.7 μm wavelength. The transfer of shortwave radiation is not as complex as longwave radiation, since the problem of the simultaneous absorption and emission of radiation from layer to layer does not occur. Thus, although not as computationally demanding as longwave schemes, shortwave schemes still require substantial computer time and approximations are needed to produce schemes that can run within operational models.

Following the discussions of Stephens (1984) and Pincus and Ackermann (2003), the shortwave irradiance is often separated into direct (*dir*) and diffuse (*dif*) components, where the direct component represents the contributions from photons that have not been scattered. Thus,

$$I(\tau, \mu, \varphi) = I_{dir}(\tau, \mu, \varphi) + I_{dif}(\tau, \mu, \varphi). \tag{8.61}$$

The direct component follows Beer's law, such that the irradiance reaching a given level z is given by

$$F_D(z, \mu_0) = \mu_0 \int_0^\infty S_\nu(\infty) \tau_\nu(z, \infty, \mu_0) \, d\nu, \tag{8.62}$$

where

$$\tau_\nu(z, \infty, \mu_0) = \exp\left(-\frac{1}{\mu_0} \int_z^\infty k_\nu \, du\right). \tag{8.63}$$

While problems similar to those found in calculating longwave radiation are present in the path and frequency integrals, the values of $S_\nu(\infty)$ are known and specified a priori. One can use a narrow- or broad-band approach to calculate the integral. The correlated-k, or k-distribution, method also can be used to improve the accuracy and speed of the calculations for shortwave radiation.

In addition, Stephens (1984) indicates that the effects of pressure and temperature on k_ν are only a complication for water vapor absorption. Thus, if desired, one can define a mean transmission function

$$\tau_{\bar{\nu}}(z, \infty, \mu_0) = \frac{1}{\Delta \nu} \int_{\Delta \nu} \exp\left(-m_r(\mu_0) \int_z^\infty k_\nu \, du\right) d\nu, \tag{8.64}$$

where $m_r(\mu_0)$ is a relative air mass factor and differs from $1/\mu_0$ only at large zenith angles greater than 75° (and for ozone – see Rodgers 1967). This could be considered a broad-band approach.

With these assumptions one can then rewrite the integral over frequency as a summation and obtain

$$F_D(z) = \mu_0 \sum_{i=1}^N S_i(\infty) \tau_{\bar{\nu}i}(u), \tag{8.65}$$

where N is the number of frequency intervals and u is the absorber path length from z to ∞ along the zenith angle.

Since the direct component of radiation follows Beer's law, and therefore is relatively easy to code, most of the literature describes the ways in which the diffuse component of shortwave radiation is handled. The azimuthally averaged monochromatic radiative transfer equation for a given frequency ν is

$$\mu \frac{dI}{d\tau} = -I(\tau, \mu) + \frac{\omega_0}{2} \int_{-1}^1 \bar{p}(\tau, \mu, \mu') I(\tau, \mu') \, d\mu' + \frac{\omega_0 S_0}{4\pi} \bar{p}(\tau, \mu, \mu_0) e^{-\tau/\mu_0}, \tag{8.66}$$

where ω_0 is the single-scattering albedo, which is the likelihood that a photon is scattered rather than absorbed at each interaction (and varies from 0 to 1), $\bar{p}$ is the scattering phase function that characterizes the angular distribution of the scattered radiation field, S_0 is the solar constant (W m^{-2}), and μ_0 is the cosine of the zenith angle of the sun as always. To obtain the irradiance, this equation must be integrated over frequency and zenith angle. The first term on the right-hand side is the diffuse radiation that enters into the layer in question. The second term represents the increase in diffuse radiation from multiple scattering, while the last term represents the increase in diffuse radiation from single scattering of direct solar radiation. Thus, it is easy to see how scattering complicates the calculations. In addition, since the intensity I appears on both sides of the equation, this an integrodifferential equation which is quite difficult to solve.

The optical thickness τ now includes contributions from scattering and absorption. The scattering phase function $\bar{p}$ describes how likely it is that radiation traveling in the (μ', ϕ') direction will be scattered, by molecules or cloud droplets, into the (μ, ϕ) direction. The value of a phase function can vary over several orders of magnitude as the scattering angle is varied and illustrates complex behavior. Analytic formulas are used to approximate the phase function and can be compared against a full Rayleigh or Mie scattering calculation for accuracy. Thankfully, the exact nature of the scattering phase function is not incredibly important, as multiple scatter tends to smooth out its peaks when hemispheric integrals are calculated. This is a key point, since the full scattering calculations show a great deal of structure in the scattering phase function. Yet it is clear that the scattering phase function approximation used does make a difference in the accuracy of the results.

The phase function can be described in terms of a scattering angle Θ between the incident and scattered radiation from a frame of reference centered on a particle. A useful parameter to use in describing the phase function is the asymmetry parameter g, such that

$$g = \frac{1}{2} \int_{-1}^{1} \cos\Theta \, \bar{p}(\Theta) \, d\cos\Theta, \qquad (8.67)$$

which varies from -1 for complete backscatter, to 0 for isotropic scatter, to $+1$ for complete forward scatter.

8.4.3 Eddington approach

As discussed earlier, the schemes are named based upon the choices made in approximating the various terms in the integrals for the shortwave radiation flux. The Eddington approximation as discussed by Pincus and Ackermann (2003) uses an expansion for both intensity and phase to first order, such that each varies linearly with μ as

$$I(\mu) = I_0 + I_1\mu$$
$$\bar{p}(\tau, \mu, \mu_0) = 1 + 3g\mu\mu'. \tag{8.68}$$

This yields for the diffuse radiative transfer equation, after evaluating the integral and rearranging some terms,

$$\mu\frac{dI_0}{d\tau} + \mu^2\frac{dI_1}{d\tau} = I_0(1 - \omega_0) + I_1\mu(1 - \omega_0 g) - \frac{\omega_0}{4}(1 - 3g\mu\mu_0)S_0e^{-\tau/\mu_0}. \tag{8.69}$$

A pair of equations for I_0 and I_1 are obtained by first integrating over μ from -1 to 1 and then multiplying by μ and again conducting this same integration. This yields

$$\frac{dI_0}{d\tau} = (1 - \omega_0 g)I_1 + \frac{3\omega_0}{4\pi}g\mu_0 S_0 e^{-\tau/\mu_0}, \tag{8.70}$$

$$\frac{dI_1}{d\tau} = 3(1 - \omega_0)I_0 - \frac{3\omega_0}{4\pi}S_0 e^{-\tau/\mu_0}. \tag{8.71}$$

These two first-order equations in I_0 and I_1 can be solved, following Shettle and Weinman (1970), providing solutions that are a sum of exponentials in τ, such as

$$I_0 = Ae^{k\tau} + Be^{-k\tau} + \phi e^{-\tau/\mu_0}, \tag{8.72}$$

where A, B, and ϕ are determined from the boundary conditions at the top and bottom of the atmosphere and from the particular solution.

The diffuse irradiances (fluxes) are then computed from I_0 and I_1 using

$$F_D(\tau, \omega_0, g, \mu_0) = 2\pi \int_0^1 (I_0 + \mu I_1)\mu \, d\mu = \pi\left(I_0 + \frac{2}{3}I_1\right), \tag{8.73}$$

$$F_U(\tau, \omega_0, g, \mu_0) = 2\pi \int_0^{-1} (I_0 + \mu I_1)\mu \, d\mu = \pi\left(I_0 - \frac{2}{3}I_1\right). \tag{8.74}$$

Since different frequencies are associated with different parameter values for ω_0, g, and τ, these expressions for F_D and F_U must be integrated over all frequencies to get the total diffuse irradiance.

8.4.4 Delta-Eddington approach

The Eddington approximation is not as accurate as many would like, and particularly has problems in optically thin atmospheres and when large absorption is involved, so Joseph *et al.* (1976) develop a modification to the phase function that performs better. The original phase function approximation, which contains a large and narrow forward peak in scattering, is replaced with a delta function in the forward direction and a smoother scaled phase function elsewhere. This results in a new approximation for the phase function, such that

$$\bar{p}(\cos \Theta) \approx 2f\partial(1 - \cos \Theta) + (1 - f)(1 + 3g' \cos \Theta), \qquad (8.75)$$

where f is the fractional scattering into the forward peak, ∂ is the Dirac delta function, and g' is the asymmetry factor of the truncated phase function (Pincus and Ackermann 2003). Joseph *et al.* (1976) require that this phase function have the same asymmetry factor (g) as the original phase function, which defines a specific relationship between g' and g. They also define $f = g^2$. When all the equations are derived, it is seen that this approach just rescales the original Eddington solutions.

One application of the delta-Eddington approach is discussed by Briegleb (1992). In this parameterization, originally developed for the NCAR Community Climate Model version 2 (CCM2), the solar spectrum is divided into 18 bands. Seven bands are for O_3, one band for the visible, seven bands are for H_2O, and the final three bands are for CO_2. The model atmosphere consists of a discrete set of horizontally homogeneous layers. The delta-Eddington solution consists of evaluating the solution for the reflectivity and transmissivity of each layer, and then combining the layers together to obtain the upward and downward spectral fluxes. This is repeated for all 18 spectral bands to accumulate broad-band fluxes, from which heating rates then can be calculated.

8.4.5 Two-stream approach

While the terminology is a little bit confusing, there is another method to solve the radiative transfer equation that is called the two-stream method (Liou

1980). This should not be confused with the general idea that most methods used to solve the radiative transfer equation consist of calculating two streams of radiation: one in the upward direction and one in the downward direction. This "two-stream" method is just one particular approach to solving the same problem. While the Eddington approach expands the intensity and phase into first order in angle, the two-stream model first averages the radiative transfer equation and phase function over each hemisphere to obtain the radiative flux, and then computes the solution. Thus, the azimuthally averaged radiative transfer equation is integrated over the hemisphere to obtain the upward and downward fluxes, leading to two separate equations

$$\bar{\mu}\frac{dF_D}{d\tau} = F_D - \omega_0(1-b)F_D + \omega_0 b F_U - \frac{\omega_0}{2\pi}[1 - b(\mu_0)]S_0 e^{-\tau/\mu_0}, \tag{8.76}$$

$$\bar{\mu}\frac{dF_U}{d\tau} = F_U - \omega_0(1-b)F_U + \omega_0 b F_D - \frac{\omega_0}{2\pi}b(\mu_0)S_0 e^{-\tau/\mu_0}, \tag{8.77}$$

where b is the backscattering coefficient (Pincus and Ackermann 2003). These are two first-order, linear coupled differential equations with constant coefficients. The solutions are found by uncoupling the equations through differentiating with respect to optical thickness, and then substituting the other equation. The solutions end up being a sum of exponentials in optical thickness, just as found for the Eddington solutions.

All two-stream methods can be generalized to a generic form, where

$$\frac{dF_U}{d\tau} = \gamma_1 F_U - \gamma_2 F_D + \frac{S_0}{4}\tilde{\omega}_0\gamma_3 e^{-\tau/\mu_0}, \tag{8.78}$$

$$\frac{dF_D}{d\tau} = \gamma_2 F_U - \gamma_1 F_D + \frac{S_0}{4}\tilde{\omega}_0\gamma_4 e^{-\tau/\mu_0}, \tag{8.79}$$

provided that an explicit assumption is made regarding the dependence of I upon μ. Values of reflectance R and transmittance T can be derived directly from these equations. The heart of these two-stream approaches is a suitable choice of τ, $\tilde{\omega}_0$, and g for each spectral interval and absorber. Note that the heating from CO_2 and O_2 absorption lines is usually added to the heating by water vapor absorption, since the contributions of both CO_2 and O_2 to solar heating are small (Stephens 1984). Meador and Weaver (1980) derive the various formulas for numerous two-stream approaches and discuss the different assumptions that go into their development.

The above discussion outlines several approaches to shortwave radiative flux calculations that apply to any given layer of the atmosphere. To incorporate these equations into a numerical weather prediction model, the values of F_U and F_D again need to be calculated at each vertical model level just as is done for the longwave flux. However, this is more complicated than the calculations for longwave flux, owing to the effects of multiple scattering that must be taken into account. The two-stream solutions discussed above apply to a single homogeneous layer with fixed values of τ, $\tilde{\omega}_0$, and g, and so the challenge is to combine several layers with these varying optical properties. Several different techniques are available that can be used for this purpose.

The adding method is one approach to calculating the shortwave radiative flux across multiple vertical levels. Take two adjacent vertical layers in the atmosphere. The upper layer has transmittance T_1 and reflectance R_1, while the lower layer has transmittance T_2 and reflectance R_2. To find out how much total flux is reflected from this combination of layers, we examine the multiple reflections. First, some of the flux that enters the upper layer is reflected immediately (R_1). However, some is also transmitted through the layer, is reflected by the lower layer, and then transmitted back through the first layer $(T_1 R_2 T_1)$. Some of this flux is reflected back from the upper layer and again reflected back from the lower layer, finally passing through the upper layer $(T_1 R_2 R_1 R_2 T_1)$. This process goes on and on (Pincus and Ackermann 2003), yielding

$$R_T = R_1 + T_1 R_2 T_1 + T_1 R_2 R_1 R_2 T_1 + T_1 R_2 R_1 R_2 R_1 R_2 T_1 + \cdots$$
$$R_T = R_1 + T_1 R_2 [1 + R_1 R_2]^{-1} T_1. \tag{8.80}$$

This equation for total reflectance is combined with its analog for transmittance with the delta-Eddington or other two-stream models to compute the transmittance and reflectance of layered atmospheres. The adding–doubling method extends this general approach to intensity (Pincus and Ackermann 2003). Stephens (1984) illustrates that an equivalent approach is to consider the atmosphere as a system of n homogeneous layers with specified values for reflectance and transmission. This allows one to construct an equivalent linear system of $3n + 3$ equations, the solution of which produces the equivalent adding algorithm. Pincus and Ackermann (2003) also discuss an eigenvector approach termed the discrete ordinate approach.

8.5 Radiative transfer data sets

Numerical weather prediction models provide the vertical profiles of temperature, pressure, and water vapor mixing ratio needed for longwave and shortwave radiative transfer calculations under clear skies, but gas and particle concentrations also are needed in these calculations. Of the gases important to radiative transfer calculations, carbon dioxide and oxygen are considered permanent constituents of the atmosphere and have nearly constant volume ratios up to 60 km (Liou 1980). Although the concentration of carbon dioxide does vary slightly throughout the annual cycle (variations of ~6 parts per million (ppm) by volume with a concentration of ~375 ppm in 2004), it also has been increasing throughout the years owing to the burning of fossil fuels and these changes are estimated for use in global climate models. In contrast, ozone is more variable in time and space, although it typically resides at levels between 15–30 km above ground. This variability requires that the ozone distribution be specified in the numerical model through reference to typical distributions (e.g. London *et al.* 1976). Total column ozone concentration also can be observed by satellite (Heath *et al.* 1975), but the radiative transfer calculations still need the vertical distribution. Jang *et al.* (2003) suggest that this can be estimated from vertical mean potential vorticity and show that the inclusion of real-time ozone data into a numerical weather prediction model can lead to improvements in forecasts. At present, operational models generally use an ozone climatology.

Finally, aerosols are important to shortwave absorption and scattering. Aerosols can scatter and absorb shortwave radiation, thereby changing the diffuse fraction of the shortwave radiative flux and influencing the terrestrial carbon cycle (Niyogi *et al.* 2004). In addition, aerosols can act as condensation nuclei for cloud droplets, thereby enhancing the amount of cloud cover, and influencing cloud lifetimes and precipitation efficiencies (Twomey 1977; Albrecht 1989). Some aerosols are created by natural processes, such as from sea spray, dust, smoke from natural forest fires, chemical reactions, boreal forests, and volcanic eruptions. Other aerosols are produced by humankind as a consequence of fuel combustion and are often commonly referred to as pollution. Obviously aerosols also can vary greatly in time and space. In general, aerosol concentrations in the troposphere are much greater than those in the stratosphere, except after volcanic eruptions. Tropospheric aerosols have lifetimes of several days, emphasizing their variability. Most models presently use typical aerosol distributions in the radiative transfer calculations, while developers are moving to include real-time total column aerosol concentrations based upon satellite data in combination with information from climatology or surface aerosol observing networks such as the Aerosol Robotic Network (AERONET; Dubovik and King 2000) to provide the distributions needed for radiative transfer calculations (see King *et al.* 1999; Al-Saadi *et al.* 2005).

8.6 Discussion

Relatively sophisticated and computationally affordable radiative transfer parameterizations have been developed for both the shortwave and longwave radiation components. These schemes provide the net surface radiation for the land surface schemes, which is partitioned into the total available energy for sensible, latent, and ground heat fluxes, and the vertical radiative flux divergence, which is used to calculate the radiative heating and cooling rates of a given atmospheric volume. Errors in the net radiation clearly influence these surface flux amounts and feed back to influence boundary layer structure and depth (Guichard *et al.* 2003; Zamora *et al.* 2003), and eventually even cloud development and precipitation. Thus, accurate radiation parameterizations are very important to the success of numerical weather prediction.

Comparisons of present operational and research models with observations generally indicate that the predicted amounts of surface incoming shortwave radiation are too large (Fig. 8.9) (Betts *et al.* 1997; Halthore *et al.* 1998; Hinkelman *et al.* 1999; Chevallier and Morcrette 2000; Barker *et al.* 2003; Marshall *et al.* 2003; Zamora *et al.* 2003). This high bias is probably due to

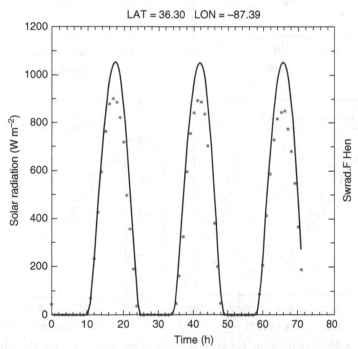

Figure 8.9. Total solar flux predicted by a numerical model (solid line) and measured at New Hendersonville, Tennessee (asterisks) between 11 and 14 June 1995. From Zamora *et al.* (2003).

either the lack of a combination of aerosol and ozone absorption (Zamora *et al.* 2003) or to excessive simplifications in the shortwave radiation parameterization (Morcrette 2002). When the aerosol optical depths and amounts of ozone are known, then Satheesh *et al.* (1999), Mlawer *et al.* (2000), and Zamora *et al.* (2003) indicate good agreement between the predicted and observed amount of surface shortwave radiation. Satheesh *et al.* (1999) show that the total shortwave flux decreases by 50–80 W m^{-2} when including aerosol radiative forcing based upon observed aerosol characteristics, with similar results from Zamora *et al.* (2005) (Fig. 8.10). Zamora *et al.* (2003) further

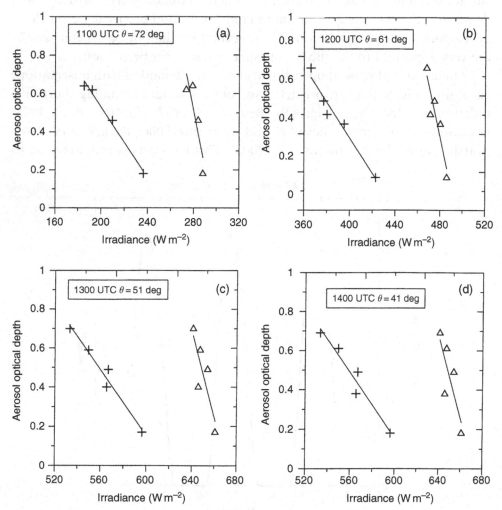

Figure 8.10. Correlation between aerosol optical depth and the observed (crosses) and Eta model (triangles) solar irradiances for zenith angles of (a) 72°, (b) 61°, (c) 51°, and (d) 41°, measured on five different cloud-free days during August 2002. From Zamora *et al.* (2005).

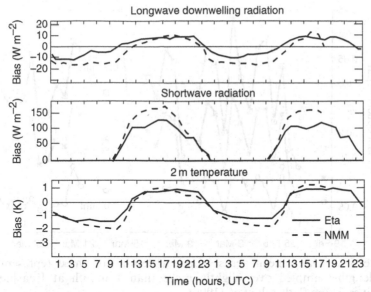

Figure 8.11. Plots of bias (model − observed) in downward longwave radiation (top), shortwave radiation (middle), and 2 m temperature (bottom) versus time (UTC) over the 48 h forecast time and averaged over 77 days. The models are the NCEP Eta model and the Non-hydrostatic Mesoscale Model (NMM). The downward longwave radiation was only observed at a single site (Plymouth, Massachusetts), whereas the solar radiation and the 2 m temperature are averages from six sites. After Stensrud *et al.* (2006).

show that stratospheric ozone alone can reduce the surface shortwave radiation by 20–30 W m^{-2}. The excessive net radiation at the surface leads to predicted daytime temperatures that are too warm (Fig. 8.11) and boundary layers that are too deep when compared with observations (Zamora *et al.* 2003).

Unfortunately, aerosol optical depths vary across a large range, from 0.02 to 0.4, and are influenced strongly by both natural and anthropogenic sources (sea salt from the oceans, mineral dust from arid land regions, sulfate and nitrate from both natural and anthropogenic sources, and organic and carbonaceous particles from burning (Satheesh *et al.* 1999)). Satheesh *et al.* (1999) show daily variations approaching 0.2 in aerosol optical depths over the Maldive Islands (Fig. 8.12). Yet aerosol optical depth is not commonly measured, reported, and ingested into models as part of our routine observational system, although this may soon change (King *et al.* 1999).

Intercomparisons of shortwave radiation parameterizations also have been undertaken and results summarized by Fouquart *et al.* (1991) and Boucher *et al.* (1998). As reported in Fouquart *et al.* (1991), even for the simplest case of pure water vapor absorption, root-mean-square (rms) differences of 1–3%

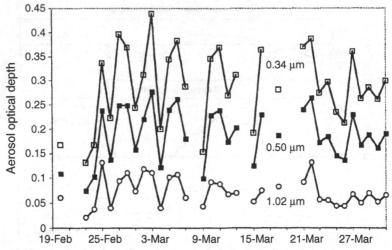

Figure 8.12. Daily mean aerosol optical depths for three representative wavelengths sampled over slightly more than 1 month at Kaashidhoo, Maldives. From Satheesh *et al.* (1999).

($\sim$10 W m^{-2}) are found for the downward flux at the surface, with larger differences of 6–11% for the total atmospheric absorption. The rms differences in downward flux at the surface increase to 4% for low elevation angles. When aerosols and/or clouds are considered, the rms differences increase even further, with clouds producing differences of up to 10% (50 W m^{-2}) and aerosols up to 21% (90 W m^{-2}). Fouquart *et al.* (1991) note two main causes of the uncertainty in the shortwave calculations. First, the calculation of water vapor absorption is critical, yet large differences exist in the various parameterizations. Fundamental inadequacies may be associated with details in the spectral lines and/or irradiances at the top of the atmosphere. Second, that the interactions between multiple scattering and molecular absorption are very difficult to handle adequately with low spectral resolution methods. This is the cause of the large rms differences when aerosols or clouds are present. Boucher *et al.* (1998) compare the direct shortwave radiative forcing by sulfate aerosols in 15 different parameterizations and find a standard deviation of the zenith-angle-averaged normalized broad-band shortwave forcing of 8%, with somewhat larger standard deviations at some zenith angles. Most of the one-dimensional parameterizations overpredict surface shortwave radiation by 15–25 W m^{-2} at overhead sun for a standard tropical atmosphere, regardless of cloud cover. However, recent results by Bush *et al.* (2000), Dutton *et al.* (2001), and Philipona (2002) suggest that much of this presumed overprediction actually may be due to an underestimation of the observed clear-sky global and diffuse solar irradiance caused by pyranometer differential cooling

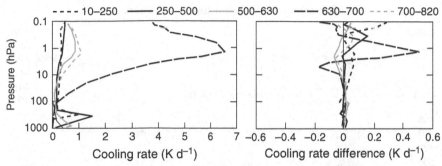

Figure 8.13. Spectrally integrated cooling rates (K day^{-1}) for each of the bands 1 through 5 (10–820 cm^{-1} as indicated in the figure) calculated by the line-by-line radiative transfer model (LBLRTM), and cooling rate differences (K day^{-1}) between RRTM and LBLRTM. Note that the cooling rate differences generally are less than 0.2 K day^{-1} and the flux differences are typically less than 0.6 W m^{-2} (not shown). After Mlawer *et al.* (1997).

when adequate ventilation and heating systems are not used. Philipona (2002) concludes that since most pyranometers in use worldwide do not have adequate ventilation and heating systems, archived global and diffuse radiation measurements largely are underestimated.

For longwave radiation parameterizations, the results are more mixed. Mlawer *et al.* (1997) show good agreement between the RRTM and a line-by-line radiative transfer model (Fig. 8.13), while Zamora *et al.* (2003) show good agreement between surface incoming longwave radiation from the RRTM and observations. However, a broad-band emissivity parameterization does not fare as well in the comparison as it overpredicts the mean incoming radiation by 80 W m^{-2} (Zamora *et al.* 2003). This overprediction leads to the model being unable to develop a nocturnal stable boundary layer and a low-level jet. Comparisons of present operational models from the USA with observations over the New England region (Stensrud *et al.* 2006) show that the model parameterizations underpredict the incoming longwave radiation by 10–20 W m^{-2}, leading to surface nighttime temperatures that are too cool (Fig. 8.11). Similar results are found in Hinkelman *et al.* (1999).

The terrain slope, orientation, and sky view also influence the amount of radiation received at the surface. Avissar and Pielke (1989) illustrate the importance that terrain slope and orientation can have on the incoming solar radiation, yet these effects are not always incorporated into models. A parameterization that includes the effects of terrain on both longwave and shortwave radiation is developed by Müller and Scherer (2005) and yields smaller 2 m air temperature errors in comparison with simulations that do not include the terrain effects.

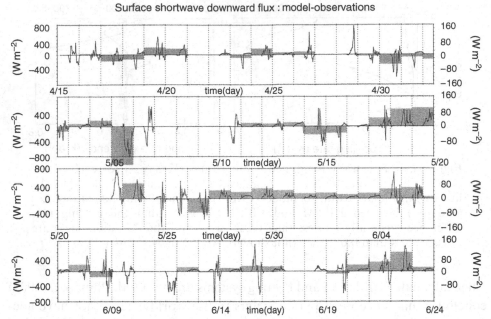

Figure 8.14. Time series of surface downward shortwave radiation differences (simulated − observed) for 15 April through 24 June 1998 over the Atmospheric Radiation Measurement Southern Great Plains site in Lamont, Oklahoma. Values are 30 minute averages. Gray shading denotes the mean daily values, using the axis on the right-hand side of the figure. From Guichard *et al.* (2003).

A comparison of amounts of surface radiation versus observations suggests that on short timescales the differences between the predicted and observed amounts can be quite large on a large number of days (Fig. 8.14). Differences of several hundred $W m^{-2}$ in incoming surface shortwave radiation and between 50 and 100 $W m^{-2}$ for incoming longwave radiation are not unusual. Undoubtedly, many of these differences are due to clouds, which are covered in the next chapter. While the daily average values of predicted radiation may be relatively close to the observations, these large short timescale differences make one wonder about how this inaccuracy in incoming surface radiation influences the model behavior. Is it good enough to get the daily average net radiation, or is it also important to get the instantaneous radiation amounts correct? This is an important question for numerical weather prediction and the answers may very well be case dependent. Certainly we have learned that small differences over even short time periods can be important when the weather situation is uncertain and a host of outcomes are possible. If this position is true, indicating that a high level of accuracy is needed in the instantaneous radiation fluxes, then we have a long way to go in developing radiation parameterizations with the needed level of accuracy for use in numerical weather prediction models.

Finally, as the grid spacing in models continues to decrease, the validity of using model data from a single vertical column (i.e., the plane-parallel assumption) for radiative transfer calculations comes into question. Clearly as horizontal grid spacings approach 1 km the use of a single vertical column to describe the radiation transfer becomes fraught with difficulties, as radiation reaching a given location on the Earth's surface clearly passes through multiple horizontal grid cells. Monte Carlo photon transport codes are available that track individual photons in three dimensions from their entry points in a column to their exit points and allow photons to move from grid cell to grid cell (e.g., Marchuk *et al.* 1980; Barker and Davies 1992; Chylek and Dobbie 1995; Marshak and Davis 2005). Another approach for three-dimensional radiative transfer is the spherical harmonic discrete ordinate method (Evans 1998). Typically these methods are used in situations where the plane-parallel assumption is poor, such as when clouds are present, and are not used in operational models owing to their computational expense. These methods are summarized and plans for their development and testing outlined in Cahalan *et al.* (2005).

8.7 Questions

The basic ideas behind most of the radiative transfer parameterization schemes are similar, so let us examine the behavior and construction of one of the schemes. Lacis and Hansen (1974) examine how ozone and water vapor influence the total solar flux. The fraction of total solar flux absorbed in the *l*th layer of the atmosphere by ozone, where $l = 1$ is the top of the atmosphere and *l* increases downward, is defined as

$$A_l = \mu_0 \{ A(x_{l+1}) - A(x_l) + \overline{R}(\mu_0) [A(x_l^*) - A(x_{l+1}^*)] \}, \qquad (8.81)$$

where $\overline{R}(\mu_0)$ is the albedo of the reflecting region, x_l is the ozone path traversed by the direct solar beam in reaching the *l*th layer, and x_l^* is the ozone path of the diffuse radiation illuminating the *l*th layer from below. Ozone absorption is modeled as a purely absorbing region sitting on top of a purely reflecting region. For our purposes, all ozone is generally assumed to lie above the troposphere, or above approximately 100 hPa. The albedo of the reflecting region is defined as

$$\overline{R}(\mu_0) = \overline{R}a(\mu_0) + [1 - \overline{R}a(\mu_0)](1 - \overline{\overline{R}}a^*)Rg/(1 - \overline{\overline{R}}a^* Rg), \qquad (8.82)$$

where Rg is the ground albedo, $\overline{\overline{R}}a^* = 0.144$, and the effective albedo of the lower atmosphere is defined as

$$\overline{Ra}(\mu_0) = \frac{0.219}{1 + 0.816\mu_0}, \tag{8.83}$$

for clear skies. They also define $x_l = u_l M$, where u_l is the amount of ozone in a vertical column above the lth layer, and

$$M = \frac{35}{(1224\mu_0^2 + 1)^{1/2}}, \tag{8.84}$$

is the magnification factor (Rodgers 1967) to account for path slant and refraction. The ozone path traversed by the diffuse radiation illuminating the lth layer from below is defined as

$$x_l^* = u_t M + \overline{M}(u_t - u_l), \tag{8.85}$$

where u_t is the total ozone above the main reflecting layer (the ground for clear skies), and $\overline{M} = 1.9$ is the effective magnification factor for diffuse radiation.

Finally, the fraction of incident solar flux absorbed by the Chappius visible band is defined as

$$A(x_l) = \frac{0.02118x_l}{1 + 0.042x_l + 0.000323x_l^2}, \tag{8.86}$$

and the absorption of the ultraviolet region is defined as

$$A(x_l) = \frac{1.082x_l}{(1 + 138.6x_l)^{0.805}} + \frac{0.0658x_l}{1 + (103.6x_l)^3}, \tag{8.87}$$

where the total absorption by ozone in the lth layer is the sum of the individual absorptions from the visible and ultraviolet regions.

If we assume a single layer of ozone above the troposphere with an ozone path x_2 above a single-scattering layer, then following Zamora *et al.* (2003) we find that

$$A = \mu_0 \{ A(x_2) + \overline{R}(\mu_0)[A[x_2(M + \overline{M})] - A[x_2 M]] \}. \tag{8.88}$$

1. Calculate the absorption of ozone for a zenith angle of 60° and for ozone path $x_2 = 0.2$, 0.3, 0.4, 0.5, and 0.6. Assuming a solar constant of $1368\,\mathrm{W\,m^{-2}}$ and a ground albedo $Rg = 0.10$, how much does the presence of ozone decrease the incoming solar radiation?
2. Now let us turn our attention to the absorption of solar radiation by water vapor. Lacis and Hansen (1974) discuss water vapor absorption further. One formula for water vapor absorption is

$$A(y) = \frac{2.9y}{(1 + 141.5y)^{0.635} + 5.925y},\tag{8.89}$$

where y is the precipitable water vapor in centimeters. The fractional absorption in the lth layer is then defined under clear skies as

$$A_l = \mu_0\{A(y_{l+1}) - A(y_l) + Rg[A(y_l^*) - A(y_{l+1}^*)]\},\tag{8.90}$$

where the effective amount of water vapor traversed by the direct solar beam is

$$y_l = \frac{M}{g} \int_0^{p_l} q\left(\frac{p}{p_0}\right)^n \left(\frac{T_0}{T}\right)^{1/2} dp,\tag{8.91}$$

in units of $\text{kg}\,\text{m}^{-2}$, and where the effective amount of water vapor traversed by the diffuse radiation reaching the lth layer from below is

$$y_l^* = \frac{M}{g} \int_0^{p_g} q\left(\frac{p}{p_0}\right)^n \left(\frac{T_0}{T}\right)^{1/2} dp + \frac{5}{3g} \int_{p_{l+1}}^{p_g} q\left(\frac{p}{p_0}\right)^n \left(\frac{T_0}{T}\right)^{1/2} dp,\tag{8.92}$$

where p_g is the ground pressure, g is the acceleration due to gravity, $p_0 = 1013\,\text{mb}$, and $T_0 = 273.15\,\text{K}$. One can convert y_l to units of cm by dividing by the density of water ($1000\,\text{kg}\,\text{m}^{-3}$) and multiplying by 100 to convert from m to cm. Assuming the atmospheric profile given below, calculate the heating rate in clear-sky conditions using $n = 0, 0.5$, and 1 to calculate the effective water vapor. Describe the influence of the pressure scaling on the resulting heating rates.

The sounding to use in calculating water vapor absorption is as follows.

p (mb)	T (°C)	q (g kg^{-1})
950.0	36.0	13.0
900.0	32.0	11.0
850.0	26.0	11.0
800.0	20.0	11.0
750.0	15.0	11.0
700.0	15.0	4.0
650.0	10.0	2.0
600.0	5.0	1.5
550.0	0.0	1.5
500.0	−8.0	1.0
450.0	−15.0	1.0
400.0	−20.0	0.4
350.0	−26.0	0.4
300.0	−30.0	0.4
250.0	−44.0	0.2
200.0	−53.0	0.1
150.0	−62.0	0.05
127.4	−67.0	0.05

9

Cloud cover and cloudy-sky radiation parameterizations

9.1 Introduction

The amount of sky covered by clouds, or cloud cover, can substantially alter the amount of radiation received at the Earth's surface, thereby influencing atmospheric circulations and climate. Clouds both reflect and absorb short-wave radiation, altering the total albedo. Under overcast skies, the reduction in the incoming shortwave radiation due to the clouds can be substantial. In addition, clouds absorb and emit longwave radiation in the atmospheric window, the wavelength band in which the atmosphere is almost transparent to longwave radiation. The ability of clouds to alter substantially both the shortwave and longwave radiative transfer highlights their importance to numerical weather prediction across a range of spatial and temporal scales.

Small-scale atmospheric circulations develop in response to horizontal differential heating as discussed in Chapter 2. Horizontal variability in cloud cover can lead to surface differential heating and the development of meso-scale thermally induced circulations between cloudy and clear areas (Koch 1984; Segal *et al.* 1986). These circulations can reach magnitudes similar to those associated with sea breezes with the potential to aid in both the development of deep convection and frontogenesis across weak or moderate cold fronts (Koch 1984; Keyser 1986; Segal *et al.* 1986, 1993; Businger *et al.* 1991). Segal *et al.* (1986) further argue that even short-lived episodes of cloudy–clear air contrasts can lead to changes in the low-level wind fields. The effects of cloud shading underneath thunderstorm anvils may be important to the evolution of thunderstorms (Markowski and Harrington 2005).

On much longer timescales, clouds play a very important role in the Earth's radiation budget. It is clear that clouds act as a radiative feedback to climate by reflecting the incoming solar radiation, and absorbing and emitting the terrestrial longwave radiation. This feedback is rather complex, with clouds generally acting to cool the surface except in polar regions where the albedo of cloudy regions can be less than the surface albedo (Schneider 1972). An increase in low- and mid-level clouds leads to cooling, owing to the increase in albedo, while an increase in high-level cirrus clouds leads to warming, owing to the associated increase in longwave absorption and emission without much change to shortwave radiation. The ability of clouds to produce both a positive and a negative climate feedback, owing to the details in the horizontal and vertical cloud characteristics, indicates both the importance and sensitivity of three-dimensional cloud cover treatment for simulations of global climate change (Stephens 2005). Uncertainties in how clouds are parameterized in models are one of the principal obstacles preventing more accurate climate change prediction (Webster and Stephens 1983). Numerous studies have shown that model simulations are sensitive to the specification of cloudiness (e.g., Meleshko and Wetherald 1981; Shukla and Sud 1981).

Cloud cover also can be modified by aerosols. Aerosols from both natural and anthropogenic sources act as condensation nuclei for cloud droplets, such that increases in the amount of aerosol can increase low-level cloud cover (Twomey 1977). However, increases in aerosols also change cloud lifetimes and precipitation efficiencies (Albrecht 1989) and some aerosols produce a reduction in cloud reflectance (Kaufman and Nakajima 1993). Thus, the interrelationships between climate, humans, and clouds are complex and highlight the challenges in global climate simulations.

The largest challenge to cloud cover parameterization is that the formation and dissipation of clouds is, in general, poorly understood. And since clouds are subgrid scale, both horizontally and vertically, there is neither a complete theory nor an observational database that can be used to relate cloud cover to the large-scale variables (Slingo 1987). Yet cloud cover is very important because it has a significant influence on incoming solar radiation and downwelling longwave radiation, not to mention the heating rates from radiative flux divergence in the atmosphere. Eliminating cloud cover in a mesoscale model can easily alter temperatures in the lowest model layer by at least 2 K in winter and 4 K in summer over the central and eastern USA (Cortinas and Stensrud 1995). The effects of cloud cover typically are even larger in desert regions.

It is important to recognize that cloud cover is three-dimensional and can vary in the vertical direction as well as the horizontal. In numerical models, cloud cover is defined as the fraction of a specified vertical column (grid cell) of air that has clouds contained within it when evaluated within selected vertical layers. Some parameterizations evaluate cloud cover over three or four fairly deep vertical layers, roughly corresponding to the depths associated with the visual groupings of clouds (low, middle, and high clouds). Other parameterizations evaluate the cloud cover for every vertical model level. Values of cloud cover range from 0 for a cloud-free sky to 1 for an overcast sky.

To understand how one might parameterize cloud cover, first consider the physical processes that lead to cloud formation. Typically, clouds occur because of interactions between vertical motions, turbulence, radiation processes, and microphysical processes. Because of the complexity of cloud formation and the lack of a sound theory to describe it, two approaches have been used to attempt to relate cloud cover to other variables (Slingo 1980).

1. A statistical or diagnostic approach, in which cloudiness is predicted empirically from observed or model variables. Note that using rawinsonde data in a statistical approach has typically not been very successful owing to problems with both the representativeness of point measurements from soundings being applied to areas the size of model grid cells and the lack of detailed information on cloud cover as a function of height (Slingo 1980).
2. A prognostic approach, in which the cloud water content has been explicitly calculated using a model. Unfortunately, until recently there has been very little available verification data to determine if the cloud water contents are reasonable and to document the fine-scale structures of the resulting clouds. Observational experiments, such as the Winter Icing and Storms Project (WISP; Rasmussen *et al.* 1992) and the Improvement of Microphysical PaRameterizations through Observational Verification Experiment (IMPROVE; Stoelinga *et al.* 2003), are beginning to fill this need. However, cloud fields are fractal (Cahalan and Joseph 1989), suggesting that some diagnostic element is always needed.

The need for a cloud cover parameterization scheme in a given numerical weather prediction model depends upon both the model grid spacing and the choices of the model user. Clearly, a cloud cover parameterization scheme is needed for grid spacing larger than 10 km, since so many clouds at that resolution are subgrid phenomena. However, as grid spacing decreases it is difficult to determine exactly when one can rely on the model to generate the appropriate clouds directly and at the right time. If the model has a microphysical parameterization scheme that includes cloud water, then certainly the model is capable of developing clouds or pseudo-clouds. However, clouds at times are very small-scale features and the dividing line between needing and not needing a cloud cover parameterization scheme as a function of grid spacing is unclear. For example, Dudhia (1989) assumes a cloud cover of either 0 or 1 for both longwave and shortwave radiation within a numerical model that uses 10 km grid spacing. Is this a wise choice? Condensation processes often occur at scales much smaller than 10 km, and many clouds have a length scale of perhaps several hundred meters. So the answer to the question is likely to depend upon the phenomena that the user wants to be able to reproduce and the user's sensitivity to errors. Certainly the errors in cloud cover should decrease as the model grid spacing decreases, since fewer and fewer clouds will be subgrid scale, but comparisons between cloud cover produced from very high-resolution model forecasts and from observations have yet to be done. We do not really know how well the smallest grid-spacing models handle clouds over large regions. In addition, it is not clear whether or not cloud cover parameterizations based upon large-scale observations are helpful as the model grid spacing becomes small. Certainly more attention is needed on this issue.

9.2 Cloud cover parameterizations

There are two main types of cloud cover parameterizations being used in numerical weather prediction models. The first type is a diagnostic cloud cover parameterization, in which cloud cover is diagnosed after each time step from the model variables and then used to modify the amounts of short-wave and longwave radiation. These parameterizations generally are very simple and computationally inexpensive. The second type is a prognostic cloud cover parameterization, in which cloud cover is added as a predicted model variable along with a second variable for cloud water. These parameterizations are more complicated and also more expensive computationally.

9.2.1 Diagnostic cloud cover

When considering how cloud cover could be statistically related to observed variables, an important consideration is what variables to explore in developing this relationship. Smagorinsky (1960) originally proposed using relative humidity to predict cloud cover, a reasonable first-order relationship. However, the relationship between model-grid-scale cloudiness and relative humidity obviously has complicating factors. While it is clear that relative humidity plays a significant role (Fig. 9.1), as when Slingo (1980) notes that in the absence of a boundary layer inversion, low clouds almost always occurred when the relative humidity exceeded 80%, there also is little doubt that other factors are also important. Thus, most studies have focused upon an examination of relative humidity, convective activity, vertical velocity, stability, wind shear, and surface fluxes. Some of these are clearly surface variables, such as surface fluxes, while others extend throughout the depth of the troposphere or have no specific vertical level assignment, such as convective activity. All of these approaches try to ascertain when clouds form and how much of a given

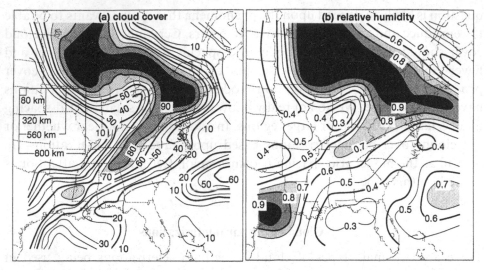

Figure 9.1. Cloud cover (a) and relative humidity (b) averaged over $(320 \text{ km})^2$ areas in the 800–730 hPa layer at 1800 UTC 23 April 1981. Cloud cover is from the United States Air Force three-dimensional cloud analysis scheme, while the relative humidity is from the interpolated observations. Squares in (a) illustrate different horizontal grid spacings. From Walcek (1994).

model grid cell contains cloud at a given time, since subgrid scale fluctuations in temperature and moisture can lead to areas where condensation occurs well before the grid mean values become saturated. While a variety of approaches can be found in the literature, several approaches are surveyed that suggest the basic ideas behind cloud cover parameterization.

The first approach examined is that of Benjamin and Carlson (1986), as also discussed in Anthes *et al.* (1987). In this approach, the cloud fraction (b), which varies from 0 to 1, is defined as

$$b = 4.0RH - 3.0,\qquad(9.1)$$

for low and middle clouds, and

$$b = 2.5RH - 1.5,\qquad(9.2)$$

for high clouds, where RH is the maximum relative humidity found within the model layers associated with each type of cloud. Low clouds are defined as those located below 800 hPa, middle clouds are defined as those between 800 and 450 hPa, and high clouds are those located above 450 hPa. Note that these relationships indicate that no low and middle clouds are found for maximum relative humidities below 75%, and no high clouds are found for maximum relative humidities below 60%. Basically, this scheme provides a linear increase in cloud cover as maximum RH increases past its layer-dependent, predetermined minimum thresholds. Once the cloud fraction b is calculated, then modifications to both the incoming solar radiation and downwelling longwave radiation due to clouds can be determined.

The second cloudiness approach examined is that of Slingo (1987), which was implemented operationally in the European Centre for Medium-range Weather Forecasting (ECMWF) medium-range forecast model in 1985. One distinct difference between this approach and that of Benjamin and Carlson (1986) is that cloud fraction is calculated for each model level in the Slingo approach, instead of only for three vertical layers in the Benjamin–Carlson approach. Yet the Slingo approach again has three basic types of clouds (low, middle, and high), but adds another level of complexity by including the effects of parameterized deep convection on cloud cover (b_{conv}), such that

$$b_{conv} = a + d\ln(P), \qquad (9.3)$$

where P is the precipitation rate from the numerical model, and a and d are empirically defined constants. The cloud fraction produced by parameterized convection may be needed in a numerical model, since some convective parameterization schemes produce little if any moistening in the upper levels when they are activated, necessitating a separate parameterization to account for the high-level cloudiness typically observed. Note that $b_{conv} < 0.80$ always.

Slingo (1987) also divides the resulting total cloudiness from b_{conv} into both low-level clouds (75%) and deep tropospheric clouds (25%). For the high- and mid-level clouds created by convective processes, the scheme has for cloud tops above 400 hPa

$$b_{high} = 2.0(b_{conv} - 0.3) \qquad \text{for} \quad b_{conv} > 0.40 \qquad (9.4)$$

and for cloud tops below 400 hPa it has

$$b_{mid} = 0.25b_{conv}. \qquad (9.5)$$

The model convective scheme provides information on the cloud base and the cloud top, and also the precipitation rate P needed to define b_{conv}.

The Slingo scheme also computes the cloud fraction based only upon the RH of each model layer, such that for extratropical and frontal cirrus clouds

$$b_{high} = \left[\max\left\{0, \frac{(RH - 0.8)}{0.2}\right\}\right]^2 \qquad (9.6)$$

and

$$b_{mid} = \left[\max\left\{0, \frac{(RH_e - 0.8)}{0.2}\right\}\right]^2, \qquad (9.7)$$

where $RH_e = RH$ $(1.0 - b_{conv})$ and it attempts to account for the so-called compensating subsidence from deep convection in the mid-levels. This scheme uses the same pressure levels to define low, middle, and high clouds as Benjamin and Carlson (1986) discussed previously.

Low-level clouds are the most troublesome to handle, since these are very dependent upon the surface fluxes and model boundary layer structures. Observational studies suggest that low-level clouds depend upon a delicate balance between cloud-top entrainment, radiative cooling, and surface fluxes. All of these factors are difficult for a model to predict accurately. In this scheme, b_{low} is divided into contributions associated with extratropical cyclones and tropical disturbances, and associated with boundary layer processes. The portion attributed to cyclones and tropical disturbances is predicted by the relationships

$$\hat{b}_{low} = \left[\max\left\{ 0, \frac{(RH_e - 0.8)}{0.2} \right\} \right]^2 \quad \text{if} \quad \omega < 0, \tag{9.8}$$

where ω is the model vertical velocity in pressure coordinates. When subsidence is indicated ($\omega \geq 0$), then

$$b_{low} = 0.0. \tag{9.9}$$

The actual diagnosed low-level cloud cover b_{low} depends directly upon $\hat{b}_{low}$, such that

$$b_{low} = \hat{b}_{low}(-10.0\omega), \tag{9.10}$$

for $\omega < -0.1 \, \text{Pa s}^{-1}$. Therefore, there is a linear relationship between the amount of cloud cover diagnosed and the model-predicted vertical motion fields.

For boundary layer clouds, the low-level cloud cover depends upon the low-level lapse rate such that

$$b_{low} = -6.67 \frac{\Delta \theta}{\Delta p} - 0.667, \tag{9.11}$$

where $\Delta\theta/\Delta p$ is the lapse rate in the most stable layer below 750 hPa. An extra check is also added to make certain that the RH is high enough to develop clouds (i.e., no clouds are diagnosed when the air is dry).

This approach is based upon an examination of the available observational studies of cloud cover as related to various observed parameters, and an in-depth analysis of the ECMWF model output and comparisons against observed cloud cover. While very empirical in nature, this approach includes a great deal of complexity in relating the model variables to cloud development.

The third scheme is from the United States National Centers for Environmental Prediction Eta Model and follows Sundqvist (1988). This scheme also calculates cloud cover for each model level, and again has three basic types of clouds. However, the pressure levels that distinguish the three layers are somewhat different from the previous two schemes. In the Eta model, the cloud cover (b) for low ($p > 800$ hPa), middle (400 hPa $< p < 800$ hPa), and high ($p < 400$ hPa) clouds is given by

$$b = 1.0 - \left[\frac{(1.0 - RH)}{(1.0 - RH_{crit})} \right]^{1/2}, \tag{9.12}$$

where $b = 0$ when $RH < RH_{crit}$ and

$$RH_{crit} = RH1 + F1(L)(0.95 - RH1)F2(t), \tag{9.13}$$

with L defined as the model level number above the ground surface. The variable $RH1$ equals 0.8 over water and equals 0.75 over land. The function $F1(L)$ varies linearly from 1 to 0 in the first ten layers above the ground. Thus, $F1(1) = 1.0$, $F1(2) = 0.9$, $F1(3) = 0.8$, and so forth, up to $F1(11) = 0.0$. This sequencing obviously causes $F1$ to increase towards the surface of the model, and causes RH_{crit} similarly to increase. Thus, it requires a larger value of RH to form clouds at the very lowest model levels. The function $F2(t)$ varies linearly from 0 to 1 in the first 24 h of the forecast and is fixed at 1 after hour 24. The maximum value of RH found in each layer is used to calculate the cloud cover.

When these three schemes are compared for low clouds at 800 hPa, it is seen that the Benjamin–Carlson scheme predicts less cloud cover than either the Slingo or Eta schemes for all relative humidity values less than 100% (Fig. 9.2). Additionally, the Slingo and Eta schemes are very similar in their shape, although the exact values for the Slingo scheme depend somewhat on the magnitude of the vertical motion. All of these types of diagnostic approaches are designed for a particular model at a particular model grid spacing, so one must be cautious when applying these schemes to other models and at other values of grid spacing. The basic relationship between relative humidity and cloud cover likely remains a reasonable first-order approximation, but the parameterization details (such as the threshold relative humidity at which clouds first begin to form) may need to be altered.

9.2.2 Prognostic cloud cover

Many numerical models include explicit microphysics parameterizations as discussed in Chapter 7. Depending upon the complexity of these schemes, the model may contain equations for cloud water, rain water, cloud ice, and snow,

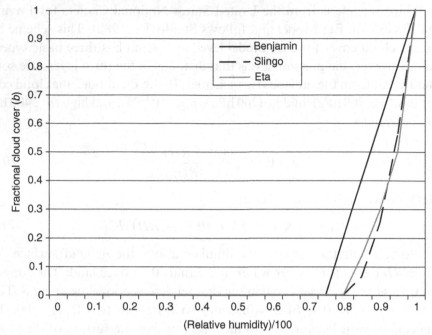

Figure 9.2. Value of cloud cover versus relative humidity scaled from 0 to 1 from the Benjamin–Carlson (solid black), Slingo (dashed), and Eta model (solid gray) cloud cover schemes for low clouds at 800 hPa. The Slingo and Eta schemes have very similar profiles as a function of relative humidity.

and perhaps even graupel and hail. When models predict these explicit cloud variables, it is important that there is consistency between the explicitly pre-dicted cloud variables and the cloud cover parameterization often used to feed needed information to the radiation parameterization. It is possible that the model-produced cloud water or cloud ice may not be consistent with the cloud cover diagnosed and used by the radiation parameterization (Sundqvist *et al.* 1989), a situation one would prefer to avoid.

For cloud-scale models and some mesoscale models with sophisticated micro-physical parameterization schemes and relative small grid spacing, a binary cloud cover can be deduced directly from the cloud water fields that the model predicts. In these cases, cloud cover is either explicitly 1 or 0, depending upon whether or not cloud water is present in a given grid cell. Typically in this case the cloud fraction is not diagnosed and there is no cloud cover parameterization; instead the predicted cloud water is used directly in the calculations for the radiative fluxes. It is important to emphasize again that there are no subgrid scale clouds in this approach. As mentioned previously, the grid spacing for which this type of approach is appropriate is likely to be less than 10 km, and may be smaller than 2 km, but needs to be determined by the user and depends upon the type of clouds the model needs to represent. For small cumulus clouds a grid spacing of a hundred meters may be needed before this type of approach is valid.

For models that use larger grid spacings, and in which cloud development needs to occur prior to grid-scale saturation, another approach is desired that allows for partial cloudiness within the grid cell. The first example of a fully prognostic approach to the parameterization of fractional cloud cover is from Sundqvist (1988) and is extended by Sundqvist *et al.* (1989). This cloud cover parameterization scheme only requires an additional predictive equation for cloud water (q_c), and so avoids the large number of additional equations required in many of the more sophisticated microphysics schemes discussed in Chapter 7. Thus, this type of scheme is much more computationally efficient than a complete microphysics parameterization, and yet contains parameterizations of many of the same physical processes – albeit at an even more simplified level.

These approaches begin by assuming that within each grid cell there are cloudy areas and cloud-free areas (Sundqvist 1988). Thus, the relative humidity U of the grid cell is a weighted average of the humidity in the cloudy area, $U_S \equiv 1$, and the humidity in the cloud-free area, U_0, such that

$$U = bU_S + (1 - b)U_0, \qquad (9.14)$$

where b is the cloud cover and varies from 0 to 1 as before. Sundqvist *et al.* (1989) further assume that the humidity of the cloud-free area depends upon both the amount of cloud cover and a basic threshold humidity U_{00} that is allowed to vary as a function of model level and temperature. These assumptions lead to the expression

$$U_0 = U_{00} + b(U_S - U_{00}), \qquad (9.15)$$

for the humidity in the cloud-free area. Finally, combining (9.14) and (9.15) yields a diagnostic equation for cloud cover in terms of U and U_{00}, such that

$$b = 1 - \left(\frac{U_S - U}{U_S - U_{00}} \right)^{1/2}, \qquad (9.16)$$

allowing cloud cover to be determined from the evolution of the grid cell relative humidity U and noting that $b = 0$ for $U < U_{00}$. This equation expresses the same relationship between relative humidity and cloud cover as used by the Eta model as described in (9.12).

The prognostic cloud cover approach hinges upon an expression for the grid-resolved latent heating rate (Q) due to condensation derived by Sundqvist (1988). Beginning with the relationship

$$q = Uq_s, \tag{9.17}$$

where q_s is the saturation specific humidity and taking the derivative with respect to time, one finds

$$\frac{\partial q}{\partial t} = \frac{\partial U}{\partial t}q_s + U\frac{\partial q_s}{\partial t}. \tag{9.18}$$

Using the approximation $q_s \cong \varepsilon e_s/p$ to replace q_s in the last term of (9.18), where e_s is the saturation vapor pressure, p is pressure, and $\varepsilon = 0.622$, yields

$$\frac{\partial q}{\partial t} = \frac{\partial U}{\partial t}q_s + \frac{U\varepsilon}{p}\frac{\partial e_s}{\partial t} - \frac{U\varepsilon e_s}{p^2}\frac{\partial p}{\partial t}. \tag{9.19}$$

Making use of the Clausius–Clapeyron equation, $de_s/dT = \varepsilon L_v e_s/R_d T^2$, where R_d is the specific gas constant for dry air, and L_v is the latent heat of vaporization, some further manipulation leads to an expression for the specific humidity tendency in terms of the tendencies of temperature, humidity, and pressure,

$$\frac{1}{q}\frac{\partial q}{\partial t} = \frac{\varepsilon L_v}{R_d T^2}\frac{\partial T}{\partial t} + \frac{1}{U}\frac{\partial U}{\partial t} - \frac{1}{p}\frac{\partial p}{\partial t}. \tag{9.20}$$

Following Sundqvist *et al.* (1989), the tendency equations for temperature, specific humidity, and the cloud water mixing ratio (q_c) in the case of stratiform condensation and evaporation are

$$\frac{\partial T}{\partial t} = A_T + \frac{L_v}{c_p}Q - \frac{L_v}{c_p}E_0, \tag{9.21}$$

$$\frac{\partial q}{\partial t} = A_q - Q + E_0, \tag{9.22}$$

$$\frac{\partial q_c}{\partial t} = A_{q_c} + Q - (P - E_r) - E_0, \tag{9.23}$$

where the A-terms represent the tendencies due to all processes other than condensation and evaporation, c_p is the specific heat of air at constant pressure, P is the rate of release of precipitation, E_r is the evaporation rate of precipitation, and E_0 is the total evaporation rate due to E_r and the evaporation of cloud water. When the environment is saturated, $U \equiv 1$ and the second term on the right-hand side of (9.20) is identically zero. If this modified version of (9.20) for saturated conditions is inserted into (9.22), and if (9.21) is then used to eliminate the temperature tendency, one arrives at

$$Q = \frac{M - q_s(\partial U/\partial t)}{1 + U\varepsilon L_v^2 q_s/R_d c_p T^2} + E_0, \tag{9.24}$$

where M is the convergence of available latent heat into the grid cell. The convergence term M is defined as

$$M = A_q - \frac{U \varepsilon L_v q_s}{R_d T^2} A_T + \frac{U q_s}{p} \frac{\partial p}{\partial t}. \tag{9.25}$$

The system of equations is closed by providing an expression for the U-tendency. Sundqvist *et al.* (1989) assume that the quantity $M + E_0$ can be divided into two parts. One part, bM, condenses in the cloudy portion of the grid cell, while the other part, $(1 - b)M + E_0$, increases the relative humidity of the cloud-free portion and hence the cloud cover of the grid cell. This assumption leads to the tendency equation

$$\frac{\partial U}{\partial t} = \frac{2(1 - b)(U_S - U_{00})[(1 - b)M + E_0]}{2 q_s (1 - b)(U_S - U_{00}) + q_c/b}. \tag{9.26}$$

Since the latent heating rate Q is incorporated into the T, q, and q_c tendency equations (9.21)–(9.23), there is a direct link between changes in cloud cover and the model tendency equations.

Since cloud cover is incorporated into the model predictive equations, it also should be incorporated into any other relevant parameterizations. For example, Sundqvist *et al.* (1989) define the autoconversion from cloud water to precipitation as

$$P_{AUTO} = c_0 q_c \left[1 - \exp \left[-\left(\frac{q_c}{b q_{c\_threshold}} \right)^2 \right] \right], \tag{9.27}$$

where c_0 and $q_{c\_threshold}$ are constants ($c_0 = 1.0 \times 10^{-4}$ s^{-1} and $q_{c\_threshold} = 3.0 \times 10^{-4}$ in Zhao and Carr (1997)). Note how the cloud cover b appearing in the exponential influences the autoconversion rate, allowing for larger autoconversion rates for smaller values of cloud cover as q_c is held constant. It is the cloud water content per cloud area that is seen to determine the rate at which precipitation is produced.

The evaporation of precipitation also depends upon the cloud cover as well as the relative humidity of the grid cell (Sundqvist *et al.* 1989). The total evaporation rate E_0 is calculated from both the evaporation of precipitation and the evaporation of cloud water that occurs when cloud water is advected into a grid cell where no condensation is taking place. In addition, a direct link between the convective parameterization scheme and the creation of cloud water is created as part of the scheme (Sundqvist *et al.* 1989). Thus, the effects of cloud cover are accounted for in all the appropriate terms in the model equations, allowing for consistency between the microphysics and cloud cover.

While Sundqvist *et al.* (1989) further include simple representations of the coalescence process, the Bergeron–Findeisen mechanism, and the microphysical properties of cirrus clouds, Zhao and Carr (1997) extend this parameterization to include cloud ice processes. The inclusion of cloud ice processes led to significant improvements in the Eta model precipitation forecasts (Zhao and Carr 1997). This type of approach is also implemented and tested by Pudykiewicz *et al.* (1992).

The prognostic representation of cloud cover is further extended by Tiedtke (1993) through the development of tendency equations for both cloud water content and cloud cover. Thus, cloud cover is a predicted variable in the numerical model just like temperature and specific humidity. For example, it is assumed that the formation of cloud water is determined by the decrease in the saturation specific humidity. In addition, condensation is assumed to occur in both existing clouds (if any) and in new clouds, leading to

$$\frac{\partial q_c}{\partial t}_{cond} = -b\frac{dq_s}{dt} - \Delta b\frac{dq_s}{dt}, \qquad \frac{dq_s}{dt} < 0, \tag{9.28}$$

where *cond* refers to the tendency from processes associated with condensation only, and Δb is the fractional cloud cover increase per time step. If moisture is distributed evenly around the mean environmental value of specific humidity, then Tiedtke (1993) finds that the change in cloud cover is given by

$$\frac{\partial b}{\partial t}_{cond} = -(1 - b)\frac{dq_s/dt}{(q_s - q)}, \qquad \frac{dq_s}{dt} < 0, \tag{9.29}$$

thereby indicating how decreases in the saturation specific humidity cause both an increase in cloud water and an increase in cloud cover. Parameterizations are also developed for the processes of evaporation, precipitation, and turbulence, and are incorporated into both the cloud water and cloud cover equations as in (9.28) and (9.29). This approach allows for cloud-related processes to be treated uniformly. The scheme further links the grid-scale variables to the convective scheme, such that the formation of cloud due to cumulus convection is tied to the detrainment of cloud mass in the updraft of the convective parameterization scheme. Tests of this scheme within the ECMWF global forecast model indicate that realistic cloud fields are produced (Tiedtke 1993). Other examples of this type of approach for mesoscale models are found in Ballard *et al.* (1991) and Bechtold *et al.* (1993), and for global climate models are found in Le Treut and Li (1991), Del Genio *et al.* (1996), Fowler *et al.* (1996), and Teixeira and Hogan (2002).

There also exists a group of schemes that often are called "statistical schemes" in which a probability density function is assigned to the total water mixing ratio (the sum of the water vapor and all liquid and ice mixing ratios) and, by defining the statistical moments of the distribution of the total water mixing ratio, one can diagnose the cloud fraction by integrating the supersaturated portion of the probability density function. Examples of these schemes are found in Smith (1990), Cusack *et al.* (1999), Lohmann *et al.* (1999), Bony and Emanuel (2001), Chaboureau and Bechtold (2002), Thompkins (2002), and Berg and Stull (2005). The Smith (1990) scheme is extended by Wilson and Ballard (1999) to include cloud ice processes. These schemes incorporate estimates of subgrid scale variability into the diagnosis of the cloud cover fraction, but still do not predict cloud cover explicitly. Chaboureau and Bechtold (2002) show that the partial cloudiness scheme reduces the biases in the shortwave and infrared spectral ranges in comparison to explicit microphysical schemes without partial cloudiness, yielding better agreement with synthetic satellite imagery.

Finally, yet another prognostic cloud cover scheme is suggested by Grabowski and Smolarkiewicz (1999) and Grabowski (2001) in which a cloud-resolving model is run within the grid cell of a model with much larger grid spacing. This scheme is called a "superparameterization" by Randall *et al.* (2003). While this approach is typically viewed as an alternative for global climate models, the approach also may be reasonable for some operational forecast models for the treatment of cumulus clouds. The basic idea is to embed a two-dimensional cloud-resolving model within the grid cell of a model that does not explicitly simulate clouds (Fig. 9.3) and to use the cloud-resolving model output to compute statistics for fractional cloudiness, cloud water, and precipitation rate for the larger-scale model's grid column. Initial results suggest that the global model simulations with this superparameterization are able to produce a vigorous Madden–Julian Oscillation (MJO), in contrast with the control run with a typical cloud parameterization

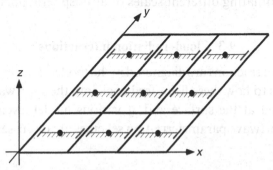

Figure 9.3. Schematic of a superparameterization of cloud effects, in which the black squares represent the large-scale model grid cells and the hatched lines represent a two-dimensional cloud-resolving model domain within each large-scale model grid cell. From Randall *et al.* (2003).

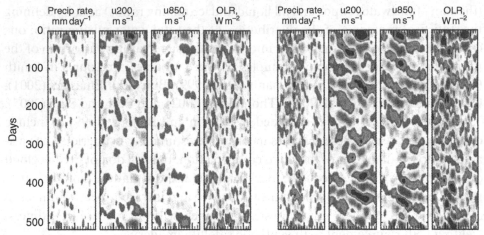

Figure 9.4. Hovmöller diagrams for the precipitation rate (mm day^{-1}), 200 hPa zonal wind (m s^{-1}), 850 hPa zonal wind (m s^{-1}), and outgoing longwave radiation (OLR, W m^{-2}) for wave periods of 20–100 days from the same global model with (left) a traditional cloud cover and convective parameterization approach and (right) a superparameterization that embeds a cloud-resolving model into each grid cell of the larger-scale model. Each subpanel spans 0° to 360° in longitude at a tropical latitude. Note that the superparameterization results indicate a vigorous Madden–Julian oscillation (illustrated by the coherent patterns) that is completely absent from the simulation with a traditional cloud cover parameterization. From Randall *et al.* (2003).

scheme that essentially has no MJO (Fig. 9.4). Although computationally much more expensive than a more traditional cloud-cover approach, the superparameterization approach is much cheaper than a full cloud-resolving model simulation and represents an interesting use of a combination of models developed for simulating different scales of atmospheric phenomena.

9.3 Cloud–radiation interactions

Now that the cloud fractions are diagnosed in the model, or cloud water is present, the question turns to how these fields feed back into the shortwave and longwave radiation predicted at the surface and at various model levels. As before, the longwave and shortwave parameterization schemes are discussed separately.

9.3.1 Longwave radiation in cloudy skies

Longwave heating rates within clouds are typically calculated by assuming that the clouds radiate as blackbodies. Stephens (1978a) shows that a cumulonimbus cloud with a liquid water content of $2.5\,\mathrm{g\,m^{-3}}$ radiates as a blackbody beyond a depth of $12\,\mathrm{m}$ into the cloud. In contrast, a thin stratus cloud with a liquid water content of $0.05\,\mathrm{g\,m^{-3}}$ radiates as a blackbody only beyond a depth of $600\,\mathrm{m}$ into the cloud, which may be more than the vertical grid increment of many of the models in use today. Cirrus clouds are probably the most difficult, because of the relatively low level of water content present in these clouds and their importance to climate (Liou 1986; Stephens *et al.* 1990). Liou and Wittman (1979) suggest that cirrus clouds should be considered gray bodies instead. Thankfully, the dominance of liquid water absorption over scattering by cloud droplets simplifies the problem and allows for the development of relatively simple approximations. This is not necessarily true for cirrus clouds when ice is present.

Stephens (1978b) shows that longwave radiative flux is most sensitive to the cloud water path (CWP), and this has been found to be true regardless of whether the cloud is composed of cloud droplets or ice crystals. The effects of rain and snow on longwave flux are two to three orders of magnitude less than the effects of cloud water and ice. This result suggests that rain and snow can be neglected to first order in quantifying the longwave radiation in cloudy skies, although approaches have been developed to incorporate the effects of rain and snow.

If the model being used has cloud water and cloud ice as predicted variables, owing to the use of a microphysical parameterization, then it is very easy to calculate the CWP. This is simply the integral

$$CWP = \int_{z_1}^{z_2} w\,dz, \qquad (9.30)$$

where z_1 and z_2 are the heights over which the CWP needs to be determined, and w is the cloud liquid water content in $\mathrm{g\,m^{-3}}$. However, if a numerical model does not include cloud water as a variable, determination of a CWP is still required to parameterize the radiative effects of subgrid scale clouds. Kiehl *et al.* (1996) prescribe a meridionally and height varying, but time-independent, cloud liquid water density profile $\rho_l(z)$ that is analytically determined. This profile represents an exponentially decaying vertical profile for in-cloud water concentration, such that

$$\rho_l(z) = \rho_{l0} e^{-z/h_l}, \qquad (9.31)$$

where the reference value ρ_{l0} is set to $0.21\,\mathrm{g\,m^{-3}}$. The cloud water scale height h_l is locally determined as a function of the vertically integrated water vapor. Thus, the CWP can be calculated even for models without explicit microphysical schemes and used in the longwave flux calculations for cloudy skies. Now that the CWP is available for the model, we turn to how the interactions between clouds and radiation occur. Much of the following discussion is based upon Stephens (1978b) and Dudhia (1989).

When dealing with the interactions between clouds and radiation, the total cloud emissivity, which can be calculated from the measurement of the long-wave radiative flux profile through the cloud and the cloud temperature, is the appropriate parameter to use (Stephens 1978b). The emissivities of very different cloud types are nearly identical if their CWP is the same. Cloud emissivity is calculated using

$$\varepsilon_{cloud} = 1 - e^{a_0 CWP}, \tag{9.32}$$

where $a_0 = 0.158$ for downward longwave flux and $a_0 = 0.130$ for upward long-wave flux. These two different values are found because the spectral compositions of the upward and downward beams incident on the cloud boundaries are so different. This total cloud emissivity is actually an "effective" emissivity, since it includes the combined effects of absorption and scattering.

Since the cloud total emissivity applies across the entire infrared spectrum, the problem of having cloud and gaseous absorption occurring in the same spectral band must be addressed. This is often accounted for using the overlap approximation, in which

$$\varepsilon_{total} = 1 - (1 - \varepsilon_{cloud})(1 - \varepsilon_{gas}), \tag{9.33}$$

where ε_{gas} can be for water vapor or another gas such as CO_2.

For ice clouds, Dudhia (1989) uses the same general approach as for cloud water, but specifies $a_0 = 0.0735 \, \mathrm{m^2 \, g^{-1}}$, a value of roughly half that for water vapor. Dudhia also develops an emissivity equation for rain and snow that can be combined with the cloud and gas emissivities using the overlap approximation.

Once the emissivities are known for each atmospheric layer in the model, the addition of clouds adds extra boundaries to the clear-sky longwave radiative flux calculations. As shown by Stephens (1978b), the flux equations above and below a cloud with cloud top height z_T and cloud base height z_b are

$$F_U(z) = \sigma T^4(z_T)[1 - \varepsilon(z, z_T)] + \int_{z_T}^{z} \sigma T^4(z') \frac{d\varepsilon(z, z')}{dz'} \, dz' \qquad z > z_T, \tag{9.34}$$

and

$$F_D(z) = \sigma T^4(z_b)[1 - \varepsilon(z_b, z)] + \int_{z_b}^{z} \sigma T^4(z') \frac{d\varepsilon(z, z')}{dz'} \, dz' \qquad z < z_b. \tag{9.35}$$

Fluxes within the cloud are defined by interposing z_T and z_b and by specifying the appropriate cloud emissivity.

Another approach is developed by Savijärvi and Räisänen (1998) in which the longwave radiation parameterization is based upon the *CWP* and the effective cloud droplet radius. In models with explicit microphysics schemes, the effective cloud droplet radius r_e can be calculated explicitly from

$$r_e = \frac{\int_0^\infty n(r)r^3 \, dr}{\int_0^\infty n(r)r^2 \, dr}, \tag{9.36}$$

where r is the droplet radius, and $n(r)$ is the droplet size distribution function. If the numerical models do not include an explicit microphysical parameterization, then Savijärvi and Räisänen (1998) provide an estimate for r_e based upon the *CWP* that depends upon whether the clouds are maritime or continental.

When ice is present in the cloud, then the cloud particles are not necessarily spherical as assumed for cloud droplets. This alters the calculations of the scattering and absorption coefficients and the asymmetry factor. Ebert and Curry (1992) and Fu *et al.* (1998) parameterize the optical properties of ice crystal clouds. Ebert and Curry (1992) use the ice water path and an effective ice crystal radius to describe the ice optical properties. They find that a change in the effective ice crystal radius alone is more effective than a change in just the ice water path in altering shortwave reflectivity and the cloud albedo feedback. Fu *et al.* (1998) use observations of 28 ice crystal distributions to relate the width of a crystal to its length and parameterize the absorbtivity, reflectivity, and transmissivity using both the ice water content and the generalized effective particle size.

Another challenge arises when a given cloud layer is only partially filled with cloud. Since models with explicit microphysics are capable of resolving clouds to some extent, these models generally assume either an entirely clear or an entirely cloudy grid cell. Thus, partially cloudy layers typically occur in models with diagnostic cloud cover schemes and convective parameterization schemes. In this case the longwave flux is determined by linearly weighting the clear-sky and cloud-sky fractions, such that

$$F_U(z) = (1 - b)F_{U_{clear}}(z) + bF_{U_{cloud}}(z), \tag{9.37}$$

where b is the cloud fraction. Harshvardham and Weinman (1982) demonstrate that this linear weighting is reasonable if the model-derived cloud fraction is replaced with

$$b' = \frac{[1 + 2a(1 + 0.15b)]b}{1 + 2ab(1 + 0.15b)}, \tag{9.38}$$

where a is the aspect ratio (depth over width) of the cloud. This result is based upon a study of black square blocks, which approximate clouds, over a radiating surface, and takes into account some of the three-dimensional effects of broken clouds.

While there are a number of other details and complexities that could be addressed, the last one to be discussed here is the problem of cloud vertical overlap in models that use a cloud cover parameterization. Most of these models diagnose cloud cover for each vertical model layer, and these values of fractional cloud cover vary with height. The concern is how the cloud cover fractions from adjacent layers are related to each other. As shown earlier, partially cloudy skies are divided into sectors in which the amount of cloud is either 0 or 1 and they are then linearly weighted to determine the resulting longwave flux. When adjacent layers both have clouds, there are two common approaches to determine how these adjacent cloud fractions are related. One approach is to assume that the cloud fractions randomly overlap (Manabe and Strickler 1964). This approach defines the amount of sky covered simultaneously by n layers by the relationship

$$1 - (1 - b_1)(1 - b_2)(1 - b_3) \cdots (1 - b_n), \tag{9.39}$$

where b_1 to b_n are the cloud fractions of the n layers. This approach assumes that clouds are distributed randomly in each layer and that the clouds in one layer are not necessarily related to the clouds in an adjacent layer. An alternative approach is a maximum/random overlap method (Geleyn and Hollingsworth 1979), where the cloud fraction from a combination of layers is given by the maximum cloud fraction determined at any one layer which is overlapped and any excess cloud is positioned randomly across the layer. This latter approach is suggested by the work of Tian and Curry (1989) to be more consistent with the distribution of observed clouds. Stephens *et al.* (2004) compare the most common cloud vertical overlap assumptions and find that two methods, the random and maximum-random overlap methods, have severe problems since they depend upon the vertical resolution of the numerical model.

It is clear that the problem of cloud–radiation interactions is a difficult and challenging one. The parameterizations examined are relatively complex, but the atmosphere is even more so. The cloud–radiation and feedback problem is one that has been highlighted as a crucial area where more work is needed in order to be able to better address global climate change and it definitely influences short-range operational weather prediction as well.

9.3.2 Shortwave radiation in cloudy skies

As with longwave radiative flux in cloudy skies, the shortwave radiative flux depends mostly on the cloud water path. The heart of the problem is to determine realistic values for the cloud optical properties of optical thickness τ, single-scattering albedo $\tilde{\omega}_0$, and asymmetry factor g. Of these properties, the optical thickness is the most important parameter. Stephens (1978b) suggests that a rough range for optical thickness is between 5 and 500. Twomey (1976) notes that if the sun's disk is not visible through a cloud layer, then the optical thickness must be equal to 10 or more. The absorption of shortwave radiation by cloud droplets cannot be neglected when compared against the absorption by water vapor and needs to be included in many models.

Stephens (1978b) provides a derivation that outlines the various steps in determining the cloud optical thickness τ. When all is said and done, the result is simply that

$$\tau \approx \frac{3}{2}\frac{CWP}{r_e}, \tag{9.40}$$

where r_e is the effective drop size from (9.36). This parameterization is based upon a specified set of calculations using eight cloud types (Stephens 1978b). Values of cloud optical thickness typically range from 1 to 500.

Further simplifications can be made by noting that liquid water absorption in clouds only occurs for $\lambda > 0.75\,\mu m$. Thus, Stephens separates the solar spectrum into two parts: 0.3–0.75 μm for which no absorption occurs and the single-scattering albedo $\tilde{\omega}_0 = 1$, and 0.75–4.0 μm for which absorption occurs and $\tilde{\omega}_0 < 1$. Scattering occurs in both of these parts of the solar spectrum. He then derives expressions for the broad-band optical thickness of cloud layers as a function of CWP only, with

$$\log_{10}(\tau) = 0.2633 + 1.7095\log_e[\log_{10}(CWP)] \qquad \tilde{\omega}_0 = 1 \tag{9.41}$$

and

$$\log_{10}(\tau) = 0.3492 + 1.6518\log_e[\log_{10}(CWP)] \qquad \tilde{\omega}_0 < 1. \tag{9.42}$$

He notes that the effective radius of a cloud droplet distribution is intuitively a function of liquid water content, and hence by extension CWP. Thus, we should not be surprised that the effective radius can be removed from these relationships.

Following Coakley and Chylek (1975), the reflection (Re) and transmission (Tr) through a cloud layer of optical thickness τ as defined in (9.41) and (9.42) is given by

$$Re(\mu_0) = \frac{\beta(\mu_0)\tau/\mu_0}{1 + \beta(\mu_0)\tau/\mu_0}, \tag{9.43}$$

$$Tr(\mu_0) = 1 - Re(\mu_0), \tag{9.44}$$

for a non-absorbing medium with $\tilde{\omega}_0 = 1$, and

$$Re(\mu_0) = (u^2 - 1)\left[\exp(\tau_{eff}) - \exp(-\tau_{eff})\right]/R, \tag{9.45}$$

$$Tr(\mu_0) = 4u/R, \tag{9.46}$$

for an absorbing medium with $\tilde{\omega}_0 < 1$ and where

$$u^2 = [1 - \bar{\omega}_0 + 2\beta(\mu_0)\bar{\omega}_0]/(1 - \bar{\omega}_0), \tag{9.47}$$

$$\tau_{eff} = \{(1 - \bar{\omega}_0)[1 - \bar{\omega}_0 + 2\beta(\mu_0)\bar{\omega}_0]\}^{1/2}\tau/\mu_0, \tag{9.48}$$

and

$$R = (u + 1)^2 \exp(\tau_{eff}) - (u - 1)^2 \exp(-\tau_{eff}). \tag{9.49}$$

Here β is the backscattered fraction of monodirectional incident radiation at the zenith angle μ_0. Stephens (1978b) tuned values of $\tilde{\omega}_0$ to match the flux values provided by more accurate numerical solutions and provides a lookup table such that one can obtain $\tilde{\omega}_0$ given the values of τ and μ_0. This approach similarly removes the dependence of $\tilde{\omega}_0$ on the effective cloud droplet radius, and is used by Dudhia (1989) among others. Thus, the gross shortwave radiative effects of clouds can be determined largely from the values of CWP and zenith angle.

A problem arises in that water vapor and cloud droplet absorption overlap, yet behave quite differently. As Stephens (1984) discusses, the absorption by liquid water is a smooth function of wavelength, whereas the absorption by water vapor varies strongly with wavelength. A method for calculating the absorption for cloudy skies that overlaps the cloud droplet and molecular absorptions is needed. Stephens discusses several options that are available.

Owing to the importance of high-level cirrus clouds to climate, Ebert and Curry (1992) and Fu (1996) develop ice cloud parameterization schemes. The single-scattering properties of solar radiation are parameterized in terms of the ice water path and the generalized effective particle size in both schemes.

When model layers are only partially cloudy, then various weightings between the clear-sky and cloudy-sky fractions are often used, as is done for longwave radiation and discussed previously. These weightings include minimum cloud overlap, maximum cloud overlap, random overlap, maximum-random overlap, and variations on linear weightings. Stephens *et al.* (2004) suggest that the maximum-random and random overlap methods depend upon the vertical resolution of the numerical model and thereby suffer severe problems. Welch *et al.* (1980) suggest that radiative transfer through the three-dimensional cloud shapes is not quite as simple as one might hope and that linear weightings are an oversimplification. The development of improved parameterizations for model-relative partial cloud (in the horizontal and vertical directions) is clearly needed.

9.4 Discussion

The parameterization of cloud cover is yet another important component in many numerical weather prediction models. While the need for this type of parameterization may indeed decrease and perhaps vanish as model grid spacings become smaller and smaller, observations suggest that even at 2 km grid spacing clouds can occur as a subgrid phenomenon. And for global models, the need for cloud cover parameterization is going to continue for some time until computers become fast enough for cloud-resolving models to be run at global scales.

We have seen that there exist a wide variety of approaches that attempt to deal with diagnosing or predicting cloud cover, ranging from simple relationships between relative humidity and cloud cover to fully prognostic schemes that include the cloud fraction as a model variable and integrate this variable with other model parameterization schemes. Comparisons between observations of relative humidity and cloud cover indicate a large scatter (Fig. 9.5). Walcek (1994) shows that, within the 800 to 730 hPa layer, at 80% relative humidity there is a nearly uniform probability of observing any cloud fraction. The standard deviation of cloud cover approaches 40% within restricted relative humidity ranges using an 80 km grid cell (Fig. 9.5c). One interpretation of these results is that the cloud cover is influenced by much more than relative humidity, which is certainly true. Another interpretation is that our ability to correctly predict cloud cover is limited, since cloud cover parameterizations based upon relative humidity are competitive with more sophisticated schemes as indicated by their continued use in operational forecast models.

Similarly, Hinkelman *et al.* (1999) compare the Eta model liquid or ice mixing ratio with observations at the Atmospheric Radiation Measurement Southern Great Plains site in Oklahoma during the first half of 1997. Their results suggest that the Eta model, in general, produces a reasonable mean representation of explicit clouds when compared against the cloud radar

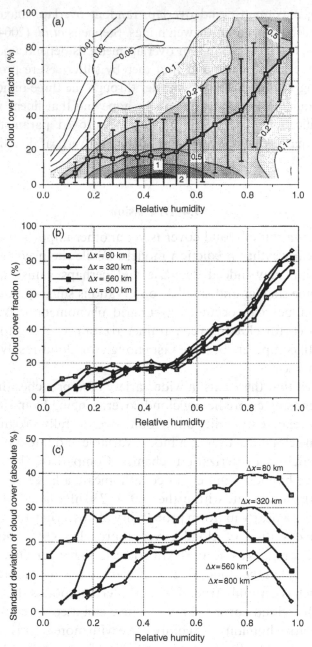

Figure 9.5. (a) A joint frequency distribution showing the probability (%) of observing various cloud cover and relative humidity combinations averaged over $(320\,km)^2$ areas in the 800–730 hPa layer from five local noon samples. The solid dark curve with error bars shows the mean and standard deviation of cloud cover. (b) The mean cloud cover as a function of relative humidity in the 800–730 hPa layer for various averaging areas. (c) The standard deviation of the cloud cover within restricted relative humidity ranges for various area sizes. From Walcek (1994).

reflectivity data. There is a mean overprediction of cirrus events in the Eta model during the first 3 months of the year and a mean underprediction of mid-level clouds in the springtime. Zhou and Cess (2000) show that the diagnosed cloud heights play a large role in the surface energy budget. Differences in the downwelling longwave radiation of $31\,\mathrm{W\,m^{-2}}$ are found when cloud heights are shifted upwards $2\,\mathrm{km}$ and downwards $2\,\mathrm{km}$ for mid-level clouds. High clouds have a smaller influence on the longwave radiation since they are cooler than the mid-level clouds. Thus, not only do we need to predict cloud cover and cloud water content accurately, but the cloud base and the cloud top as well. Barker *et al.* (2003) compare results from numerous solar radiation parameterizations with observations and conclude that none of the schemes do well under all cloud cover conditions.

Part of the problem in predicting cloud cover is that the observational data set for cloud cover has a number of uncertainties, including the differences in the perspectives of the data sets. Clouds can be observed above a point using ceilometer measurements for cloud base heights of up to ~3600 m, cloud cover can be estimated visually by surface and aircraft observers, and the cloud top and the cloud fraction can be estimated from satellite data (Schreiner *et al.* 1993). Buriez *et al.* (1988) and Chang and Coakley (1993) find that different methods of diagnosing the cloud cover fraction from satellite imagery yield differences of 16–30%, which is very similar to the standard deviations of cloud cover when compared to relative humidity measurements (Fig. 9.5c). In contrast to the observations, cloud cover in a numerical model is determined for a specific horizontal area and height that change as the model grid spacing changes. Applying the satellite cloud amount to models requires the use of a cloud overlap assumption, and Zhou and Cess (2000) indicate that the use of different overlap assumptions can produce differences in the incoming surface longwave radiation of $20\,\mathrm{W\,m^{-2}}$. They conclude that more information on cloud overlap and more accurate cloud profiling are needed. Thus, uncertainties in the observed cloud fractions are likely to contribute to the difficulty in developing good parameterizations for cloud cover.

However, even assuming there exists a near-perfect parameterization for the cloud cover fraction, challenges still exist in calculating the absorption of shortwave and longwave radiation by clouds. Stephens and Tsay (1990) summarize the disagreements between the theoretical and observed values of both cloud absorption and reflection of solar radiation. They show that a number of studies have measured the mean atmospheric absorption by clouds to be as high as 15–40%, whereas theoretical values are typically in the 5–10% range for the cloud types considered. One possible explanation for this disagreement is the effect of cloud heterogeneity on cloud absorption and reflection, although no clear explanation to account for the discrepancy between observation and theory has emerged.

Another challenge that has emerged is the importance of the cloud microphysics variables to cloud–radiation interactions. Results from a single-column model suggest that the effective cloud droplet radius influences the model radiative fluxes (Iacobellis and Somerville 2000). Similarly, Liu *et al.* (2003) show that the assumed ice crystal habits can have a significant influence on the evolution of cirrus clouds and, therefore, on both the solar and infrared radiative heating rates for these clouds. Results from Gu and Liou (2000) indicate that the model simulated ice crystal size distribution within cirrus clouds depends strongly on the simulated radiative heating profiles. Within mesoscale convective systems, the strength of the rear inflow jet is influenced by the interactions between longwave cooling and ice microphysics (Chen and Cotton 1988). Yet it is not at all clear that the radiation and microphysics parameterization schemes in most models have identical specifications for the effective cloud droplet radius or ice crystal size distributions and also are able to pass this information from one parameterization to the other as needed.

Since the magnitudes of downwelling longwave radiation are not as large as the incoming solar radiation, errors in this parameterization may have a smaller influence on the model forecast accuracy. However, errors in cloud representation can lead to non-trivial differences in the downwelling radiation (Fig. 9.6) that may influence local predictions. In addition, cloud geometry effects play a role in longwave scattering under broken cloudiness conditions (Takara and Ellingson 2000), further complicating the situation.

As mentioned in the previous chapter, if the instantaneous values of observed and predicted surface shortwave or longwave radiation are examined, large differences are seen (e.g., Fig. 9.6). Many if not all of these larger differences are due to clouds passing over the observation site, which obviously are not accurately predicted by the numerical model. It may be that these differences, which often persist for relatively short periods of time, do not matter and the model forecasts do not suffer from these inaccuracies. But it also may be true that under some situations these differences do matter and result in unexpected consequences, such as the initiation of deep convection from growing thermals. Thus, the way one views the accuracy of parameterizations in general depends upon what types of events are important and what processes influence these events. It remains an open question as to how important these differences are to accurate and useful numerical weather prediction.

Finally, three-dimensional effects can be very important for radiation–cloud interactions. All the one-dimensional radiative transfer parameterizations discussed are challenged in cloudy atmospheres and as the horizontal grid

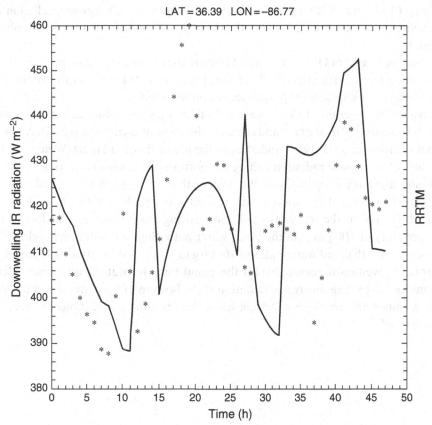

Figure 9.6. Downwelling infrared radiative flux predicted by the RRTM (solid line) from a mesoscale model simulation and measured (asterisks) at Cornelia Fort Airpark, Tennessee, as a function of time for 3–4 July 1995. From Zamora *et al.* (2003).

spacing in the model decreases, making the plane-parallel assumption dubious. The transfer of photons between different parts of a cloud or between neighboring grid cells cannot be accounted for in one-dimensional radiative transfer parameterizations. Unfortunately, while three-dimensional radiative transfer approaches have been developed, they are not yet affordable for use in operational models (Marshak and Davis 2005; Cahalan *et al.* 2005).

9.5 Questions

1. Fill in the details in the derivation of the Sundqvist cloud cover parameterization approach. Starting with (9.17) derive (9.24).
2. Let us examine the effects of clouds on longwave radiation. Assume a very simple atmosphere with only one layer, with $F_{Dclear} = 315 \, \mathrm{W \, m^{-2}}$ and $F_{Dcloud} = 350 \, \mathrm{W \, m^{-2}}$.

Using (9.37) and (9.38), calculate the downwelling surface longwave radiation as a function of cloud cover. Also vary the cloud aspect ratio from 1 to 5. Describe the results.

3. Using the results of Question 2, how large are the errors in the downwelling surface radiation for uncertainties of 30% in the cloud cover? What does this say about the importance of cloud cover parameterization in models?

4. Turning our attention to shortwave radiation, vary the values of the cloud water path from 10 to 10 000 $\mathrm{g\,m}^{-2}$ and calculate the transmissivity using (9.42). Assuming that the incoming shortwave radiation at the top of the cloud is 800 $\mathrm{W\,m}^{-2}$, plot the values of shortwave radiation exiting the bottom of the cloud as a function of the cloud water path using (9.43)–(9.49). Assume that $\mu_0 = 0.87$, $\beta = 0.06$, and $\tilde{\omega}_0 = 0.8$.

5. Using the results of Question 4 with a cloud water path of 1000 $\mathrm{g\,m}^{-2}$, determine the difference in the shortwave radiation exiting the bottom of the cloud for an uncertainty of 100 $\mathrm{g\,m}^{-2}$ in the cloud water path. Since cumulonimbus clouds are associated with cloud water paths of 10 000 $\mathrm{g\,m}^{-2}$, an error of 10% or 1000 $\mathrm{g\,m}^{-2}$ is probably even more reasonable for this cloud type. Repeat the difference calculations for the exiting shortwave radiation at the bottom of the cumulonimbus cloud. How important are these uncertainties to the resulting calculations of shortwave radiation?

10

Orographic drag parameterizations

10.1 Introduction

Mountains are one of the most visually stunning features of the landscape. Created by the collision of tectonic plates and the eruption of volcanoes, and then sculpted by rain and winds, mountains are a source of both inspiration and wonder. The Greek gods were believed to dwell on top of Mount Olympus, while folklore suggests that wise men and women seek mountain tops to find solitude and ponder the fate of the universe. Yet in reality, mountains are obstacles to most forms of transportation and acted as nearly impenetrable barriers to many early human communities. In a similar manner, mountains also influence the atmosphere by acting as obstacles to air flow and their effects are included in parameterization schemes for many numerical weather prediction and climate simulation models.

The flow of water in a fast-moving mountain stream provides a useful analogy to the influences of mountains on the atmosphere. The surface of the water in such a stream is often far from uniform, and clearly shows the influence of submerged obstacles, such as large rocks, that act to perturb the water surface both slightly ahead of and downstream from the obstacles. Mountains act in a similar fashion to perturb the atmospheric flow. This occurs because a stable atmospheric stratification creates buoyancy forces that act to return vertically displaced parcels to their equilibrium levels and because slight ascent often leads to saturation of the atmosphere and cloud formation (Smith 1979).

Analyses from the output of medium-range numerical weather prediction models and general circulation models in the 1970s and 1980s during the northern hemisphere cold season indicate that the zonally averaged upper-tropospheric flow was too strong in the midlatitudes after several days of prediction time (Lilly 1972; Palmer *et al.* 1986; Kim *et al.* 2003). This systematic error influenced the predicted values of sea-level pressure, low-level winds,

373

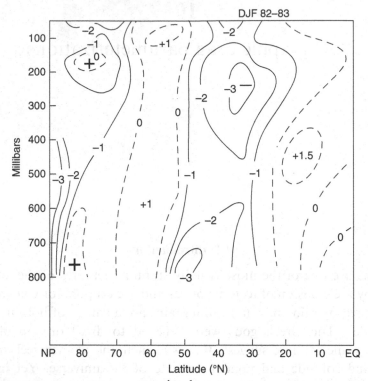

Figure 10.1. Residual term $(m\,s^{-1}\,d^{-1})$ from the zonal mean momentum budget calculated using uninitialized global analyses from December 1982 to February 1983 and attributable to subgrid-scale motions. Note the vertical coherence in the drag between 30° and 50° N and the occurrence of both low-level and upper-level minima. From Palmer *et al.* (1986).

geopotential heights, and the evolution of extratropical cyclones (Palmer *et al.* 1986). Since the model-produced zonally averaged wind speeds agreed in many other respects with observations, this overprediction of the midlatitude westerly subtropical jet suggested that an important physical process was not included (Lilly 1972). Zonal mean momentum budget calculations from uninitialized forecast model analyses also indicated that the residual terms, representing the effects of subgrid-scale motions, produce a drag in midlatitudes during the cool season (Fig. 10.1) further strengthening the argument that some subgrid-scale process was not incorporated in the models (Palmer *et al.* 1986).

Further studies indicate that this cold season wind speed error is due to the lack of sufficient drag from rugged mountain ranges. At low levels, mountains can produce flow blocking under stable atmospheric stratifications, changing the effective mountain height and influencing the stationary waves in the atmosphere (Wallace *et al.* 1983). However, as described by Palmer *et al.*

(1986), in stratified flow a drag force can be imposed on the atmosphere via internal gravity waves. These waves have horizontal wavelengths of ~6 km for flow over a hill of width 1 km and propagate vertically into the atmosphere. Wave breaking occurs when the waves reach a critical level; therefore, these internal gravity waves are able to create an upper-level drag on the atmospheric flow (Lindzen 1981). If the model grid spacing is greater than 5–10 km, then there can be a significant underestimation of the drag force exerted by rugged mountain ranges since these waves and the terrain features that generate them are not resolved adequately by the model (Clark and Miller 1991). This systematic model wind speed error is much smaller or negligible during the warm season, when the atmospheric stratification and surface winds are weaker. It also is negligible when the elevated terrain varies smoothly, such as over plateaus, where there is little subgrid variation in terrain height.

10.2 Simple theory

The response of the atmosphere to changes in terrain, such as mountain ranges and more isolated hills, is complicated because it depends upon the ever evolving atmospheric stratification. One measure of atmospheric stability is the Brunt-Väisälä frequency N (s^{-1}), where

$$N^2 = \frac{g}{\theta}\frac{\partial \theta}{\partial z},$$ (10.1)

g is the acceleration due to gravity, and θ is the atmospheric potential temperature (K). This is the frequency at which a vertically displaced air parcel oscillates within a statically stable environment ($N^2 > 0$). Air easily flows over any obstacles to the flow for unstable ($N^2 < 0$) or neutral ($N^2 = 0$) stratification. This is not the case for moderately stable stratification when buoyancy forces are important to the flow over the obstacles and internal gravity waves can be produced (Carruthers and Hunt 1990). As the stratification increases even further, the buoyancy forces are strong enough that vertical motion is suppressed below a certain height and the air flows around the obstacles (Fig. 10.2). Although the height and length of the hill influence the exact response (Carruthers and Hunt 1990), the atmospheric stratification plays the dominant role.

One way to determine whether or not the air flows over or around an obstacle of height h is to examine the inverse Froude number (Fr). This number is defined as

$$Fr = \frac{Nh}{U},$$ (10.2)

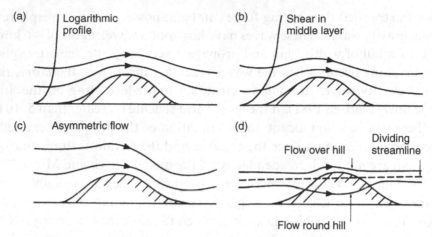

Figure 10.2. Atmospheric flow patterns over a three-dimensional hill for: (a) neutral stratification, (b) weak stratification, (c) moderate stratification, and (d) strong stratification. From Carruthers and Hunt (1990).

where U is the wind speed of the incident flow (Gill 1982). When $Fr \gg 1$ the air flows around the obstacle since the stratification is strong. In contrast, when $Fr \ll 1$ the air flows over the obstacle. For values of $Fr \sim 1$ the air generally flows both around and over the obstacle. As seen in (10.2), for a fixed stratification ($N =$ constant) the value of Fr decreases directly in proportion to increases in the incident wind speed. Thus, both stratification and the incident wind speed are important to determining the influence of the obstacle on the atmospheric flow. The height h of the obstacle also plays a role, as larger heights increase the value of Fr, thereby increasing the likelihood that the flow goes around the obstacle. The amount of drag produced by the obstacle on the atmosphere changes if the flow goes over the obstacle, goes around the obstacle, or if the obstacle generates internal gravity waves, or some combination of these flow responses (Fig. 10.3).

Another important parameter to consider when examining the atmospheric response to obstacles is the Richardson number (Ri), the dimensionless ratio of the buoyant suppression of turbulence to the generation of turbulence by vertical wind shear. It is defined as

$$Ri = \frac{(g/\theta)\,\partial\theta/\partial z}{(\partial U/\partial z)^2} = \frac{N^2}{(\partial U/\partial z)^2}, \tag{10.3}$$

where $U(z)$ is the mean horizontal wind speed. When the Richardson number falls below 0.25, the commonly accepted value of the critical Richardson number, the air becomes unstable and turbulence is generated.

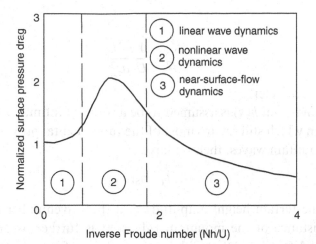

Figure 10.3. Schematic of the normalized surface pressure drag as a function of *Fr* for constant values of N and U that outlines the three flow regimes identified from the numerical simulations. The pressure drag is normalized by the value expected by linear theory due to freely propagating gravity waves. From Scinocca and McFarlane (2000).

The response of a stably stratified atmosphere to mountains can be investigated analytically following Durran (1986, 1990). For stationary, two-dimensional airflow over small-amplitude mountains the inviscid Boussinesq equations contain the essential physics governing the flow. If these equations are linearized about a horizontally uniform basic state with a mean wind $U(z)$, the result is

$$U\frac{\partial u}{\partial x} + w\frac{\partial U}{\partial z} + c_p\theta_0\frac{\partial \pi}{\partial x} = 0, \tag{10.4}$$

$$U\frac{\partial w}{\partial x} + c_p\theta_0\frac{\partial \pi}{\partial z} = b, \tag{10.5}$$

$$U\frac{\partial b}{\partial x} + N^2 w = 0, \tag{10.6}$$

$$\frac{\partial u}{\partial x} + \frac{\partial w}{\partial z} = 0, \tag{10.7}$$

where $b = g\theta/\theta_0$ and π is the Exner function, which is equal to $(p/p_0)^{R/c_p}$. These equations can be combined to form a single equation in the velocity component w, such that

$$\frac{\partial^2 w}{\partial x^2} + \frac{\partial^2 w}{\partial z^2} + l^2 w = 0, \tag{10.8}$$

in which

$$l^2 = \frac{N^2}{U^2} - \frac{1}{U}\frac{d^2 U}{dz^2} \tag{10.9}$$

is the Scorer parameter.

If the terrain height $h(x)$ is assumed to be a series of infinite periodic ridges, an assumption which still retains most of the fundamental properties of small-amplitude mountain waves, then we have

$$h(x) = h_0 \cos(kx), \tag{10.10}$$

where h_0 is the terrain height amplitude and the wavenumber k defines the separation distance of the mountain ridges. It is further assumed that the values of N and U are constant with height and time. Since the terrain surface is fixed and impenetrable, the velocity normal to the topography must vanish. This lower boundary condition requires that

$$w(x, z = 0) = (U + u)\frac{\partial h}{\partial x} \approx U\frac{\partial h}{\partial x} = -U h_0 k \sin(kx). \tag{10.11}$$

Thus, solutions to (10.8) may be written in the form

$$w(x, z) = \hat{w}_1(z) \cos(kx) + \hat{w}_2(z) \sin(kx). \tag{10.12}$$

When this solution is substituted into (10.8), then an equation governing the vertical structure of the perturbation vertical velocity is found, such that both $\hat{w}_1$ and $\hat{w}_2$ satisfy

$$\frac{\partial^2 \hat{w}_i}{\partial z^2} + (l^2 - k^2)\hat{w}_i = 0, \qquad i = 1, 2. \tag{10.13}$$

Since both N and U are assumed to be constant, $l^2 - k^2 = m^2$ is also a constant and the solutions to (10.13) are

$$\hat{w}_i(z) = \begin{cases} A_i e^{\mu z} + B_i e^{-\mu z} & k > l \\ C_i \cos(mz) + D_i \sin(mz) & k < l, \end{cases} \tag{10.14}$$

where $\mu^2 = -m^2$ (Durran 1986). The coefficients A_i, B_i, C_i, and D_i are determined by the boundary conditions at the surface and as z approaches infinity. Since the wave amplitude cannot grow exponentially without bound, it is required that $A_i = 0$. The lower boundary condition (10.11) then defines $B_1 = 0$ and $B_2 = -U h_0 k$, yielding stationary waves that decrease exponentially with height for $k > l$. For the other case when $k < l$, then trigonometric identities can be used to rewrite the solution for w as

$$w(x, z) = E_1 \sin(kx + mz) + E_2 \sin(kx - mz) + E_3 \cos(kx + mz)$$
$$+ E_4 \cos(kx - mz), \tag{10.15}$$

where $m > 0$ and $k > 0$. The lower boundary condition leads to $E_1 + E_2 = -Uh_0 k$ and $E_3 + E_4 = 0$.

As discussed by Durran (1986), the terms in (10.15) with $(kx + mz)$ correspond to waves in which the lines of constant phase tilt upstream $(kx + mz = \text{constant})$. These waves transport energy upward and momentum downward, which is the desired behavior since the mountain acts as an energy source and the waves should thus transport energy away from the mountain. Note that in this case the horizontal wavenumber k is generally much smaller than the vertical wavenumber m, yielding $m = N/U$ from (10.9). In contrast, the terms with $(kx - mz)$ transport energy downward and towards the mountain – a situation that makes no physical sense. Thus, the upper boundary condition requires that $E_1 = -Uh_0 k$, and that $E_2 = E_3 = E_4 = 0$. Durran (1986, 1990) provides further explanation for this choice of upper boundary condition and why it is the correct one.

Finally, the perturbation vertical velocity field for waves forced by a sinusoidal terrain field defined by (10.10) is given as

$$w(x, z) = \begin{cases} -Uh_0 k e^{-\mu z} \sin(kx) & k > l \\ -Uh_0 k \sin(kx + mz) & k < l. \end{cases} \tag{10.16}$$

These two wave structures are depicted in Fig. 10.4, where the waves with $k > l$ are evanescent waves that decay exponentially with height and the waves with $k < l$ are waves that propagate vertically without loss of amplitude. If ϕ is defined as the angle between the vertical and the slanting parcel trajectories for the vertically propagating waves, then Durran (1986) shows that this angle can be determined from

$$\cos(\phi) = \frac{Uk}{N}, \tag{10.17}$$

as long as $Uk < N$. For $Uk > N$ the waves decay since there is no way for buoyancy forces to support the oscillation.

For more realistic terrain shapes, the terrain profiles are constructed from periodic functions using Fourier transforms (see Durran 1990). Results show that the perturbation vertical velocity over an isolated bell-shaped ridge behaves very similarly to the vertical velocity over the infinite series of sinusoidal ridges. If the mountain quarter width is specified by L, then for a narrow mountain with $N \ll U/L$ the mountain primarily forces evanescent waves. In contrast, for a wide mountain with $N \gg U/L$ the mountain primarily forces waves that propagate vertically and their lines of constant phase tilt upstream.

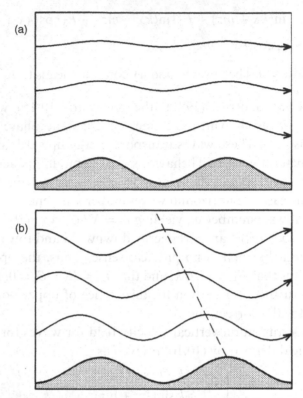

Figure 10.4. Two-dimensional streamlines over an infinite series of sinusoidal ridges for: (a) evanescent waves and (b) vertically propagating gravity waves. The dashed line in (b) indicates the vertical upstream tilt of the lines of constant phase. The airflow is from left to right. From Durran (1990).

When U and N vary in the vertical direction, another type of mountain wave – the trapped lee wave – can occur (Fig. 10.5). In this case the wave activity is confined to the lower troposphere on the lee side of the mountain (Scorer 1949; Durran 1986). If one assumes two vertical atmospheric layers with different values of N in each layer, a necessary condition for the existence of trapped waves can be determined that depends upon the value of the Scorer parameter in each layer and the depth of the lower layer (Scorer 1949). These trapped waves propagate vertically in the lower layer and then decay exponentially with height in the upper layer. The wave energy is repeatedly reflected from the flat ground downstream of the mountain and from the upper atmospheric layer, producing waves with no vertical tilt (Durran 1990). Thus, trapped lee waves are referred to as resonant waves when these conditions are met.

Internal gravity waves are important to the large-scale atmosphere because they can produce drag. If the variables in the horizontal equation of motion

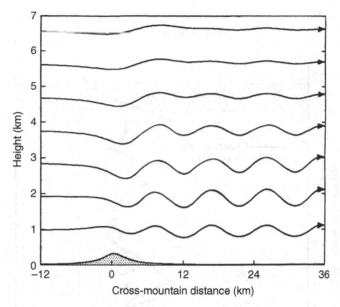

Figure 10.5. Two-dimensional streamlines in steady airflow over a single mountain ridge when the Scorer parameter varies in the vertical allowing trapped lee waves. The airflow is from left to right. From Durran (1990).

written in z coordinates are separated into mean (bar) and perturbation (prime) quantities, then it can be shown that

$$\bar{\rho}\frac{d\bar{V}_H}{dt} + \nabla_H\bar{p} + fk \times \bar{V}_H = -\frac{\partial}{\partial z}\left(\overline{\rho w' V_H'}\right) + \cdots,$$ (10.18)

where the first term on the right-hand side is the vertical eddy stress convergence term due to gravity waves (Lilly 1972). In a stably stratified atmosphere, gravity waves are able to transport momentum substantial distances between their sources and sinks without affecting the intervening vertical layers (Bretherton 1969). In essence, $\overline{\rho w' V_H'}$ is non-zero when small-amplitude waves are present, but its vertical derivative is zero (Eliassen and Palm 1961). This also implies that the momentum flux for vertically propagating waves at any vertical level is equal to their momentum flux, or stress, at the ground surface in the absence of dissipation (Eliassen and Palm 1961). Drag is produced when the momentum flux changes with height, which occurs during wave breaking and for trapped waves (Durran 1990).

Wave breaking occurs when the gravity waves grow to large amplitude, in part due to the decrease of density with height, and overturn. This occurs because as the amplitude of the waves increases, the influence of the wave circulations on the local environment may lead to the local Richardson number (Ri) dropping below its critical value of 0.25 (Lindzen 1981). Turbulence is then generated,

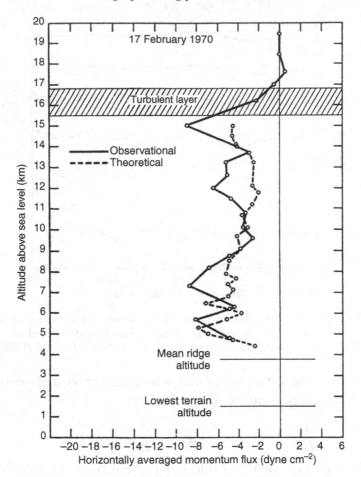

Figure 10.6. Mean observed profile of momentum flux obtained by averaging values obtained from three different analysis methods (solid black line) and from theory (dashed line). Note how the momentum flux is approximately constant with height below the turbulent layer, and then decreases to zero within the turbulent layer. From Lilly and Kennedy (1973).

producing an exchange of momentum between the waves and the environment and hence drag. Analyses of aircraft observations of mountain wave momentum flux indicate that the momentum flux is fairly constant below the turbulent layer (Fig. 10.6) in agreement with theory (Lilly and Kennedy 1973).

Trapped lee waves can also produce momentum flux that varies with height, and produce drag as seen in both model simulations (Durran 1990) and observations (Georgelin and Lott 2001). Although the horizontal average of the momentum flux over one phase of a resonant wave is zero, it is the disturbance nearest the mountain that generates the vertically varying non-zero momentum flux (Durran 1990). Although likely not as important as wave breaking, the drag

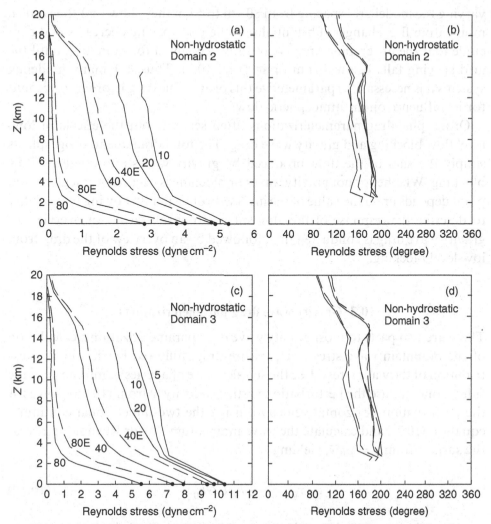

Figure 10.7. Vertical profiles of Reynolds stress for various model grid spacings over the Alps region in northern Italy. Numbers 80, 40, 20, 10, and 5 denote the model grid spacing used, while the letter E refers to a model simulation with envelope orography. The magnitude of the Reynolds stress is on the left, while the direction of the stress is on the right. It is clear that the Reynolds stress at 80 km grid spacing is an order of magnitude smaller above 3 km than the stress determined using 5 km grid spacing, illustrating the need for orographic drag parameterizations. From Clark and Miller (1991).

due to trapped lee waves is a potentially important source of orographic wave drag, especially for models with smaller grid spacing.

The dependence of this vertical momentum flux on model grid spacing is examined by Clark and Miller (1991) for a simulation of flow over the Alps. By varying the model grid spacing from 80 km to 5 km they show that the vertical momentum flux changes substantially as the grid spacing decreases and conclude that the orographic drag is not fully accounted for in models until the grid spacing falls below 10 km or so (Fig. 10.7). Thus, for many modeling systems it is necessary to parameterize this orographic drag in order to account for its influence on the atmospheric flow.

Orographic drag parameterizations often separate out the effects of low-level flow blocking and gravity wave drag. The total drag due to orography is simply the sum of the drag produced by gravity waves and low-level flow blocking. Whether or not gravity waves or blocking is occurring at a given grid point depends upon the value of the inverse Froude number owing to its ability to describe the response of the atmosphere to an obstacle. An overview of gravity wave drag is conducted first, followed by an overview of the drag from low-level blocking.

10.3 Gravity wave drag parameterizations

There are two parts to most gravity wave drag parameterizations: calculation of the mountain wave stress, or pressure drag, followed by the vertical distribution of the wave stress. For the simplest case of a series of infinite periodic ridges, one can use the perturbation vertical velocity w from (10.16), solve for the perturbation horizontal velocity u using the two-dimensional continuity equation (10.7), and calculate the wave momentum flux or Reynolds stress at the surface from $\tau = \rho \overline{uw}$, yielding

$$\tau_{sfc} = -\frac{1}{2} k \rho U N h_0^2. \tag{10.19}$$

When used in numerical models the stress is often defined similarly as

$$\tau_{sfc} = -B \rho_0 U_0 N_0 h^2, \tag{10.20}$$

where U_0 is the low-level wind speed, N_0 is the low-level Brunt-Väisälä frequency, ρ_0 is the surface density, h^2 is the variance of the subgrid scale orography, and B is a tunable constant that depends upon the characteristics of the orography (Phillips 1984; Palmer *et al.* 1986). For example, Boer *et al.* (1984) define $B \propto 1/d_0$, where d_0 is the typical separation distance between the important topographic features in the grid cell. Kim and Arakawa (1995) show that this type of approach to defining the surface stress due to the mountain waves is fairly common in gravity wave drag parameterizations (e.g., Chouinard *et al.* 1986; McFarlane 1987; Iwasaki *et al.* 1989).

A slightly different formula is developed by Pierrehumbert (1986) in which

$$\tau_{sfc} = -B \frac{\rho_0 U_0^3}{N_0} \left(\frac{Fr^2}{Fr^2 + 1} \right). \tag{10.21}$$

Assuming that B is constant, then when $Fr \gg 1$ and air flows around the obstacle the surface stress is at its maximum. When $Fr \sim 1$ and air flows both around and over the obstacle, the surface stress is roughly half its potential maximum value. And when $Fr \ll 0$ and air easily flows over the obstacle there is little surface stress. This type of surface stress formulation is used, for example, by Stern *et al.* (1987), Alpert *et al.* (1988), Kim and Arakawa (1995), Gregory *et al.* (1997), and Kim and Doyle (2005).

However, the value of B in (10.20) and (10.21) is not a constant and can vary from gridpoint to gridpoint. The expression for B often is a complicated function of both the subgrid orography and the surface wind direction. For example, Lott and Miller (1997) use estimates of the height variation in the along-ridge and cross-ridge directions to specify the value of B. Estimates of the sharpness, slope, width, and profile of the orography based upon the surface wind direction are used in Kim and Doyle (2005) to calculate B. The formulation in Kim and Arakawa (1995) is based upon results from over 100 two-dimensional mountain wave simulations with various shapes and sizes of mountain. The need for all of these simulations illustrates the sensitivity of the gravity wave drag parameterizations to the subgrid orography and the surface wind direction.

Once the surface stress is determined, the vertical profile of wave stress can be calculated. This is accomplished by estimating the influence of the orographic gravity waves on the local static stability and vertical wind shear and then determining a minimum Richardson number that represents the smallest value the Richardson number can obtain under the influence of the gravity waves. When this minimum Richardson number falls below the critical value for the onset of turbulence, then wave breaking occurs. This minimum Richardson number is defined as

$$Ri_{min} = Ri \frac{1 - (N\delta h / U)}{[1 + Ri^{1/2}(N\delta h / U)]^2}, \tag{10.22}$$

where δh is the vertical displacement of an isentropic surface due to the gravity waves, N is the local value of the Brunt-Väisälä frequency, and U is the local wind speed in the direction of the surface wind direction (Palmer *et al.* 1986).

Beginning at the surface, the calculations to determine the gravity wave drag are made by moving upward in the grid cell. The required values of Ri, N, and U are determined directly from the model data for any given vertical level. The value of δh is determined by assuming that the momentum flux at any level in the atmosphere equals the surface momentum flux unless wave breaking occurs and some of the momentum is transferred to the environment (Eliassen and Palm 1961). Thus, one solves for δh using (10.20) knowing the surface value of the momentum flux (or the momentum flux at the next lowest model level) and replacing the surface values of ρ, N, and U with their values at the given model level. This yields

$$\delta h_i = \left(-\frac{\tau_i}{B\rho_i U_i N_i} \right)^{1/2}, \tag{10.23}$$

where τ_i is the momentum flux reaching to vertical level i and the value of U_i is the component of the wind parallel to the surface wind (Kim and Arakawa 1995). Once the value of δh is determined, the minimum Richardson number is calculated from (10.22). If $Ri_{min} \geq 0.25$, the critical value for the onset of turbulence, then wave breaking does not occur, the vertical wave momentum flux is unchanged, and the next highest vertical level is evaluated. When $Ri_{min} < 0.25$ at a given vertical level in the grid cell, wave breaking occurs and some of the wave momentum flux is transferred to the environment, producing drag.

The most common approach to specifying the amount of momentum flux transferred to the environment is based upon the saturation hypothesis of Lindzen (1981), in which dissipation processes are assumed to limit the wave amplitudes and to produce drag. When wave breaking occurs, one can calculate the value of δh needed to reset the value of Ri_{min} to 0.25 in (10.22) and thus cause the turbulence to cease. The momentum flux that remains after wave breaking is determined using this modified value of δh in (10.20). The difference between the original surface momentum flux and the momentum flux left after wave breaking represents the amount of momentum transferred to the atmosphere at the model level height. Several caveats and practical points regarding these calculations are discussed by Palmer *et al.* (1986).

A comparison of results from a gravity wave drag parameterization that is based upon the saturation hypothesis with results from two-dimensional mountain wave simulations suggests that the magnitude of the parameterized drag is underestimated in low-level downstream regions with wave breaking or lee wave trapping (Kim and Arakawa 1995). Further exploration of this problem by Kim and Arakawa (1995) suggests that the vertical gradient of the Scorer parameter is a more useful approach to estimating the change of wave momentum flux with height at low levels. Thus, the change in wave momentum flux is defined using

$$\frac{\tau_i}{\tau_{i-1}} = \min\left(C_i \frac{l_i^2}{l_{i-1}^2}, 1 \right), \tag{10.24}$$

where $C_i = 1$ (a tunable constant), i defines the vertical model layer and increases upwards, and l is again the Scorer parameter (Kim and Arakawa 1995; Kim and Doyle 2005). Kim and Arakawa (1995) use (10.24) to determine the change in wave momentum flux below 10 km and use the saturation hypothesis (10.22) above 10 km.

The effects of low-level wave breaking within the first vertical wavelength above the surface also can be approximated by examining the results from fully three-dimensional simulations of breaking waves. Results indicate that the surface stress is amplified by up to three times above expectations from linear theory for flow over obstacles, and this stress amplification depends upon the value of Fr (see Fig. 10.3). Thus, Scinocca and McFarlane (2000) develop a simple empirical relationship to increase the surface stress for the appropriate range of Fr values for which low-level breaking is likely. This stress decreases linearly between the surface and the height of wave breaking. Kim and Doyle (2005) include an enhancement factor in their calculation of low-level drag to account for low-level wave breaking and wave trapping, which is also tied to the value of Fr.

Most gravity wave drag parameterizations assume only one vertically propagating gravity wave is generated by the subgrid orographic mountains. When the model winds turn with height and become normal to the surface wind direction (i.e., the two-dimensional wave orientation) a critical level is created and all of the remaining wave momentum is transferred to the atmosphere at this one vertical level. This can result in excessive momentum deposition at these critical levels (Scinocca and McFarlane 2000). To help alleviate this problem, Scinocca and McFarlane (2000) assume that there exist two vertically propagating gravity waves within each grid cell with different orientations, while Gregory *et al.* (1997) assume a finite spectrum of gravity waves. Most models that focus upon the evolution of the atmosphere in the troposphere assume a single gravity wave, whereas models that also need to predict the evolution of the stratosphere tend to allow for a spectrum of waves (Kim *et al.* 2003).

10.4 Low-level blocking drag parameterizations

The blocking of low-level flow due to subgrid-scale mountains also can produce an extra source of drag for the atmosphere. Initial attempts to account for this blocking effect focused upon modifying the model-resolved orography. For example, the original model terrain heights (i.e., the average terrain height within a model grid cell) are increased in proportion to the standard deviation of the subgrid terrain heights when applying envelope orography (Wallace *et al.* 1983). A similar approach is used by Mesinger *et al.* (1988) in a study of lee cyclogenesis in which the model terrain heights are defined by the tallest peaks and by Tibaldi (1986) in studying the maintenance of quasi-stationary waves. However, while these approaches increase the drag, since the terrain is taller and interacts with a deeper layer of the atmosphere, they also appear to have undesirable consequences. Envelope orography makes the assimilation of low-level observations more difficult, while also leading to excessive precipitation over the enhanced orography (Lott and Miller 1997).

The problems with envelope orography and other modifications to the original average model terrain field led to the development of low-level blocking parameterizations. These approaches use the value of the inverse Froude number to estimate the depth of the flow that is blocked by the subgrid mountains. Scale analysis and results from analytic studies of flow over objects of various shapes allow for an estimation of the surface form drag due to subgrid orographic features. A typical form of the surface stress due to low-level blocking is

$$\tau_{sfc} = -\frac{\Lambda}{2L^2}\rho_0 C h_b l_b U_0 |U_0|, \tag{10.25}$$

where L^2 is the area of the grid cell, Λ is an estimate of the number of ellipses needed to represent the unresolved topographic features, C is a bulk drag coefficient, h_b is the height of the blocked layer, and l_b is the width of the blocked layer (Scinocca and McFarlane 2000). An alternative formulation to (10.25) is found in Kim and Doyle (2005) that includes a unique set of parameters that take into consideration the details of the orographic configurations. The height of the blocked layer is related to the inverse Froude number, such that

$$h_b = \frac{U_0}{N_0}[Fr - Fr_c], \tag{10.26}$$

where Fr_c is the critical inverse Froude number that specifies when low-level wave breaking occurs and has a value of ~ 1 (Scinocca and McFarlane 2000). Over the depth of the blocked flow region the stress is typically held constant or decreased linearly with height (Lott and Miller 1997; Scinocca and McFarlane 2000).

10.5 Discussion

Most orographic drag parameterizations used within operational forecast models today are based upon a combination of low-level blocking and gravity wave drag diagnostic calculations. We have seen that these calculations depend upon the values of N, U, and Fr, and the variation of the orography within the model grid cell. The amount of drag imparted to the atmosphere also depends upon the assumptions used for determining where and when wave breaking occurs. While most of the early orographic drag parameterizations were evaluated only by their ability to improve the model forecasts, most studies today also try to evaluate the parameterizations off-line in comparisons with observations from a handful of field campaigns and wave-resolving model simulations (Kim *et al.* 2003). For example, the orographic drag parameterization of Kim and Doyle (2005) is compared with simulations that explicitly resolve the momentum flux over complex orography. Results suggest good agreement between the parameterization and the explicit momentum flux calculations (Fig. 10.8). However, the objective evaluation of orographic drag parameterizations remains difficult and many of the parameters used within these schemes are not constrained and so must be selected carefully.

The need for gravity wave drag also is not limited to flow over mountain ranges. Deep convection also produces gravity waves that propagate vertically and may need to be parameterized (Lindzen 1984; Kershaw 1995; Gregory *et al.* 1997; Bosseut *et al.* 1998; Chun and Baik 1998, 2002). It appears that organized convective regions, such as mesoscale convective systems, may be important generators of gravity waves that influence atmospheric circulations in the Tropics (Kim *et al.* 2003). In particular, Alexander and Holton (1997) hypothesize that convectively generated gravity waves are a possible mechanism for the quasi-biennial oscillation. Unfortunately, developing parameterizations for convectively generated gravity waves is even more challenging than for the orographically generated gravity waves, in part because the generated waves are largely non-stationary and there are a number of mechanisms that can account for the wave generation.

One of the challenges to orographic drag parameterizations is the ever changing grid spacing of the models. For large horizontal grid spacings the gravity waves forced by convection or terrain likely remain contained within the model grid cell to a first approximation. However, as the grid spacing is reduced it is possible and perhaps even likely that the gravity waves actually propagate out of the model vertical grid column to influence neighboring grid points. Methods to approximate this wave propagation into neighboring grid cells have been developed, but are presently too computationally expensive for widespread use (Kim and Doyle 2005). The vertical grid spacing is also likely to be an important consideration as it influences the predicted values of U and N.

The length of the model simulation also is an important consideration when deciding if an orographic drag parameterization is needed. Certaintly for model forecasts beyond a few days the potential influences of orographic drag are large enough to warrant inclusion of the parameterization. However, the strong mixing and drag produced by wave breaking associated with downslope wind storms can be substantial even for shorter time periods. Thus, these types of parameterizations may need to be included even for

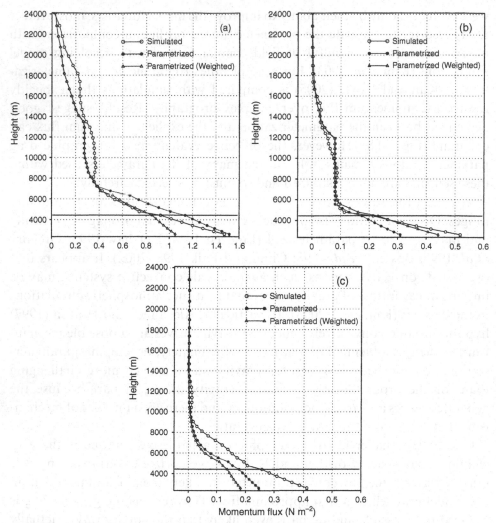

Figure 10.8. Momentum flux from an orographic drag parameterization compared with the flux calculated from an explicit 2 km grid spacing simulation of the gravity wave event produced by flow over mountains. The sign of the momentum flux is reversed. Results from three different cases are shown. In (a) the momentum flux does not reach 0 below 24 km. In (b) the momentum flux goes to zero between 8 and 18 km, whereas in (c) the momentum flux goes to zero below 14 km. From Kim and Doyle (2005).

short-range forecasts. The point at which the horizontal grid spacing begins to fully resolve the gravity waves, such that no orographic drag parameterization is needed, also remains an open question (see Kim *et al.* 2003). The effects of moisture on the development of trapped mountain waves also may be important, as the wave response can be amplified or damped due to the presence of moisture (Durran and Klemp 1982; Kim and Doyle 2005).

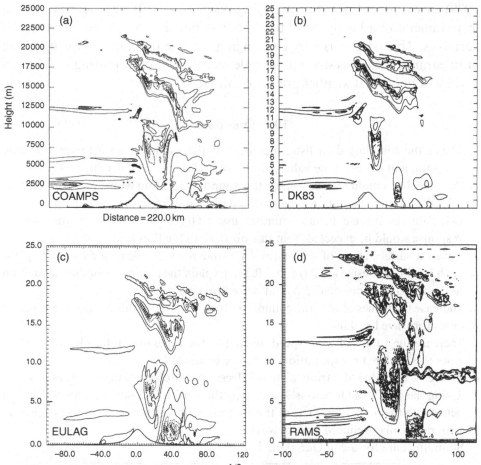

Figure 10.9. Isolines of sign(Ri) $|Ri|^{1/2}$ at the 3 h time from idealized simulations of the 11 January 1972 Boulder, Colorado, downslope windstorm using 4 different non-hydrostatic models. All simulations are two-dimensional and use an idealized mountain profile. While the general regions of wave breaking are very similar, the details are very different. From Doyle *et al.* (2000).

Explicit two-dimensional simulations of gravity wave breaking over an analytic orographic profile from 11 different numerical models show many similarities (Doyle *et al.* 2000). All the runs produce wave breaking in approximately the same location in the stratosphere and produce downslope winds in the lee of the mountain. However, there also are notable differences in the details of the strength of the downslope winds, the depth of the wave breaking region, and the detailed structures within the breaking region (Fig. 10.9). So while these results are encouraging in that all the models produce wave breaking, they also raise some concerns about the ability of explicit gravity wave simulations to produce accurate reproductions of all aspects of the wave breaking. Since wave breaking is a very non-linear process (Peltier and Clark

1979) this is probably not a surprising result. The difficulty becomes trying to parameterize a highly non-linear process based upon the resolved model variables. This process is very challenging, yet the developers of orographic drag parameterizations strive to provide realistic estimates of drag for use in a variety of numerical weather prediction models.

10.6 Questions

1. Using the sounding data listed below, plot the sounding on a thermodynamic diagram, and calculate the values of N using (10.1) for each height level.
2. Assume that a mountain exists near this sounding location with a ridge orientation of $180°$. Vary the mountain height between 100 and 500 m in 50 m increments, and calculate the inverse Froude number using (10.2). What type of atmospheric response would be expected from this obstacle to the flow?
3. Given a surface stress of $-1.0 \, kg \, m^{-1} s^{-2}$ and $B=0.2$, calculate δh and Ri_{min} for each height level using (10.22) and (10.23). Explain the mechanics of this calculation and any assumptions made. Would wave breaking occur?
4. Increase the values of N in the column by 10%. How does this change the expectations for wave breaking?
5. Increase the values of wind speed normal to the ridge orientation by 10%. How does this change the expectations for wave breaking?
6. Decrease the value of B from 0.2 to 0.1. Describe the changes to the values of Ri_{min}.
7. Using the results of Questions 3–6, discuss the sensitivities of the parameterization scheme to the model forecasts of the environment that the schemes use to predict orographically produced gravity waves.

Sounding data for use in Questions 1–7.

P (hPa)	H (m)	T (°C)	TD (°C)	U (m s^{-1})
861	1 475	−2.90	−6.40	2.00
850	1 573	0.00	−8.00	8.00
813	1 930	0.20	−13.80	6.00
762	2 438	−1.80	−10.73	0.70
726	2 831	−1.90	−8.90	3.00
700	3 120	−4.30	−11.30	7.00
653	3 658	−6.20	−15.94	8.00
604	4 267	−9.90	−21.35	14.00
500	5 710	−19.10	−39.10	16.00
400	7 320	−32.30	−48.30	20.00
300	9 300	−45.30	−55.30	25.00
250	10 500	−52.50	−61.50	41.00
200	11 920	−55.30	−71.30	25.00
150	13 750	−56.70	−80.70	20.00
100	16 270	−64.90	−87.90	15.00

11

Thoughts on the future

11.1 Introduction

Throughout the previous chapters we have examined a number of the most commonly used types of parameterization schemes within numerical weather prediction models. While individual parameterization schemes are constantly undergoing revision and new schemes appear in both the literature and operational models routinely, the underlying need for the parameterization of fundamental atmospheric processes has not changed. Indeed, the number of processes that are being parameterized has increased over the past 20 years to allow for more realism in both forecasts and climate simulations. These additional parameterizations may not be crucial to the model forecasts at all times and places, but they can make a significant difference regarding a particular event of importance to a specific user or community.

This evolution of parameterization highlights the fact that numerical models are becoming more capable (see Roebber *et al.* 2004). Models now can reproduce many of the phenomena that are observed in the atmosphere. As simple examples, moderate- and high-resolution models of today can reproduce mesoscale convective systems with their leading line of deep convection and trailing stratiform precipitation region as well as sea breezes and mountain-valley flows, while coupled ocean–atmosphere models can reproduce El Niño–Southern Oscillation (ENSO) events. These phenomena could not be reproduced by any of the operational models (or even many research models) in use back in the 1970s, owing in part to their large grid spacing and in part to the parameterization schemes in use. This increased model capability often leads to higher expectations as well as higher perceived confidence in the forecasts. It can be very difficult for a human forecaster to challenge the prediction from a numerical model forecast at 1 km grid spacing that indicates the development of a severe thunderstorm, or from a seasonal climate model that predicts an ENSO event. These model forecasts often provide details that cannot be observed, while the structures and behaviors produced by the model appear very realistic.

Yet it is hoped that some degree of uncertainty or doubt has been created as the various parameterization schemes have been examined and their sensitivities explored. Parameterization schemes generally develop from a reasonable theoretical foundation, but require a number of simplifying assumptions and are tested using a limited data set over a specific range of environmental conditions. However, when incorporated into numerical models, these schemes are then used to make predictions over the entire globe and in environments for which they likely have never been tested fully. In addition, the behavior of an individual parameterization often depends upon the behavior of other parameterization schemes and interactions between schemes often occur. For example, planetary boundary layer schemes depend upon the behavior of the soil–vegetation–atmosphere transfer scheme that predicts ground temperature and soil moisture, yet the boundary layer scheme also influences the net radiation reaching the surface through the vertical mixing of moisture. Empirical tuning also comes into play as model developers and some users attempt to optimize model skill (however defined) for particular problems or scenarios. This empirical tuning often does not occur through a systematic approach and may instead be guided by case studies of important events or model intercomparison tests with a small sample size. The truly remarkable aspect of all this is that the resulting numerical predictions have value! And this value appears to be increasing over time, in part because the model parameterization schemes are becoming more and more accurate and realistic. This trend in the ever increasing realism in parameterization schemes is expected to continue. But this situation also should cause us to pause and think about how best to use these valuable, yet flawed, tools that we create.

It is important to recognize that it is impossible to test a parameterization scheme for all atmospheric conditions that may occur. The observational data do not exist for such a test, and the time needed to conduct such tests would be prohibitive. This is not to suggest that parameterization schemes should not be tested on large data sets. Indeed, many parameterization schemes are already tested on fairly large data sets and this testing is important to the improvement and validation of schemes. However, we need to recognize and appreciate that a truly comprehensive test of a parameterization scheme is impossible. And even if a comprehensive test indicated a weakness of a scheme under specific conditions, this does not imply that the scheme should be discarded. It is just that the scheme has limitations, which is true for all parameterization schemes. Numerical models are imperfect, so the key to success is how we deal with these imperfections.

We move at this point from the realm of what is known into a realm that mixes knowledge with conjecture and opinion. This is perhaps a bit unusual for a meteorological textbook, but it seems appropriate to discuss what the future may hold to stimulate discussions and debate in this important enterprise called numerical weather prediction. Topics briefly touched upon are ensemble predictions, ensembles and high-resolution single forecasts, statistical postprocessing, and the road forward.

11.2 Ensemble predictions

Ever since the pioneering study of Lorenz (1963), it has been recognized that small errors can grow rapidly in non-linear models. However, even if the model is perfect, there is a finite time limit to the predictability of the atmosphere, since it is impossible to observe the atmosphere perfectly due to both sampling and instrument errors. This predictability time limit becomes shorter as the scales of interest become smaller (Lorenz 1969). Model simulations starting from ever so slightly different initial conditions diverge and eventually have little relationship to one another (Fig. 11.1). Since the true atmospheric state at any point in time can only be known approximately, the atmosphere prediction problem needs to be formulated in terms of the time evolution of a probability distribution function (PDF) for the atmosphere. This realization that the atmosphere is chaotic and has this sensitive dependence upon initial conditions led to the development of ensemble forecasting systems that explicitly attempt to predict the evolution of the atmospheric PDF (see the historical review by Lewis 2005). Since their first operational use in the early 1990s, ensembles have become a critical component of both operational numerical weather prediction and climate studies, and remain an important research topic. Ensembles are now used for climate, seasonal, medium-range, and short-range forecasting. As the models used within ensemble forecasting systems improve, the ensembles improve as well. Thus, model and ensemble forecast improvements go hand-in-hand.

Ensembles are simply groups of forecasts that are valid over the identical time period. Typically, each forecast member of the ensemble differs in its initial conditions in order to provide an initial sample of the atmospheric PDF. Differences in model characteristics may also be included as part of the ensemble to account for model error. A numerical weather prediction model is used to provide a forecast from each of these different initial conditions, and the properties of the atmospheric PDF are assumed to be determined by the statistics calculated from the ensemble members at any selected forecast time (Epstein 1969; Leith 1974; Molteni *et al.* 1996). The ensemble statistics

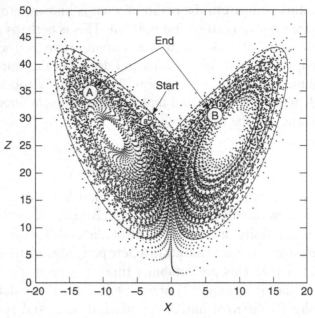

Figure 11.1. Depiction of the Lorenz (1963) attractor (dots), with trajectories *a* and *b* indicated that start from an initial small difference and evolve into a large difference at their end points *A* and *B*.

approximate the true atmospheric PDF closely if the initial perturbations accurately represent the PDF of analysis errors and if the numerical model provides a very good approximation of the atmosphere. Thus, instead of a single (deterministic) forecast that only provides one picture of the evolution of the atmosphere, an ensemble of forecasts is created and the forecast is now inherently probabilistic in nature. Murphy and Winkler (1979) strongly argue that forecasts cannot be used to their best advantage unless the forecast uncertainty is quantified and expressed in a useful manner to the end users. Ensemble forecasting systems are one way to express forecast uncertainty (Fig. 11.2).

Ensembles initially consisted of the same model with the same parameterization schemes, but using different initial conditions for each ensemble member. A number of different techniques have been developed to perturb the model initial conditions around a control, or best estimate, analysis of the atmospheric state in order to sample the analysis error. One Monte Carlo approach mimics the differences between the global analyses of different operational centers, which is an estimate of analysis error, and so draws the different initial conditions from this specified distribution (Errico and Baumhefner 1987; Mullen and Baumhefner 1994). However, it takes a very

COM 500mb ht(m) 5820m Spgt 39H fcst from 09Z 22 AUG 2005
verified time: 00z, 08/24/2005

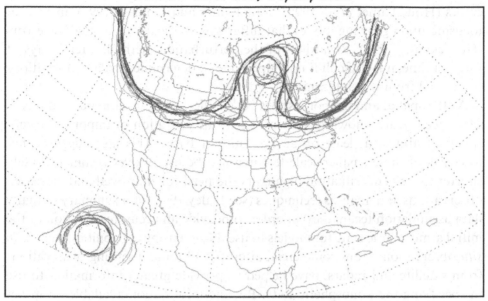

Figure 11.2. Spaghetti diagram of 5820 m, 500 hPa geopotential height isolines from the NCEP short-range ensemble forecasting system started at 0900 UTC 22 August 2005. Geopotential height isolines valid at the 39 h forecast time, or 0000 UTC 24 August 2005. Notice the variability over central North America, with some members indicating that a ridge or anticyclone will develop and others indicating zonal flow across the United States. Courtesy of Dr. Jun Du of the National Centers for Environmental Prediction.

large number of ensemble members to randomly sample the analysis error well and it is prohibitively expensive at this time to produce forecasts from hundreds of ensemble members. Thus, several initial condition perturbation techniques sample the analysis uncertainty space more strategically. Both singular vectors (Buizza and Palmer 1995; Molteni *et al.* 1996; Buizza 1997) and the breeding of growing modes (Toth and Kalnay 1993, 1997) select perturbations to the control analysis that are the most unstable and grow the fastest. It is hoped that these fastest growing perturbations lead to a reasonable sampling of the true atmospheric PDF with fewer ensemble members, since they should dominate the ensemble variability. Results further indicate that these two perturbation techniques, initially developed for medium-range ensemble systems, focus upon perturbations to synoptic-scale weather systems over the midlatitudes that are associated with baroclinic instability (Toth and Kalnay 1997). This temporal and spatial scale is well suited for numerical weather prediction, since baroclinic instability is inherent within the equations of motion. Yet another perturbation approach uses perturbed observations in data assimilation systems to produce a set of representative analysis errors (Houtekamer *et al*. 1996; Houtekamer and Lefaivre 1997), instead of focusing upon growing modes that differ systematically from analysis errors (Houtekamer 1995). Regardless of the perturbation method, after 5 days of forecast time, it is difficult if not impossible to determine which perturbations originated from which technique.

As the spatial and temporal scales get smaller, more instabilities and physical processes are known to play a role in the evolution of important atmospheric features, at least on an intermittent basis. This also suggests that forecasts of these smaller-scale features will be useful over commensurately shorter times. This situation highlights the need for improved and increased observations and analysis techniques (see Daley 1991 for a summary of many data assimilation techniques) in order to provide an accurate depiction of the initial atmospheric state for models to use. New sensing capabilities, such as *in situ* observations from commercial aircraft and remote sensing observations from satellite and radars, have helped to provide greater information to use in specifying the atmospheric state. But the information available on meso-scale and cloud-scale atmospheric features is still woefully inadequate. Sophisticated data assimilation techniques, such as three- and four-dimensional variational assimilation (e.g., Derber 1989; Županski and Mesinger 1995; Gauthier *et al*. 1999; Lorenc *et al*. 2000; Rabier *et al*. 2000; Barker *et al*. 2004; Županski *et al*. 2005) and the ensemble Kalman filter (Evensen 1994, 1997; Mitchell and Houtekamer 2000; Houtekamer and Mitchell 2001; Snyder and Zhang 2003; Dowell *et al*. 2004) help to make the most use of the available observations. Yet it is clear that many uncertainties remain in the initial conditions provided to models.

The uncertainties present in specifying the atmospheric state at any given time also influence the perturbation strategies designed for creating the ensemble members. For example, perturbation techniques designed for the medium-range forecast problem may not be well suited to the short-range forecast problem. While baroclinic instability is still important at short ranges, many initial state uncertainties and short-range forecast concerns have little to do with synoptic-scale features. The use of optimization periods of 12 h to 2 days for error growth may define the perturbation types that are generated and these perturbations may not be especially meaningful for the short range. Alternative approaches to generating initial condition perturbations that use input from human forecasters appear worthy of further exploration for predicting unlikely events (e.g., Xu *et al*. 2001; Homar *et al*. 2006). Yet it may be that combining various approaches, such as Monte

Carlo and the breeding of growing models, yields the best results. More research is needed on how to generate perturbations for short-range ensembles. The same can probably be said of ensembles used for seasonal and climate simulations and forecasts.

We have seen in earlier chapters that parameterization schemes play a large role in forecasts of sensible weather – low-level temperatures, dewpoints, winds, and rainfall about which the public is most concerned. And, therefore, one would think that parameterization schemes may contribute to forecast sensitivities and to important forecast errors (Fig. 11.3). This realization that model imperfections may contribute substantially to forecast error has led to the inclusion of different models or different physical process parameterization schemes or stochastic errors within ensemble forecast systems for both the medium and short range (Houtekamer *et al.* 1996; Atger 1999; Buizza *et al.* 1999; Harrison *et al.* 1999; Stensrud *et al.* 1999, 2000; Fritsch *et al.* 2000; Evans *et al.* 2000; Ziehmann 2000; Wandishin *et al.* 2001; Hou *et al.* 2001; Stensrud and Weiss 2002). Results from these studies clearly indicate that ensembles containing different models or different parameterization schemes are more skillful than ensembles that do not contain some aspect of model uncertainty. It is anticipated that as we explore the optimum balance of ensemble member grid spacing and the number of ensemble members, the probabilistic forecasts from multimodel ensemble systems will only further improve.

The value of multimodel ensembles also is seen in seasonal to interannual climate prediction. Nine-member ensembles are created from each of seven different coupled ocean–atmosphere models that use nearly the same control initial conditions and the results from all nine ensembles pooled into one large multimodel ensemble as part of the Development of a European Multimodel Ensemble system for seasonal-to-inTERannual prediction (DEMETER; Palmer *et al.* 2004) project. Results indicate that coupled ocean–atmosphere multimodel ensembles are more reliable, have better resolution, and have consistently better performance than single-model ensembles (Hagedorn *et al.* 2005).

For climate simulations, Stainforth *et al.* (2005) use widely distributed desktop computers to conduct over 2500 simulations of future climate in a world with doubled CO_2 levels. These simulations differ in that parameters within the model physical process parameterization schemes are varied over a range of values deemed plausible by the scheme developers, and in that they have variations in the initial conditions. The ensemble mean forecasts are in agreement with other predictions, indicating a global warming of nearly 3.4 K for doubled CO_2 conditions. However, the range of warming is much greater than seen in previous studies, with some simulations indicating warming of

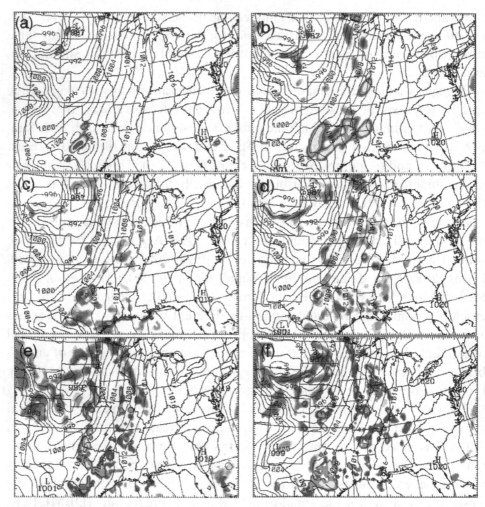

Figure 11.3. Sea-level pressure (contoured every 2 hPa) and 3 h accumulated convective rainfall (mm shaded) at the 24 h forecast time from six different mesoscale short-range ensemble members with parameterization scheme diversity valid at 0000 UTC 4 May 1999. While the sea-level pressure patterns are very similar in the six members, the rainfall patterns and amounts are very different. This highlights the importance of model parameterizations to the resulting forecasts and the need for parameterization diversity in ensemble forecasting systems. From Stensrud and Weiss (2002).

over 10 K is possible and others indicating very little warming at all. This result highlights the importance of the model physical process parameterizations to climate modeling, the need for improved parameterizations, and the need for a better understanding of how to create ensembles to respond to the ever increasing demand for more accurate and skillful weather information across a range of temporal scales.

In retrospect, this situation should not be surprising. We have seen that parameterization schemes can provide very different answers under the same environmental conditions. It is also possible that some parameterization schemes are entirely locked out of some environmental conditions and will never be able to reproduce the observed atmospheric behaviors in these environments. One example of this situation is the Betts–Miller convective parameterization being unable to develop convection in "loaded-gun" soundings (Fawbush and Miller 1954) that are typical of severe storm environments. This situation occurs because loaded-gun soundings have very dry mid-levels, such that the scheme fails to activate and hence is not able to produce rainfall. This is not to say that the Betts–Miller scheme is without value; it does an incredibly good job in many locations, but on the relatively rare occasion when these severe storm environments occur it is not able to activate. One can easily think of other examples in microphysical parameterizations, where some interactions between the various microphysics species are not included in the scheme, thereby locking out some behaviors. Another example is radiation parameterizations that do not allow for partly cloudy skies.

With the realization that parameterization schemes are imperfect and may not even function realistically in some environments, the idea of ensembles with model or model parameterization diversity in addition to initial condition uncertainty becomes very appealing and intuitive. The value of this type of approach is that the ensemble is more likely to be capable of reproducing the observed atmospheric phenomena over a broader range of environmental conditions, owing to the variety of parameterization schemes being used. It maximizes the chances that at least one of the parameterization schemes is capable of producing a realistic forecast for a given environment. It may be that not all members are equally plausible in a given environment, but over the entire range of environments visited by the atmosphere many parameterizations are equally skillful. The only argument for ensembles without model or model physics diversity is that all the parameterization schemes function well in all environmental conditions, a hypothesis that appears questionable at best.

11.3 Ensembles and high-resolution single forecasts

Computer resources are a finite quantity. Even with the incredibly rapid increases in computer processor speed, model developers and users are able to consume all the available processor cycles. This is particularly true in operational centers, where one also wants to make the best use of these precious resources. This situation has led occasionally to a perceived competition between high-resolution single deterministic forecasts and ensembles,

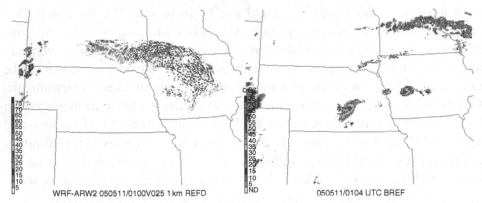

WRF-ARW2 050511/0100V025 1 km REFD   ND   050511/0104 UTC BREF

Figure 11.4. Cloud-scale model prediction (left) and observed (right) radar reflectivity fields valid 0100 UTC 11 May 2005. While the cloud-scale model clearly captures the typical types of reflectivity structures seen in the observations, the model has clearly missed the thunderstorms active in Nebraska and Iowa and misplaced the convection in Minnesota (in the upper right portion of the figure) southwestward. Image created as part of the Storm Prediction Center 2005 spring forecasting experiment.

since both ensembles and high-resolution (i.e., small grid spacing) single forecasts are very computationally demanding. The argument for high-resolution deterministic forecasts derives from the desire to have forecast models that are capable of predicting the observed atmospheric phenomena that are deemed most important. Without argument, high-resolution forecasts may be quite valuable for forecasting commonly observed, small-scale features with large societal or economic impact, such as severe thunderstorms and sea breezes. However, participation in several spring forecasting experiments that evaluated high-resolution (2 and 4 km grid spacing) operational forecasts of deep convection suggests that the lack of uncertainty information hampers the best of these forecasts (Fig. 11.4). It is not clear that some of the behaviors that are explicitly seen in high-resolution forecasts cannot be anticipated from the environmental conditions of lower-resolution forecasts with just as much skill (see discussions of Brooks *et al.* 1992 and Roebber *et al.* 2004). It is generally easier to disagree with lower-resolution forecasts than it is to disagree with forecasts that provide data on the same scales as the best of our observational systems.

In forecasts over western Washington state, Colle *et al.* (2000) show significant forecast improvement as the model grid spacing is decreased from 36 to 12 km. However, little forecast improvement is seen as the grid spacing is further decreased from 12 to 4 km. Over the northeastern United States, Colle *et al.* (2003) show little improvement at all when going from 36 to 12 km grid spacing, likely owing to the less sharp terrain features in the northeastern as compared to the northwestern United States. Gallus (1999) further indicates little or no increase in forecast skill when reducing from 30 to 10 km grid spacing in simulations of several midwestern convective systems. These results clearly highlight that model grid spacing by itself is not necessarily the answer to forecast improvements. While the optimal grid spacing for a given model likely depends upon a host of factors, including data assimilation systems, observational density, and model parameterization schemes, any general assumption that a reduction in the grid spacing automatically leads to improved forecasts must be suspect.

Instead of a competition between ensembles and high-resolution deterministic forecasts, there may be ways to merge high-resolution forecasts that are capable of reproducing the smaller-scale features of significant societal and economic interest with an ensemble forecasting system. Stensrud *et al.* (2000b) and Leslie and Speer (2000) discuss this situation, and suggest a combined forecast system in which an ensemble using larger grid spacing is run first and the results evaluated using clustering methods to define the most likely forecast scenarios of the day (e.g., Alhamed *et al.* 2002; Yussouf *et al.* 2004). These most likely forecast scenarios are then used to provide boundary conditions and a first-guess field for assimilating data into a small grid spacing forecast system. Using this type of approach, ensembles are used to ensure that the high-resolution forecast is actually a likely scenario and not one that is outside of the set of ensemble solutions. With a bit more ingenuity, many other possible ways to merge ensembles with high-resolution forecasts are certainly possible to provide both detailed forecasts and information on forecast uncertainty. As the forecast time increases past a few days, the value of single (deterministic) forecasts decreases rapidly and ensembles or other statistical approaches become the only useful prediction approach.

11.4 Statistical postprocessing

Arguably one of the most overlooked aspects of numerical weather prediction is the postprocessing of the forecast data to remove or reduce obvious and persistent errors. Most of the original operational postprocessing schemes used multivariate linear regression to relate the model forecast variables to observations (Glahn and Lowry 1972; Jacks *et al.* 1990). These techniques have provided improved forecast guidance to human forecasters for many decades. The downside to this type of approach is that it requires a lengthy data archive of both observations and an unchanged model, making it difficult to use this type of approach when models are changing frequently. Modifications to this approach to allow for updates have been developed

(Ross 1989; Wilson and Vallée 2002). Other approaches are also possible, such as using a Kalman filter (Homleid 1995; Roeger *et al.* 2003) and other regression techniques (Mao *et al.* 1999; Hart *et al.* 2004).

With the advent of ensembles, other approaches to postprocessing have been developed. Krishnamurti *et al.* (2000) show how a simple bias correction approach can improve the precipitation forecasts in a global ensemble. Stensrud and Yussouf (2003, 2005) and Woodcock and Engel (2005) show that a bias correction approach when applied to near-surface variables yields results that improve upon model output statistics (MOS) and also provide reliable probability forecasts for the short-range predictions of sensible weather (Fig. 11.5). A different approach, called reforecasting, produces an ensemble of retrospective reforecasts from a fixed model over a long time period (15–25 years) in order to diagnose the operational model bias and to provide improved precipitation forecasts using a regression approach (Hamill

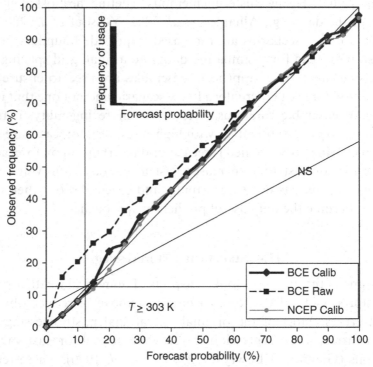

Figure 11.5. Attribute diagram for 2 m temperatures equal to or exceeding 303 K created from a 31-member ensemble over North America. Dashed line indicates results from the raw ensemble, while solid lines indicate results from two versions of postprocessed ensemble data with the bias removed from each ensemble member. Note the improved reliability in the postprocessed ensemble data. From Stensrud and Yussouf (2005).

et al. 2004, 2006). Although producing these reforecasts takes months of computer time, the process is not difficult and leads to significantly improved operational forecasts (Hamill *et al.* 2006). These types of efforts maximize the benefits of the numerical prediction systems and often lead to forecast improvements that are equivalent to decades worth of model improvement alone.

11.5 The road forward

A recurring theme in this final chapter is that there is a lot of uncertainty in how to best use our numerical weather prediction models and their associated computational resources. The meteorology community has spent over 50 years learning how to develop deterministic numerical weather prediction models. The successes arising from this investment are clear and evident every day. Numerical models handle a wide variety of weather scenarios on a vast range of spatial and temporal scales very well and influence operational and investment decisions in a large number of industries. Numerical models and parameterizations will continue to improve as new observations, theories, and ideas are converted into algorithms and studied. Without doubt, numerical weather prediction is an incredibly active and exciting field that will continue to yield forecast improvements for years to come, and one key component of these forecast improvements will continue to be the physical process parameterization schemes.

The improvement of many parameterization schemes often requires collaborations with scientists outside of traditional meteorology and atmospheric science. Not only are meteorologists working with computer scientists, but also with biologists, plant physiologists, remote sensing scientists, foresters, engineers, hydrologists, statisticians, ecologists, economists, and oceanographers. It is clear that the atmospheric sciences are already a multidisciplinary effort and this trend is only going to continue and probably accelerate. We must continue to learn the languages of other sciences in order to learn from their expertise and continue to improve model forecasts.

Beyond the models themselves, we also need to be concerned with how best to use and support our computational and human resources. Some scientists strive to improve finite-differencing or parameterization schemes, while others are working in data assimilation or basic research. Few live in the transition zone between research and operations, and fewer still have the time to step back and look more broadly at how we use these tools we build. For example, not many meteorologists examine how these numerical tools are used by human forecasters to provide guidance products to the public, or to produce forecasts tailored to specific end user communities. Many studies show that human forecasters routinely improve upon numerical guidance in a variety of ways (Olson *et al.* 1995; Leftwich *et al.* 1998; McCarthy *et al.* 1998; Reynolds 2003), yet it is not clear that we are providing the forecast data to forecasters in ways that would allow for the best use of this information or that we are even providing the correct information for all forecast concerns. It also may be that the output from weather and climate models can provide sufficient information to improve human and environmental conditions in ways we never thought possible. An increased emphasis on collaboration between model developers and a broad spectrum of the model user community could be very beneficial.

It is particularly encouraging that seasonal numerical weather prediction systems are starting to be linked directly to models that predict specific human impacts such as malaria incidence (Morse *et al.* 2005) and crop yield (Cantelaube and Terres 2005; Challinor *et al.* 2005; Marletto *et al.* 2005). These types of linkages could have a substantial impact on planning activities that lead to disease prevention and crop selection. Connections between the predictions from global climate models and ecosystem responses are also being explored (Higgins and Vellinga 2004). In short-range forecasting, numerical weather prediction models are being linked to emission and chemistry models to predict air quality (Otte *et al.* 2005), and are used to alert the public to poor air quality conditions that affect human and ecosystem health. These types of activities need to continue and to increase over time to link weather and climate predictions to other quality of life and quality of environment concerns, and may end up profoundly changing the way in which weather and climate model predictions are valued and used by the public.

A need also exists to determine more accurately the economic value of forecasts (e.g., Morss *et al.* 2005) in order to strengthen support for these activities within government budgeting agencies often looking to cut programs (Doswell and Brooks 1998). While there are a number of case studies illustrating the value of weather forecasts (e.g., Katz and Murphy 1997), there has yet to be a sector-wide evaluation of the value of forecasts. This information will be difficult to obtain, but the need for this activity increases from year to year.

Perhaps we have come to the point where the success of numerical weather prediction and the forecast enterprise often is taken for granted by many outside of meteorology. Forecast failures are highlighted, while forecast successes are simply viewed as routine. The success of the numerical weather prediction enterprise breeds increased expectations that model forecasts can provide information on not only the large-scale weather pattern, but also the sensible local weather elements (2 m temperatures, rainfall, precipitation type, and turbulence) that are much more difficult to predict. There also is increased expectation that models can provide information on unlikely events, such as damaging windstorms, tornadoes, tropical cyclones and floods, in order to help with emergency management activities and disaster planning. On longer timescales, the global society is looking to the atmospheric sciences for guidance regarding the best crops to plant for the upcoming growing season and help in understanding how our societies influence global climate change and the outcomes of any mitigation activities. At this point in time, it should not be surprising that there is a need to take the time both to learn how to use these numerical tools most effectively and perhaps even to defend our desires for further forecast improvement. In some ways, parts of numerical weather prediction are a victim of their own success. Continued improvements will occur, but we need to be more careful to illustrate and quantify the value of numerical model improvements to our constituents – the public – and educate them in how to use the output from these tools to their best advantage. If this education occurs, then the future of numerical weather prediction will be brighter than ever.

References

Adamec, D. and R. L. Elsberry (1985). Numerical simulations of the response of intense ocean currents to atmospheric forcing. *J. Phys. Oceanogr.*, **15**, 273–287.

Adlerman, E. J. and K. K. Droegemeier (2002). The sensitivity of numerically simulated cyclic mesocyclogenesis to variations in model physical and computational parameters. *Mon. Wea. Rev.*, **130**, 2671–2691.

Agee, E. and A. Gluhovsky (1999). LES model sensitivities to domain, grids, and large-eddy timescales. *J. Atmos. Sci.*, **56**, 599–604.

Al Nakshabandi, G. and H. Kohnke (1965). Thermal conductivity and diffusivity of soils as related to moisture tension and other physical properties. *Agric. Meteor.*, **2**, 271–279.

Al-Saadi, J., J. Szykman, R. B. Pierce, *et al.* (2005). Improving national air quality forecasts with satellite aerosol observations. *Bull. Amer. Meteor. Soc.*, **86**, 1249–1261.

Albrecht, B. A. (1989). Aerosols, cloud microphysics, and fractional cloudiness. *Science*, **245**, 1227–1230.

Albrecht, B. A., V. Ramanathan, and B. A. Boville (1986). The effects of cumulus moisture transports on the simulation of climate with a general circulation model. *J. Atmos. Sci.*, **43**, 2443–2462.

Alexander, M. J. and J. R. Holton (1997). A model study of zonal forcing in the equatorial stratosphere by convectively induced gravity waves. *J. Atmos. Sci.*, **54**, 408–419.

Alhamed, A., S. Lakshmivarahan, and D. J. Stensrud (2002). Cluster analysis of multi-model ensemble data from SAMEX. *Mon. Wea. Rev.*, **130**, 226–256.

Alpaty, K., J. E. Pleim, S. Raman, D. S. Niyogi, and D. W. Byun (1997). Simulation of atmospheric boundary layer processes using local- and nonlocal-closure schemes. *J. Appl. Meteor.*, **36**, 214–233.

Alpert, J. C., M. Kanamitsu, P. M. Caplan, *et al.* (1988). Mountain induced gravity wave drag parameterization in the NMC medium-range forecast model. In *Eighth Conf. on Numerical Weather Prediction*, Baltimore, MD. American Meteorological Society, pp. 726–733.

Andreas, E. L. (1992). Sea spray and the turbulent air–sea heat fluxes. *J. Geophys. Res.*, **97**, 11 429–11 441.

(1998). A new sea spray generation function for wind speed up to $32\,\mathrm{m\,s}^{-1}$. *J. Phys. Oceanogr.*, **28**, 2175–2184.

Andreas, E. L. and J. DeCosmo (1999). Sea spray production and influence on air–sea heat and moisture fluxes over the open ocean. In *Air–Sea Exchange: Physics, Chemistry, and Dynamics*, ed. G. L. Geernaert. Kluwer Academic Publishers, pp. 327–362.

Andreas, E. L. and K. A. Emanuel (2001). Effects of sea spray on tropical cyclone intensity. *J. Atmos. Sci.*, **58**, 3741–3751.

Andreas, E. L., J. B. Elson, E. C. Monahan, M. P. Rouault, and S. D. Smith (1995). The sea spray contribution to net evaporation from the sea: a review of recent progress. *Bound.-Layer Meteor.*, **72**, 3–52.

Anthes, R. A. (1977). A cumulus parameterization scheme utilizing a one-dimensional cloud model. *Mon. Wea. Rev.*, **105**, 270–286.

(1982). *Tropical Cyclones, Their Evolution, Structure and Effects. Meteorology Monographs*, No. 41. American Meteorological Society.

(1984). Enhancement of convective precipitation by mesoscale variations in vegetative covering in semiarid regions. *J. Clim. Appl. Meteor.*, **23**, 541–554.

Anthes, R. A. and T. T. Warner (1978). Development of hydrodynamic models suitable for air pollution and other mesometeorological studies. *Mon. Wea. Rev.*, **106**, 1045–1078.

Anthes, R. A., E.-Y. Hsie, and Y.-H. Kuo (1987). Description of the Penn State/NCAR mesoscale model version 4 (MM4). NCAR Tech. Note NCAR/TN-282+STR.

Arakawa, A. (1993). Closure assumptions in the cumulus parameterization problem. In *The Representation of Cumulus Convection in Numerical Models of the Atmosphere*, ed. K. A. Emanuel and D. J. Raymond. *Meteorology Monographs*, No. 46. American Meteorological Society, pp. 1–15.

Arakawa, A. and W. H. Schubert (1974). Interaction of a cumulus cloud ensemble with the large-scale environment. Part I. *J. Atmos. Sci.*, **31**, 674–701.

Arpe, K., L. Dümenil, and M. A. Giorgetta (1998). Variability of the Indian monsoon in the CHAM3 model: sensitivity to sea surface temperature, soil moisture, and the stratospheric quasi-biennial oscillation. *J. Climate*, **8**, 1837–1858.

Artaz, M.-A. and J.-C. Andre (1980). Similarity studies of entrainment in convective mixed layers. *Bound.-Layer Meteor.*, **19**, 51–66.

Asai, T. (1965). A numerical study of the air-mass transformation over the Japan Sea in winter. *J. Meteor. Soc. Japan*, **43**, 1–15.

Atger, F. (1999). The skill of ensemble prediction systems. *Mon. Wea. Rev.*, **127**, 1941–1953.

Atkinson, B. W. (1981). *Meso-scale Atmospheric Circulations*. Academic Press.

Atwater, M. A. and J. T. Ball (1981). A surface solar radiation model for cloudy atmospheres. *Mon. Wea. Rev.*, **109**, 878–888.

Avissar, R. and R. A. Pielke (1989). A parameterization of heterogeneous land surfaces for atmospheric numerical models and its impact on regional meteorology. *Mon. Wea. Rev.*, **117**, 2113–2136.

Avissar, R., P. Avissar, Y. Mahrer, and B. Bravdo (1985). A model to simulate response of plant stomata to environmental conditions. *Agric. For. Meteor.*, **34**, 21–29.

Ayotte, K. W., P. P. Sullivan, A. Andren, *et al.* (1996). An evaluation of neutral and convective planetary boundary-layer parameterizations relative to large eddy simulations. *Bound.-Layer Meteor.*, **79**, 131–175.

Bailey, M. and J. Hallett (2004). Growth rates and habits of ice crystals between $-20°$ and $-70°C$. *J. Atmos. Sci.*, **61**, 514–544.

Baldocchi, D. and K. S. Rao (1995). Intra-field variability of scalar flux densities across a transition between a desert and an irrigated potato field. *Bound.-Layer Meteor.*, **76**, 109–121.

Baldwin, M. E., J. S. Kain, and M. P. Kay (2002). Properties of the convection scheme in NCEP's Eta Model that affect forecast sounding interpretation. *Wea. Forecasting*, **17**, 1063–1079.

Ball, J., I. Woodrow, and J. Berry (1987). A model predicting stomatal conductance and its contribution to the control of photosynthesis under different environmental conditions. In *Progress in Photosynthesis Research*, Vol. IV. Martinus Nijhoff, pp. 221–224.

Ballard, S. P., B. Golding, and R. N. B. Smith (1991). Mesoscale model experimental forecasts of the haar of northeast Scotland. *Mon. Wea. Rev.*, **119**, 2107–2123.

Bane, J. M. and K. E. Osgood (1989). Wintertime air–sea interaction processes across the Gulf Stream. *J. Geophys. Res.*, **94**, 10 755–10 772.

Bao, J.-W., J. M. Wilczak, J.-K. Choi, and L. H. Kantha (2000). Numerical simulations of air–sea interaction under high wind conditions using a coupled model: a study of hurricane development. *Mon. Wea. Rev.*, **128**, 2109–2210.

Bao, J.-W., S. A. Michelson, P. J. Neiman, F. M. Ralph, and J. M. Wilczak (2006). Interpretation of enhanced integrated water vapor bands associated with extratropical cyclones: their formation and connection to tropical moisture. *Mon. Wea. Rev.*, **134**, 1063–1080.

Barker, D. M., W. Huang, Y.-R. Guo, A. J. Bourgeois and Q. N. Xiao (2004). A three-dimensional variational data assimilation system for MM5: implementation and initial results. *Mon. Wea. Rev.*, **132**, 897–914.

Barker, H. W. and J. A. Davies (1992). Solar radiative fluxes for stochastic, scaling invariant broken cloud field. *J. Atmos. Sci.*, **49**, 1115–1126.

Barker, H. W., G. L. Stephens, P. T. Partain, *et al.* (2003). Assessing 1D atmospheric solar radiative transfer models: interpretation and handling of unresolved clouds. *J. Climate*, **16**, 2676–2699.

Bartels, D. L. and R. A. Maddox (1991). Midlevel cyclonic vortices generated by mesoscale convective systems. *Mon. Wea. Rev.*, **119**, 104–118.

Barton, I. J. (1979). A parameterization of the evaporation from nonsaturated surface. *J. Appl. Meteor.*, **18**, 43–47.

Basara, J. B. (2000). The value of point-scale measurements of soil moisture in planetary boundary layer simulations. Ph.D. Dissertation, University of Oklahoma.

Basara, J. B. and K. C. Crawford (2002). Linear relationships between root-zone soil moisture and atmospheric processes in the planetary boundary layer. *J. Geophys. Res.*, **107**, 10.1029/2001JD000633.

Bazzaz, F. A. and E. D. Fajer (1992). Plant life in a CO_2-rich world. *Sci. Amer.*, **266**(1), 68–74.

Beard, K. V. and H. R. Pruppacher (1971). A wind tunnel investigation of the rate of evaporation of small water drops falling at terminal velocity in the air. *J. Atmos. Sci.*, **28**, 1455–1464.

Bechtold, P., J. P. Pinty, and P. Mascart (1993). The use of partial cloudiness in a warm-rain parameterization: a subgrid-scale precipitation scheme. *Mon. Wea. Rev.*, **121**, 3301–3311.

Beljaars, A. C. M. and A. A. M. Holtslag (1991). Flux parameterization over land surfaces for atmospheric models. *J. Appl. Meteor.*, **30**, 327–341.

Beljaars, A. C. M., P. Viterbo, M. J. Miller, and A. K. Betts (1996). Anomalous rainfall over the United States during July 1993: sensitivity to land surface parameterization and soil moisture anomalies. *Mon. Wea. Rev.*, **124**, 362–383.

Benjamin, S. G. (1983). Some effects of surface heating and topography on the regional severe storm environment. Ph.D. Thesis, Department of Meteorology, The Pennsylvania State University, University Park, PA.

Benjamin, S. G. and T. N. Carlson (1986). Some effects of surface heating and topography on the regional severe storm environment. Part I: Three-dimensional simulation. *Mon. Wea. Rev.*, **114**, 307–329.

Berg, L. K. and R. B. Stull (2005). A simple parameteriztion coupling the convective daytime boundary layer and fair-weather cumuli. *J. Atmos. Sci.*, **62**, 1976–1988.

Berry, E. X. (1965). Cloud droplet growth by condensation. Ph.D. Thesis, University of Nevada.

(1967). Cloud droplet growth by collection. *J. Atmos. Sci.*, **24**, 688–701.

(1968). Modification of the warm rain process. In *Proc. Ist National Conf. on Weather Modification*, Albany, NY. American Meteorological Society, pp. 81–88.

Berry, E. X. and R. L. Reinhardt (1974a). An analysis of cloud drop growth by collection. Part I: Double distributions. *J. Atmos. Sci.*, **31**, 1814–1824.

(1974b). An analysis of cloud drop growth by collection. Part II: Single initial distributions. *J. Atmos. Sci.*, **31**, 2127–2135.

Betts, A. K. (1973). Non-precipitating cumulus convection and its parameterization. *Quart. J. Roy. Meteor. Soc.*, **99**, 178–196.

(1983). Thermodynamics of mixed stratocumulus layers: saturation point budgets. *J. Atmos. Sci.*, **40**, 2655–2670.

(1985). Mixing line analysis of clouds and cloudy boundary layers. *J. Atmos Sci.*, **42**, 2751–2763.

(1986). A new convective adjustment scheme. Part I: Observational and theoretical basis. *Quart. J. Roy. Meteor. Soc.*, **112**, 677–691.

(1992). FIFE atmospheric boundary layer budget methods. *J. Geophys. Res.*, **97**(D17), 18 523–18 531.

Betts, A. K. and M. J. Miller (1986). A new convective adjustment scheme. Part II: Single column tests using GATE wave, BOMEX, ATEX and arctic air-mass data sets. *Quart. J. Roy. Meteor. Soc.*, **112**, 693–709.

(1993). The Betts–Miller scheme. In *The Representation of Cumulus Convection in Numerical Models of the Atmosphere*, ed. K. A. Emanuel and D. J. Raymond. *Meteorology Monographs*, No. 46. American Meteorological Society, pp. 107–121.

Betts, A. K. and J. H. Ball (1997). Albedo over the boreal forest. *J. Geophys. Res.*, **102**, 28 901–28 909.

Betts, A. K., R. L. Desjardins, J. I. MacPherson, and R. D. Kelly (1990). Boundary layer heat and moisture budgets from FIFE. *Bound.-Layer Meteor.*, **50**, 109–137.

Betts, A. K., R. L. Desjardins, and J. I. MacPherson (1992). Budget analysis of the boundary layer grid flights during FIFE 1987. *J. Geophys. Res.*, **97**(D17), 18 533–18 546.

Betts, A. K., F. Chen, K. Mitchell, and Z. I. Janjic (1997). Assessment of the land surface and boundary layer models in two operational versions of the NCEP Eta Model using FIFE data. *Mon. Wea. Rev.*, **125**, 2896–2916.

Bhumralkar, C. M. (1975). Numerical experiments on the computation of ground surface temperature in an atmospheric general circulation model. *J. Appl. Meteor.*, **14**, 1246–1258.

Bitz, C. M. and W. H. Lipscomb (1999). An energy-conserving thermodynamic model of sea ice. *J. Geophys. Res.*, **104**, 15 669–15 677.

Black, T. L. (1994). The new NMC mesoscale Eta Model: description and forecast examples. *Wea. Forecasting*, **9**, 265–284.

Blackadar, A. K. (1957). Boundary layer wind maxima and their significance for the growth of nocturnal inversions. *Bull. Amer. Meteor. Soc.*, **38**, 283–290.

(1976). Modeling the nocturnal boundary layer. In *Proceedings 3rd Symposium on Atmospheric Turbulence, Diffusion and Air Quality*. American Meteorological Society, pp. 46–49.

(1978). Modeling pollutant transfer during daytime convection. Preprints. In *Fourth Symp. on Atmos. Turbulence, Diffusion and Air Quality*, Reno, NV. American Meteorological Society, pp. 443–447.

(1979). High-resolution models of the planetary boundary layer. In *Advances in Environmental Science and Engineering*, vol. 1, ed. J. R. Pfafflin and E. N. Ziegler. Gordon and Breach Science Publishers, pp. 50–85.

Blyth, E. M. and A. J. Dolman (1995). The roughness length for heat of sparse vegetation. *J. Appl. Meteor.*, **54**, 583–585.

Boer, G. J., N. A. McFarlane, R. Laprise, J. D. Henderson, and J.-P. Blanchet (1984). The Canadian Climate Centre spectral atmospheric general circulation model. *Atmos.-Ocean*, **22**, 397–429.

Bonan, G. B., S. Levis, S. Sitch, M. Vertenstein, and K. W. Oleson (2003). A dynamic global vegetation model for use with climate models: concepts and description of simulated vegetation. *Global Change Biol.*, **9**, 1543–1566.

Bony, S. and K. A. Emanuel (2001). A parameterization of the cloudiness associated with cumulus convection; evaluation using TOGA COARE data. *J. Atmos. Sci.*, **58**, 3158–3183.

Bosart, L. F. (1999). Observed cyclone life cycles. In *The Life Cycles of Extratropical Cyclones*, ed. M. A. Shapiro and S. Grønås. American Meteorological Society, pp. 187–213.

(2003). Whither the weather analysis and forecasting process? *Wea. Forecasting*, **18**, 520–529.

Bosseut, C., M. Déqué, and D. Cariolle (1998). Impact of a simple parameterization of convective gravity-wave drag in a stratosphere–troposphere general circulation model and its sensitivity to vertical resolution. *Ann. Geophysicae*, **16**, 238–239.

Boucher, O., S. E. Schwartz, T. P. Ackerman, *et al.* (1998). Intercomparison of models representing direct shortwave radiative forcing by sulfate aerosols. *J. Geophys. Res.*, **103**, 16 979–16 998.

Bourassa, M. A., D. G. Vincent, and W. L. Wood (1999). A flux parameterization including the effects of capillary waves and sea state. *J. Atmos. Sci.*, **56**, 1123–1139.

(2001). A sea state parameterization with nonarbitrary wave age applicable to low and moderate wind speeds. *J. Phys. Oceanogr.*, **31**, 2840–2851.

Bowdle, D. A., P. V. Hobbs, and L. F. Radke (1985). Particles in the lower troposphere over the High Plains of the United States. Part III: Ice nuclei. *J. Clim. Appl. Meteor.*, **24**, 1370–1376.

Bowen, I. S. (1926). The ratio of heat losses by conduction and by evaporation from any water surface. *Phys. Rev.*, **27**, 779–787.

Bresch, J. F., R. J. Reed, and M. D. Albright (1997). A polar-low development over the Bering Sea: analysis, numerical simulation, and sensitivity experiments. *Mon. Wea. Rev.*, **125**, 3109–3130.

Bretherton, C. S., J. R. McCaa, and H. Grenier (2004). A new parameterization for shallow cumulus convection and its application to marine subtropical cloud-topped boundary layers. Part I: Description and 1D results. *Mon. Wea. Rev.*, **132**, 864–882.

Bretherton, F. P. (1969). Momentum transport by gravity waves. *Quart. J. Roy. Meteor. Soc.*, **95**, 213–243.

Briegleb, B. P. (1992). Delta-Eddington approximation for solar radiation in the NCAR Community Climate Model. *J. Geophys. Res.*, **97**, 7603–7612.

Bright, D. R. and S. L. Mullen (2002). The sensitivity of the numerical simulation of the southwest monsoon boundary layer to the choice of PBL turbulence parameterization in MM5. *Wea. Forecasting*, **17**, 99–114.

Brooks, D. A. (1983). The wake of Hurricane Allen in the western Gulf of Mexico. *J. Phys. Oceanogr.*, **13**, 117–129.

Brooks, H. E., C. A. Doswell III, and R. A. Maddox (1992). On the use of mesoscale and cloud-scale models in operational forecasting. *Wea. Forecasting*, **7**, 120–132.

Brooks, H. E., C. A. Doswell III, and J. Cooper (1994b). On the environments of tornadic and nontornadic mesocyclones. *Wea. Forecasting*, **9**, 606–618.

Brooks, H. E., C. A. Doswell III, and R. B. Wilhelmson (1994a). On the role of midtropospheric winds in the evolution and maintenance of low-level mesocyclones. *Mon. Wea. Rev.*, **122**, 126–136.

Brown, R. A. (1980). Longitudinal instabilities and secondary flows in the planetary boundary layer: a review. *Rev. Geophys. Space Phys.*, **18**, 683–697.

Brutsaert, W. (1982). *Evaporation into the Atmosphere. Theory, History, and Applications*. Reidel.

Bryan, G. H., J. C. Wyngaard, and J. M. Fritsch (2003). Resolution requirements for the simulation of deep moist convection. *Mon. Wea. Rev.*, **131**, 2394–2416.

Buermann, W., J. Dong, X. Zeng, R. B. Myneni, and R. E. Dickinson (2001). Evaluation of the utility of satellite-based vegetation leaf area index data for climate simulations. *J. Climate*, **14**, 3536–3550.

Buirez, J.-C., B. Bonnel, Y. Fouquart, J.-F. Geleyn, and J.-J. Morcrette (1988). Comparison of model-generated and satellite-derived cloud cover and radiation budget. *J. Geophys. Res.*, **93**, 3705–3719.

Buizza, R. (1997). Potential forecast skill of ensemble prediction and spread and skill distributions of the ECMWF ensemble prediction system. *Mon. Wea. Rev.*, **125**, 99–119.

Buizza, R. and T. N. Palmer (1995). The singular-vector structure of the atmospheric general circulation. *J. Atmos. Sci.*, **52**, 1434–1456.

Buizza, R., M. Miller, and T. N. Palmer (1999). Stochastic representation of model uncertainties in the ECMWF ensemble prediction system. *Quart. J. Roy. Meteor. Soc.*, **125**, 2887–2908.

Burnash, R. J., R. L. Ferral, and R. A. McGuire (1973). A generalized streamflow simulation system: conceptual modeling for digital computers. Technical Report, Joint Fed. and State River Forecast Center, Sacramento, CA.

Bush, B. C., F. P. J. Valero, A. S. Simpson, and L. Bignone (2000). Characterization of thermal effects in pyranometers: a data correction algorithm for improved measurement of surface insolation. *J. Atmos. Oceanic Technol.*, **17**, 165–175.

Businger, S., W. H. Bauman III, and G. F. Watson (1991). The development of the Piedment front and associated outbreak of severe weather on 13 March 1986. *Mon. Wea. Rev.*, **119**, 2224–2251.

Businger, J. A., J. C. Wyngaard, Y. Izumi, and E. F. Bradley (1971). Flux profile relationships in the atmospheric surface layer. *J. Atmos. Sci.*, **28**, 181–189.

Byers, H. R. (1965). *Elements of Cloud Physics*. The University of Chicago Press.

Cahalan, R. F. and J. H. Joseph (1989). Fractal statistics of cloud fields. *Mon. Wea. Rev.*, **117**, 261–272.

Cahalan, R. F., L. Oreopoulos, A. Marshak, *et al.* (2005). The I3RC: bringing together the most advanced radiative transfer tools for cloudy atmospheres. *Bull. Amer. Meteor. Soc.*, **86**, 1275–1293.

Camacho-B, S. E., A. E. Hall, and M. R. Kaufmann (1974). Efficiency and regulation of water transport in some woody and herbaceous species. *Plant Physiol.*, **54**, 169–172.

Campbell, G. S. (1974). A simple method for determining unsaturated conductivity from moisture retention data. *Water Resources Res.*, **12**, 1118–1124.

Canny, M. J. (1998). Transporting water in plants. *Amer. Scientist*, **86**, 152–160.

Cantelaube, P. and J.-M. Terres (2005). Seasonal weather forecasts for crop yield modelling in Europe. *Tellus*, **57A**, 476–487.

Capehart, W. J. and T. N. Carlson (1994). Estimating near-surface moisture availability using a meteorologically driven soil–water profile model. *J. Hydrol.*, **160**, 1–20.

  (1997). Decoupling of surface and near-surface soil water content: a remote sensing perspective. *Water Resources Res.*, **33**, 1383–1395.

Carlson, T. N. and F. E. Boland (1978). Analysis of urban–rural canopy using a surface heat flux/temperature model. *J. Appl. Meteor.*, **17**, 998–1013.

Carlson, T. N. and D. A. Ripley (1997). On the relation between NDVI, fractional vegetation cover, and leaf area index. *Remote Sens. Environ.*, **62**, 241–252.

Carlson, T. N., S. G. Benjamin, and G. S. Forbes (1983). Elevated mixed layers in the regional severe storm environment: conceptual model and case studies. *Mon. Wea. Rev.*, **111**, 1453–1474.

Carruthers, D. J. and J. C. R. Hunt (1990). Fluid mechanics of airflow over hills: turbulence, fluxes, and waves in the boundary layer. In *Atmospheric Processes over Complex Terrain*, ed. W. Blumen. *Meteorology Monographs*, No. 45. American Meteorological Society, pp. 83–107.

Carson, D. J. (1973). The development of a dry inversion-capped convectively unstable boundary layer. *Quart. J. Roy. Meteor. Soc.*, **99**, 450–467.

Carson, R. B. (1950). The Gulf Stream front: a cause of stratus on the lower Atlantic coast. *Mon. Wea. Rev.*, **78**, 91–101.

Cary, J. W. and H. F. Maryland (1972). Salt and water movement in unsaturated frozen soil. *Soil Sci. Soc. Amer. Proc.*, **36**, 549–555.

Cavalieri, D. J., *et al.* (1991). Aircraft active and passive microwave validation of sea-ice concentration from the Defense Meteorological Satellite Program (DMSP) Special Sensor Microwave Imager. *J. G. R. Oceans*, **96**(C12), 21 989–22 008.

Cess, R. D., G. L. Potter, J. P. Blanchet, *et al.* (1990). Intercomparison and interpretation of climate feedback processes in 19 atmospheric general circulation models. *J. Geophys. Res.*, **95**, 16 601–16 615.

Chaboureau, J.-P. and P. Bechtold (2002). A simple cloud parameterization derived from cloud resolving model data: diagnostic and prognostic applications. *J. Atmos. Sci.*, **59**, 2362–2372.

Challinor, A. J., J. M. Slingo, T. R. Wheeler, and F. J. Doblas-Reyes (2005). Probabilistic simulations of crop yield over western Indian using the DEMETER seasonal hindcast ensembles. *Tellus*, **57A**, 498–512.

Chang, F.-L. and J. A. Coakley (1993). Estimating errors in fractional cloud cover obtained with infrared threshold methods. *J. Geophys. Res.*, **98**, 8825–8839.

Chang, J.-T. and P. J. Wetzel (1991). Effects of spatial variations of soil moisture and vegetation to the evolution of a prestorm environment: a numerical case study. *Mon. Wea. Rev.*, **119**, 1368–1390.

Charlock, T. and B. Herman (1976). Discussion of the Elsasser formulation for infrared fluxes. *J. Appl. Meteor.*, **15**, 657–661.

Charney, J. G., R. Fjørtoft, and J. von Neuman (1950). Numerical integration of the barotropic vorticity equation. *Tellus*, **2**, 237–254.

Charnock, H. (1955). Wind stress on a water surface. *Quart. J. Roy. Meteor. Soc.*, **81**, 639–640.

Chase, T. N., R. A. Pielke, Sr., T. G. F. Kittel, *et al.* (2001). Relative climatic effects of landcover change and elevated carbon dioxide combined with aerosols: a comparison of model results and observations. *J. Geophys. Res.*, **106**, 31 685–31 691.

Chelton, D. B. and F. J. Wentz (2005). Global microwave satellite observations of sea surface temperature for numerical weather prediction and climate research. *Bull. Amer. Meteor. Soc.*, **86**, 1097–1115.

Chen, F. and J. Dudhia (2001). Coupling an advanced land surface-hydrology model with the Penn State-NCAR MM5 modeling system. Part I: Model implementation and sensitivity. *Mon. Wea. Rev.*, **129**, 569–585.

Chen, F. and K. Mitchell (1999). Using GEWEX/ISLSCP forcing data to simulate global soil moisture fields and hydrological cycle for 1987–1988. *J. Meteor. Soc. Japan*, **77**, 1–16.

Chen, F., K. Mitchell, J. Schaake, *et al.* (1996). Modeling of land surface evaporation by four schemes and comparison with FIFE observations. *J. Geophys. Res.*, **101**, 7251–7268.

Chen, S. and W. R. Cotton (1988). The sensitivity of a simulated extratropical mesoscale convective system to longwave radiation and ice phase microphysics. *J. Atmos. Sci.*, **45**, 3897–3910.

Chen, T. H., A. Henderson-Sellers, P. C. D. Milly, *et al.* (1997). Cabauw experimental results from the Project for Intercomparison of Land-Surface Parameterization Schemes. *J. Climate*, **10**, 1194–1215.

Cheng, C.-P. and R. A. Houze, Jr. (1979). The distribution of convective and mesoscale precipitation in GATE radar echo patterns. *Mon. Wea. Rev.*, **107**, 1370–1381.

Chevallier, F. and J.-J. Morcrette (2000). Comparison of model fluxes with surface and top-of-the-atmosphere observations. *Mon. Wea. Rev.*, **128**, 3839–3852.

Chouinard, C., M. Beland, and N. McFarlane (1986). A simple gravity-wave drag parameterization for use in medium-range weather forecast models. *Atmos.-Ocean*, **24**, 91–110.

Chun, H.-Y. and J.-J. Baik (1998). Momentum flux by thermally induced internal gravity waves and its approximation for large-scale models. *J. Atmos. Sci.*, **55**, 3299–3310.

   (2002). An updated parameterization of convectively forced gravity wave drag for use in large-scale models. *J. Atmos. Sci.*, **59**, 1006–1017.

Chylek, P. and J. S. Dobbie (1995). Radiative properties of finite inhomogeneous cirrus clouds: Monte Carlo simulations. *J. Atmos. Sci.*, **52**, 3512–3522.

Clapp, R. B. and G. M. Hornberger (1978). Empirical equations for some soil hydraulic properties. *Water Resources Res.*, **14**, 601–604.

Clark, T. (1973). Numerical modeling of the dynamics and microphysics of warm cumulus convection. *J. Atmos. Sci.*, **30**, 857–878.

(1974). A study in cloud phase parameterization using the gamma distribution. *J. Atmos. Sci.*, **31**, 142–155.

(1976). Use of log-normal distributions for numerical calculations of condensation and collection. *J. Atmos. Sci.*, **33**, 810–821.

Clark, T. and M. J. Miller (1991). Pressure drag and momentum fluxes due to the Alps. II: Representation in large-scale atmospheric models. *Quart. J. Roy. Meteor. Soc.*, **117**, 527–552.

Coakley, J. A. and P. Chylek (1975). The two stream approximation in radiative transfer: including the angle of incident radiation. *J. Atmos. Sci.*, **32**, 409–418.

Collatz, G. J., J. T. Ball, C. Grivet, and J. A. Berry (1991). Physiological and environmental regulation of stomatal conductance, photosynthesis and transpiration: a model that includes a laminar boundary layer. *Agric. For. Meteor.*, **54**, 107–136.

Colle, B. A., C. F. Mass, and K. J. Westrick (2000). MM5 precipitation verification over the Pacific Northwest during the 1997–99 cool seasons. *Wea. Forecasting*, **15**, 730–744.

Colle, B. A., J. B. Olson, and J. S. Tongue (2003). Multiseason verification of the MM5. Part I: Comparison with the Eta model over the central and eastern United States and impact of MM5 resolution. *Wea. Forecasting*, **18**, 431–457.

Colle, B. A., M. F. Garvert, J. B. Wolfe, *et al.* (2005b). The 13–14 December 2001 IMPROVE-2 event. Part III: Simulated microphysical budgets and sensitivity studies. *J. Atmos. Sci.*, **62**, 3535–3558.

Colle, B. A., J. B. Wolfe, W. J. Steenburgh, *et al.* (2005a). High-resolution simulations and microphysical validation of an orographic precipitation event over the Wasatch Mountains during IPEX IOP3. *Mon. Wea. Rev.*, **133**, 2947–2971.

Coniglio, M. C. and D. J. Stensrud (2001). Simulation of a progressive derecho using composite initial conditions. *Mon. Wea. Rev.*, **129**, 1593–1616.

Cooper, W. A. (1986). Ice initiation in natural clouds. In *Precipitation Enhancement – A Scientific Challenge. Meteorology Monographs*, No. 43. American Meteorological Society, pp. 29–32.

Cortinas, J. V., Jr. and D. J. Stensrud (1995). The importance of understanding mesoscale model parameterization schemes for weather forecasting. *Wea. Forecasting*, **10**, 716–740.

Cosby, B. J., G. M. Hornberger, R. B. Clapp, and T. R. Ginn (1984). A statistical exploration of the relationships of soil moisture characteristics to the physical properties of soils. *Water Resources Res.*, **20**, 682–690.

Cotton, W. R. (1972). Numerical simulation of precipitation development in supercooled cumuli – Part I. *Mon. Wea. Rev.*, **100**, 757–763.

Cotton, W. R. and R. A. Anthes (1989). *Storm and Cloud Dynamics*. Academic Press.

Cotton, W. R., G. Thompson, and P. W. Mielke (1994). Realtime mesoscale prediction on workstations. *Bull. Amer. Meteor. Soc.*, **75**, 349–362.

Cox, P. M., C. Huntingford, and R. J. Harding (1998). A canopy conductance and photosynthesis model for use in a GCM land surface scheme. *J. Hydrol.*, **212–213**, 79–94.

Cox, S. K. and K. T. Griffith (1979). Estimates of radiative divergence during Phase III of the GARP Atlantic Tropical Experiment. Part II: Analysis of Phase III results. *J. Atmos. Sci.*, **36**, 586–601.

Crawford, T. M., D. J. Stensrud, T. N. Carlson, and W. J. Capehart (2000). Using a soil hydrology model to obtain regionally averaged soil moisture values. *J. Hydrometeor.*, **1**, 353–363.

Csanady, G. T. (2001). *Air–Sea Interaction.* Cambridge University Press.

Curtis, A. R. (1952). Discussion of "A statistical model for water vapour absorption" by R. M. Goody. *Quart. J. Roy. Meteor. Soc.,* **78**, 638.

Cusack, S., J. M. Edwards, and R. Kershaw (1999). Estimating the subgrid variance of saturation, and its parameterization for use in a GCM cloud scheme. *Quart. J. Roy. Meteor. Soc.,* **125**, 3057–3076.

Dai, Y., R. E. Dickinson, and Y.-P. Wang (2004). A two-big-leaf model for canopy temperature, photosynthesis, and stomatal conductance. *J. Climate,* **17**, 2281–2299.

Dai, Y., X. Zeng, R. E. Dickinson, *et al.* (2003). The common land model. *Bull. Amer. Meteor. Soc.,* **84**, 1013–1023.

Daley, R. (1991). *Atmospheric Data Analyses.* Cambridge University Press.

Daly, E., A. Porporato, and I. Rodriguez-Iturbe (2004). Coupled dynamics of photosynthesis, transpiration, and soil water balance. Part I: Upscaling from hourly to daily level. *J. Hydrometeor.,* **5**, 546–558.

Davies, J. A. and C. D. Allen (1973). Equilibrium, potential and actual evaporation from cropped surfaces in southern Ontario. *J. Appl. Meteor.,* **12**, 649–657.

Davis, C. A., K. W. Manning, R. E. Carbone, S. B. Trier, and J. D. Tuttle (2003). Coherence of warm-season continental rainfall in numerical weather prediction models. *Mon. Wea. Rev.,* **131**, 2667–2679.

Davis, M. H. and J. D. Sartor (1967). Theoretical collision efficiencies for small cloud droplets in Stokes flow. *Nature,* **215**, 1371–1372.

Deardorff, J. W. (1966). The counter-gradient heat flux in the lower atmosphere and in the laboratory. *J. Atmos. Sci.,* **23**, 503–506.

(1972). Theoretical expression for the countergradient vertical heat flux. *J. Geophys. Res.,* **77**, 5900–5904.

(1978). Efficient prediction of ground surface temperature and moisture, with inclusion of a layer of vegetation. *J. Geophys. Res.,* **83**, 1889–1903.

(1979). Prediction of convective mixed-layer entrainment for realistic capping inversion structure. *J. Atmos. Sci.,* **36**, 424–436.

Deardorff, J. W., G. E. Willis, and D. K. Lilly (1969). Laboratory investigation of non-steady penetrative convection. *J. Fluid. Mech.,* **35**, 7–31.

DeFries, R. S. and J. R. G. Townshend (1994). NDVI derived land cover classifications at a global scale. *Int. J. Remote Sens.,* **5**, 3567–3586.

Del Genio, A. D., M. S. Yao, W. Kovari, and K. W. W. Lo (1996). A prognostic cloud water parameterization for global climate models. *J. Climate,* **9**, 270–304.

DeMott, P. J., M. P. Meyers, and W. R. Cotton (1994). Parameterization and impact of ice initiation processes relevant to numerical model simulations of cirrus clouds. *J. Atmos. Sci.,* **51**, 77–90.

Deng, A. and D. R. Stauffer (2006). On improving 4-km mesoscale model simulations. *J. Appl. Meteor. Climatology,* **45**, 361–381.

Deng, A., N. L. Seaman, and J. S. Kain (2003). A shallow-convection parameterization for mesoscale models. Part I: Submodel description and preliminary applications. *J. Atmos. Sci.,* **60**, 34–56.

Dennis, A. S., P. L. Smith, Jr., G. A. P. Peterson, and R. D. McNeil (1971). Hailstone size distributions and equivalent radar reflectivity factors computed from hailstone momentum records. *J. Appl. Meteor.,* **10**, 79–85.

Derber, J. C. (1989). A variational continuous assimilation technique. *Mon. Wea. Rev.,* **117**, 2437–2446.

Desborough, C. E. (1997). The impact of root weighting on the response of transpiration to moisture stress in land surface schemes. *Mon. Wea. Rev.*, **125**, 1920–1930.

Dickinson, R. E. (1983). Land surface processes and climate-surface albedos and energy balance. In *Advances in Geophysics*, vol. **25**. Academic Press, pp. 305–353.

(1984). Modeling evapotranspiration for three-dimensional global climate models. In *Climate Processes and Climate Sensitivity*, ed. J. E. Hansen and T. Takahashi. American Geophysical Union, pp. 58–72.

Ding, P. and D. A. Randall (1998). A cumulus parameterization with multiple cloud base levels. *J. Geophys. Res.*, **103**, 11 341–11 354.

Donner, L. J. (1993). A cumulus parameterization including mass fluxes, vertical momentum dynamics, and mesoscale effects. *J. Atmos. Sci.*, **50**, 889–906.

Dopplick, T. G. (1972). Radiative heating of the global atmosphere. *J. Atmos. Sci.*, **29**, 1278–1294.

Doswell, C. A. III and H. E. Brooks (1998). Budget cutting and the value of weather services. *Wea. Forecasting*, **13**, 206–212.

Douglas, M. W. (1993). Current research into the monsoon. *Sonorensis, Arizona Sonora Desert Museum Newsletter*, **13**(3), 10–11.

Douville, H. (2003). Assessing the influence of soil moisture on seasonal climate variability with AGCMs. *J. Hydrometeor.*, **4**, 1044–1066.

Dowell, D. C., F. Zhang, L. J. Wicker, C. Snyder, and N. A. Crook (2004). Wind and temperature retrievals in the 17 May 1981 Arcadia, Oklahoma supercell: ensemble Kalman filter experiments. *Mon. Wea. Rev.*, **132**, 1982–2005.

Doyle, J. D. (1995). Coupled ocean wave/atmosphere mesoscale model simulations of cyclogenesis. *Tellus*, **47A**, 766–788.

(2002). Coupled atmosphere–ocean wave simulations under high wind conditions. *Mon. Wea. Rev.*, **130**, 3087–3099.

Doyle, J. D. and T. T. Warner (1990). Mesoscale processes during GALE IOP 2. *Mon. Wea. Rev.*, **118**, 283–308.

Doyle, J. D., D. R. Durran, C. Chen, *et al.* (2000). An intercomparison of model-predicted wave breaking for the 11 January 1972 Boulder windstorm. *Mon. Wea. Rev.*, **128**, 901–914.

Drake, V. A. (1985). Radar observations of moths migrating in a nocturnal low-level jet. *Ecol. Entomol.*, **10**, 259–265.

Dubosclard, G. (1980). A comparison between observed and predicted values for the entrainment coefficient in the planetary boundary layer. *Bound.-Layer Meteor.*, **18**, 473–483.

Dubovik, O. and M. D. King (2000). A flexible inversion algorithm for retrieval of aerosol optical properties from Sun and sky radiance measurements. *J. Geophys. Res.*, **105**, 20 673–20 696.

Dudhia, J. (1989). Numerical study of convection observed during the winter monsoon experiment using a mesoscale two-dimensional model. *J. Atmos. Sci.*, **46**, 3077–3107.

Durran, D. R. (1986). Mountain waves. In *Mesoscale Meteorology and Forecasting*, ed. P. Ray. American Meteorological Society, pp. 472–492.

(1990). Mountain waves and downslope winds. In *Atmospheric Processes over Complex Terrain*, ed. W. Blumen. *Meteorology Monographs*, No. 45. American Meteorological Society, pp. 59–81.

(1999). *Numerical Methods for Wave Equations in Geophysical Fluid Dynamics*. Springer-Verlag.

Durran, D. R. and J. B. Klemp (1982). The effects of moisture on trapped mountain lee waves. *J. Atmos. Sci.*, **39**, 2490–2506.

Dusek, U., G. P. Frank, L. Hildebrandt, *et al.* (2006). Size matters more than chemistry for cloud-nucleating ability of aerosol particles. *Science*, **312**, 1375–1378.

Dutton, E. G., J. J. Michalsky, T. Stoffel, *et al.* (2001). Measurements of broadband diffuse solar irradiance using current commercial instrumentation with a correction for thermal offset errors. *J. Atmos. Oceanic Technol.*, **18**, 297–314.

Dutton, J. A. (1976). *The Ceaseless Wind: an Introduction to the Theory of Atmospheric Motion*. McGraw-Hill.

Dutton, J. A. and G. H. Fichtl (1969). Approximate equations of motion for gases and liquids. *J. Atmos. Sci.*, **26**, 241–254.

Dyer, A. J. (1963). The adjustment of profiles and eddy fluxes. *Quart. J. Roy. Meteor. Soc.*, **89**, 276–280.

(1974). A review of flux-profile relations. *Bound.-Layer Meteor.*, **1**, 363–372.

Ebert, E. E. and J. A. Curry (1992). A parameterization of ice cloud optical properties for climate models, *J. Geophys. Res.*, **97**, 3831–3836.

(1993). An intermediate one-dimensional thermodynamic sea ice model for investigating ice–atmosphere interactions. *J. Geophys. Res.*, **98**, 10 085–10 109.

Ebert, E. E., U. Schumann, and R. B. Stull (1989). Nonlocal turbulent mixing in the convective boundary layer evaluated from large-eddy simulation. *J. Atmos. Sci.*, **46**, 2178–2207.

Edwards, J. M. and A. Slingo (1996). Studies with a flexible new radiation code. I: Choosing a configuration for a large-scale model. *Quart. J. Roy. Meteor. Soc.*, **122**, 689–719.

Ek, M. B., K. E. Mitchell, Y. Lin, *et al.* (2003). Implementation of Noah land surface model advances in the National Centers for Environmental Prediction operational mesoscale Eta model. *J. Geophys. Res.*, **108**, 8851, doi: 10.1029/2002JD003296.

Eliassen, A. and E. Palm (1961). On the transfer of energy in stationary mountain waves. *Geofys. Publ.*, **22**, 1–23.

Ellingston, R. G., J. Ellis, and S. Fels (1991). Intercomparison of radiation codes used in climate models: long wave results. *J. Geophys. Res.*, **96**, 8929–8953.

Elsasser, W. M. and M. F. Culbertson (1960). *Atmospheric Radiation Tables*. *Meteorology Monographs*, No. 4. American Meteorological Society.

Emanuel, K. A. (1986). An air–sea interaction theory for tropical cyclones. Part I: Steady-state maintenance. *J. Atmos. Sci.*, **43**, 585–605.

(1991) A scheme for representing cumulus convection in large-scale models. *J. Atmos. Sci.*, **48**, 2313–2335.

(1994) *Atmospheric Convection*. Oxford University Press.

Emanuel, K. A. and D. J. Raymond (1993). *The Representation of Cumulus Convection in Numerical Models*, ed. K. A. Emanuel and D. J. Raymond. *Meteorology Monographs*, No. 46. American Meteorological Society.

Emanuel, K. A. and M. Zivkovic-Rothman (1999). Development and evaluation of a convection scheme for use in climate models. *J. Atmos. Sci.*, **56**, 1766–1782.

Epstein, E. S. (1969). Stochastic dynamic prediction. *Tellus*, **21**, 739–759.

Errico, R. and D. P. Baumhefner (1987). Predictability experiments using a high-resolution limited-area model. *Mon. Wea. Rev.*, **115**, 488–504.

Estoque, M. A. (1968). Vertical mixing due to penetrative convection. *J. Atmos. Sci.*, **25**, 1046–1051.

Eumas, D. and P. G. Jarvis (1989). The direct effects of increase in the global atmospheric CO_2 concentration on natural and commercial temperate trees and forests. *Adv. Ecol. Res.*, **19**, 1–55.

Evans, K. F. (1998). The spherical harmonics discrete ordinate method for three-dimensional atmospheric radiative transfer. *J. Atmos. Sci.*, **55**, 429–446.

Evans, R. E., M. S. J. Harrison, R. J. Graham, and K. R. Mylne (2000). Joint medium-range ensembles from the Met. Office and ECMWF systems. *Mon. Wea. Rev.*, **128**, 3104–3127.

Evensen, G. (1994). Sequential data assimilation with a nonlinear quasi-geostrophic model using Monte Carlo methods to forecast error statistics. *J. Geophys. Res.*, **99**, 10 143–10 162.

  (1997). Advanced data assimilation for strongly nonlinear dynamics. *Mon. Wea. Rev.*, **125**, 1342–1354.

Fairall, C. W., J. D. Kepert, and G. J. Holland (1994). The effect of sea spray on surface energy transports over the ocean. *Global Atmos. Ocean Syst.*, **2**, 121–142.

Fairall, C. W., E. F. Bradley, J. E. Hare, A. A. Grachev, and J. B. Edson (2003). Bulk parameterization of air–sea fluxes: updates and verification for the COARE algorithm. *J. Climate*, **16**, 571–591.

Fawbush, E. J. and R. C. Miller (1954). The types of air masses in which North American tornadoes form. *Bull. Amer. Meteor. Soc.*, **4**, 154–165.

Feddema, J. J., K. W. Oleson, G. B. Bonan, *et al.* (2005a). The importance of land-cover change in simulating future climates. *Science*, **310**, 1674–1678.

Feddema, J. J., K. W. Oleson, G. B. Bonan, *et al.* (2005b). A comparison of a GCM response to historical anthropogenic land cover change and model sensitivity to uncertainty in present-day land cover representations. *Clim. Dyn.*, **25**, 581–609.

Federer, B. and A. Waldvogel (1975). Hail and raindrop size distributions from a Swiss multicell storm. *J. Appl. Meteor.*, **14**, 91–97.

Ferrier, B. S. (1994). A double-moment multiple-phase four-class bulk ice scheme. Part I: Description. *J. Atmos. Sci.*, **51**, 249–280.

Ferrier, B. S., W.-K. Tao, and J. Simpson (1995). A double-moment multiple-phase four-class bulk ice scheme. Part II: Simulations of convective storms in different large-scale environments and comparisons with other bulk parameterizations. *J. Atmos. Sci.*, **52**, 1001–1033.

Fels, S. and M. D. Schwarzkopf (1975). The simplified exchange approximation. A new method for radiative transfer calculations. *J. Atmos. Sci.*, **32**, 1475–1488.

Fisher, E. L. (1958). Hurricane and the sea surface temperature field. *J. Meteor.*, **15**, 328–333.

Flerchinger, G. N. and K. E. Saxton (1989). Simultaneous heat and water model of a freezing snow–residue-soil system. I: Theory and development. *Trans. Amer. Soc. Agric. Eng.*, **32**, 565–571.

Fletcher, N. H. (1962). *The Physics of Rain Clouds*. Cambridge University Press.

Foukal, P. (1994). Study of solar irradiance variations holds key to climate questions. *Eos*, **75**(33), 377–383.

Fouquart, Y., B. Bonnel, and V. Ramaswamy (1991). Intercomparing shortwave radiation codes for climate studies. *J. Geophys. Res.*, **96**, 8955–8968.

Fowler, L. D., D. A. Randall, and S. A. Rutledge (1996). Liquid and ice cloud microphysics in the CSU general circulation model. Part I: Model description and simulated microphysical processes. *J. Climate*, **9**, 489–529.

Frank, W. M. and C. Cohen (1987). Simulation of tropical convective systems. Part I: A cumulus parameterization. *J. Atmos. Sci.*, **44**, 3787–3799.

Frank, W. M. and J. Molinari (1993). Convective adjustment. In *The Representation of Cumulus Convection in Numerical Models of the Atmosphere*, ed. K. A. Emanuel and D. J. Raymond. *Meteorology Monographs*, No. 46. American Meteorological Society, pp. 101–106.

Frey, H., M. Latif, and T. Stockdale (1997). The coupled GCM ECHO-2. Part I: The tropical Pacific. *Mon. Wea. Rev.*, **125**, 703–720.

Friedl, M. A., D. K. McIver, J. C. F. Hodges, *et al.* (2002). Global land cover mapping from MODIS: algorithms and early results. *Remote Sens. Environ.*, **83**, 287–302.

Fritsch, J. M. and C. F. Chappell (1980). Numerical prediction of convectively driven mesoscale pressure systems. Part I: Convective parameterization. *J. Atmos. Sci.*, **37**, 1722–1733.

Fritsch, J. M. and R. A. Maddox (1981). Convectively driven mesoscale pressure systems aloft. Part I: Observations. *J. Appl. Meteor.*, **20**, 9–19.

Fritsch, J. M. and R. E. Carbone (2004). Improving quantitative precipitation forecasts in the warm season: A USWRP research and development strategy. *Bull. Amer. Meteor. Soc.*, **85**, 955–965.

Fritsch, J. M., C. F. Chappell, and L. R. Hoxit (1976). The use of large-scale budgets for convective parameterization. *Mon. Wea. Rev.*, **104**, 1408–1418.

Fritsch, J. M., J. Hilliker, J. Ross, and R. L. Vislocky (2000). Model consensus. *Wea. Forecasting*, **15**, 571–582.

Fu, Q. (1996). An accurate parameterization of the solar radiative properties of cirrus clouds for climate models. *J. Climate*, **9**, 2058–2082.

Fu, Q. and K.-N. Liou (1992). On the correlated-k distribution method for radiative transfer in non-homogeneous atmospheres. *J. Atmos. Sci.*, **49**, 2139–2156.

Fu, Q., P. Yang, and W. B. Sun (1998). An accurate parameterization of the infrared properties of cirrus clouds for climate models. *J. Climate*, **11**, 2223–2237.

Gallus, W. A., Jr. (1999). Eta simulations of three extreme precipitation events: sensitivity to resolution and convective parameterization. *Wea. Forecasting*, **14**, 405–426.

Gallus, W. A., Jr. and R. H. Johnson (1991). Heat and moisture budgets of an intense midlatitude squall line. *J. Atmos. Sci.*, **48**, 122–146.

Gardiner, M. J. (1982). Use of retrieval and global soils data for global modeling. In *Land Surface Processes in Atmospheric General Circulation Models*, ed. P. S. Eagleson. Cambridge University Press.

Garratt, J. R. (1992). *The Atmospheric Boundary Layer*. Cambridge University Press.

Garratt, J. R. and B. B. Hicks (1973). Momentum, heat and water vapour transfer to and from natural and artificial surfaces. *Quart. J. Roy. Meteor. Soc.*, **99**, 680–687.

(1990). Micrometeorological and PBL experiments in Australia. *Bound.-Layer Meteor.*, **50**, 11–29.

Garvert, M. F., C. P. Woods, B. A. Colle, *et al.* (2005). The 13–14 December 2001 IMPROVE-2 event. Part II: Comparisons of MM5 model simulations of clouds and precipitation with observations. *J. Atmos. Sci.*, **62**, 3520–3534.

Gates, L. W., J. S. Boyle, C. Covey, *et al.* (1999). An overview of the results of the atmospheric model intercomparison project (AMIP I). *Bull. Amer. Meteor. Soc.*, **80**, 229–55.

Gauthier, P., C. Charette, L. Fillion, P. Koclas, and S. Laroche (1999). Implementation of a 3D variational data assimilation system at the Canadian Meteorological Centre. Part I: The global analysis. *Atmos.-Ocean*, **37**, 103–156.

Gedney, N. and P. M. Cox (2003). The sensitivity of global climate model simulations to the representation of soil moisture heterogeneity. *J. Hydrometeor.*, **4**, 1265–1275.

Geleyn, J. P. and A. Hollingsworth (1979). An economical analytical method for the computation of the interaction between scattering and line absorption of radiation. *Contrib. Atmos. Phys.*, **52**, 1–16.

Georgelin, M. and F. Lott (2001). On the transfer of momentum by trapped lee waves: case of the IOP3 of PYREX. *J. Atmos. Sci.*, **58**, 3563–3580.

GEWEX Cloud System Science Team (1993). The GEWEX cloud system study (GCSS). *Bull. Amer. Meteor. Soc.*, **74**, 387–399.

Gill, A. E. (1982). *Atmosphere–Ocean Dynamics*. Academic Press.

Gilmore, M. S., J. M. Straka, and E. N. Rasmussen (2004a). Precipitation uncertainty due to variations in precipitation particle parameters within a simple microphysics scheme. *Mon. Wea. Rev.*, **132**, 2610–2627.

(2004b). Precipitation and evolution sensitivity in simulated deep convective storms: comparisons between liquid-only and simple ice and liquid phase microphysics. *Mon. Wea. Rev.*, **132**, 1897–1916.

Giorgi, F., M. R. Marinucci, G. T. Bates, and G. De Canio (1993). Development of a second-generation regional climate model (RegCM2). Part II: Convective processes and assimilation of lateral boundary conditions. *Mon. Wea. Rev.*, **121**, 2814–2832.

Glahn, H. R. and D. A. Lowry (1972). The use of model output statistics (MOS) in objective weather forecasting. *J. Appl. Meteor.*, **11**, 1203–1211.

Godfrey, C. M., D. J. Stensrud, and L. M. Leslie (2005). The influence of improved land surface and soil data to mesoscale model predictions. Preprints. In *19th Conf. on Hydrology*, San Diego, CA, Paper 4.7. American Meteorological Society.

Godson, W. L. (1954). Spectral models and the properties of transmission functions. In *Proc. Toronto Meteor. Conf., 1953*. Royal Meteorological Society, pp. 35–42.

Goldman, A. and T. G. Kyle (1968). A comparison between statistical model and line calculation with application to the 9.6 μm ozone and the 2.7 μm water vapor bands. *Appl. Opt.*, **7**, 1167–1177.

Goody, R. M. (1964). *Atmospheric Radiation I: Theoretical Basis*. Clarendon Press.

Goody, R. M. and Y. L. Yung (1995). *Atmospheric Radiation: Theoretical Basis*. Oxford University Press.

Grabowski, W. W. (2001). Coupling cloud processes with the large-scale dynamics using cloud-resolving convection parameterization (CRCP). *J. Atmos. Sci.*, **58**, 978–997.

Grabowski, W. W. and P. K. Smolarkiewicz (1999). CRCP: a cloud resolving convection parameterization for modeling the tropical convective atmosphere. *Physica D*, **133**, 171–178.

Grabowski, W. W., P. Bechtold, A. Cheng, *et al.* (2006). Daytime convective development over land: a model intercomparison based on LBA observations. *Quart. J. Roy. Meteor. Soc.*, **132**, 317–344.

Gregory, D. and P. R. Rowntree (1990). A mass flux convection scheme with representation of cloud ensemble characteristics and stability-dependent closure. *Mon. Wea. Rev.*, **118**, 1483–1506.

Gregory, D., R. Kershaw, and P. M. Inness (1997). Parametrization of momentum transport by convection. II: Tests in single-column and general circulation models. *Quart. J. Roy. Meteor. Soc.*, **123**, 1153–1183.

Gregory, D., J.-J. Morcrette, C. Jakob, A. C. M. Beljaars, and T. Stockdale (2000). Revision of convection, radiation, and cloud schemes in the ECMWF Integrated Forecasting System. *Quart. J. Roy. Meteor. Soc.*, **126**, 1685–1710.

Grell, G. and D. Devenyi (2002). A generalized approach to parameterizing convection combining ensemble and data assimilation techniques. *Geophys. Res. Lett.*, **29**, doi: 10.1029/2002GL015311.

Grell, G., Y.-H. Kuo, and R. J. Pasch (1991). Semiprognostic tests of three cumulus parameterization schemes in middle latitudes. *Mon. Wea. Rev.*, **119**, 5–31.

Grubisic, V. and M. W. Moncrieff (2000). Parameterization of convective momentum transport in highly baroclinic conditions. *J. Atmos. Sci.*, **57**, 3035–3049.

Gu, Y. and K. N. Liou (2000). Interactions of radiation, microphysics, and turbulence in the evolution of cirrus clouds. *J. Atmos. Sci.*, **57**, 2463–2479.

Guichard, F., D. B. Parsons, J. Dudhia, and J. Bresch (2003). Evaluating mesoscale model predictions of clouds and radiation with SGP ARM data over a seasonal timescale. *Mon. Wea. Rev.*, **131**, 926–944.

Gunn, R. and G. D. Kinzer (1949). The terminal velocity of fall for water drops in stagnant air. *J. Meteor.*, **6**, 243–248.

Gutman, G. and A. Ignatov (1998). The derivation of the green vegetation fraction from NOAA/AVHRR data for use in numerical weather prediction models. *Int. J. Remote Sens.*, **19**, 1533–1543.

Hagedorn, R., F. J. Doblas-Reyes, and T. N. Palmer (2005). The rationale behind the success of multi-model ensembles in seasonal forecasting – I. Basic concept. *Tellus*, **57A**, 219–233.

Hallett, J. and S. C. Mossop (1974). Production of secondary ice particles during the riming process. *Nature*, **249**, 26–28.

Halthore, R. N., S. Nemesure, S. E. Schwartz, *et al.* (1998). Models overestimate diffuse clear-sky irradiance: a case for excess atmospheric absorption. *Geophys. Res. Lett.*, **25**, 3591–3594.

Haltiner, G. J. and R. T. Williams (1980). *Numerical Prediction and Dynamic Meteorology*. John Wiley and Sons.

Hamill, T. M., J. S. Whitaker, and X. Wei (2004). Ensemble reforecasting: improving medium-range forecast skill using retrospective forecasts. *Mon. Wea. Rev.*, **132**, 1434–1447.

Hamill, T. M., J. S. Whitaker, and S. L. Mullen (2006). Reforecasts: an important dataset for improving weather predictions. *Bull. Amer. Meteor. Soc.*, **86**, 33–46.

Han, J. and H.-L. Pan (2006). Sensitivity of hurricane intensity to convective momentum transport parameterization. *Mon. Wea. Rev.*, **134**, 664–674.

Hanks, R. J. and G. L. Ashcroft (1986). *Applied Soil Physics*. Springer-Verlag.

Hansen, M. C., R. S. DeFries, J. R. G. Townshend, and R. Sohlberg (2000). Global land cover classification at 1 km spatial resolution using a classification tree approach. *Int. J. Remote Sens.*, **21**, 1331–1364.

Harrison, M. S. J., T. N. Palmer, D. S. Richardson, and R. Buizza (1999). Analysis and model dependencies in medium-range ensembles: two transplant case studies. *Quart. J. Roy. Meteor. Soc.*, **125**, 2487–2515.

Harshvardhan and J. Weinman (1982). Infrared radiative transfer through a regular array of cuboidal clouds. *J. Atmos. Sci.*, **39**, 431–439.

Harshvardhan, R. Davies, D. A. Randall, and T. G. Corsetti (1987). A fast radiation parameterization for atmospheric circulation models. *J. Geophys. Res.*, **92**, 1009–1015.

Hart, K. A., W. J. Steenburgh, D. J. Onton, and A. J. Siffert (2004). An evaluation of mesoscale-model-based model output statistics (MOS) during the 2002 Olympic and Paralympic winter games. *Wea. Forecasting*, **19**, 200–218.

Hastenrath, S. L. (1966). The flux of atmospheric water vapor over the Caribbean Sea and the Gulf of Mexico. *J. Appl. Meteor.*, **5**, 778–788.

Heath, D. F., A. J. Krueger, A. J. Roder, and B. D. Henderson (1975). The solar scatter ultraviolet and total ozone mapping spectrometer (SBUV/TOMS) for Nimbus G. *Opt. Eng.*, **14**, 323–331.

Heidt, F. D. (1977). The growth of the mixed layer in a stratified fluid due to penetrative convection. *Bound.-Layer Meteor.*, **12**, 439–461.

Henderson-Sellers, A., Z.-L. Yang, and R. E. Dickinson (1993). The project for intercomparison of land–surface parameterization schemes. *Bull. Amer. Meteor. Soc.*, **74**, 1335–1350.

Henderson-Sellers, A., K. McGuffie, and C. Gross (1995). Sensitivity of global climate model simulations to increased stomatal resistance and CO_2 increases. *J. Climate*, **8**, 1738–1756.

Henry, W. K. and A. H. Thompson (1976). An example of polar air modification over the Gulf of Mexico. *Mon. Wea. Rev.*, **104**, 1324–1327.

Higgins, P. A. T. and M. Vellinga (2004). Ecosystem responses to abrupt climate change: teleconnections, scale and the hydrological cycle. *Climatic Change*, **64**, 127–142.

Hinkelman, L. M., T. P. Ackerman, and R. T. Marchand (1999). An evaluation of NCEP Eta model predictions of surface energy budget and cloud properties by comparison with measured ARM data. *J. Geophys. Res.*, **104**, 19 535–19 549.

Hobbs, P. V. (1987). The Gulf Stream rainband. *Geophys. Res. Lett.*, **14**, 1142–1145.

Hobbs, P. V. and A. L. Rangno (1985). Ice particle concentrations in clouds. *J. Atmos. Sci.*, **42**, 2523–2549.

Hocking, L. M. (1959). The collision efficiency of small drops. *Quart. J. Roy. Meteor. Soc.*, **85**, 44–50.

Hocking, L. M. and P. R. Jonas (1970). The collision efficiency of small drops. *Quart. J. Roy. Meteor. Soc.*, **96**, 722–729.

Hodur, R. M. (1997). The Naval Research Laboratory's coupled ocean/atmosphere mesoscale prediction system (COAMPS). *Mon. Wea. Rev.*, **125**, 1414–1430.

Holt, T. R. and S. Raman (1992). Three-dimensional mean and turbulence structure of a coastal front influenced by the Gulf Stream. *Mon. Wea. Rev.*, **120**, 17–39.

Holtslag, A. A. and A. C. M. Beljaars (1989). Surface flux parameterization schemes, developments and experiences at KNMI. In *Parameterizations of Fluxes over Land Surface, ECMWF Workshop Proceedings*, October 1988, Reading, UK, pp. 121–147.

Holtslag, A. A. and C.-H. Moeng (1991). Eddy diffusivity and countergradient transport in the convective atmospheric boundary layer. *J. Atmos. Sci.*, **48**, 1690–1698.

Holtslag, A. A. and B. A. Boville (1993). Local versus nonlocal boundary layer diffusion in a global climate model. *J. Climate*, **6**, 1825–1842.

Holtslag, A. A., I. F. Bruijn, and H.-L. Pan (1990). A high resolution air mass transformation model for short-range weather forecasting. *Mon. Wea. Rev.*, **118**, 1561–1575.

Homar, V., R. Romero, D. J. Stensrud, C. Ramis, and S. Alonso (2003). Numerical diagnosis of a small, quasi-tropical cyclone over the western Mediterranean: dynamical vs. boundary factors. *Quart. J. Royal Meteor. Soc.*, **129**, 1469–1490.

Homar, V., D. J. Stensrud, J. J. Levit, and D. R. Bright (2006). Value of human-generated pertubations in short-range ensemble forecasts of severe weather. *Wea. Forecasting*, **21**, 347–363.

Homleid, M. (1995). Diurnal corrections of short-term temperature forecasts using the Kalman filter. *Wea. Forecasting*, **10**, 689–707.

Hong, S.-Y. and H.-L. Pan (1996). Nonlocal boundary layer vertical diffusion in a medium-range forecast model. *Mon. Wea. Rev.*, **124**, 2322–2339.

Hong, X., S. W. Chang, S. Raman, L. K. Shay, and R. Hodur (2000). The interaction between Hurricane Opal (1995) and a warm core ring in the Gulf of Mexico. *Mon. Wea. Rev.*, **128**, 1347–1365.

Hopwood, W. P. (1995). Surface transfer of heat and momentum over an inhomogeneous vegetated land. *Quart. J. Roy. Meteor. Soc.*, **121**, 1549–1574.

Hou, D., E. Kalnay, and K. Drogemeier (2001). Objective verification of the SAMEX98 ensemble forecasts. *Mon. Wea. Rev.*, **129**, 73–91.

Houser, P. R. (2003). Infiltration and soil moisture processes, ch. 27. In *Handbook of Weather, Climate, and Water: Atmospheric Chemistry, Hydrology, and Societal Impacts*, ed. T. D. Potter and B. R. Colman. John Wiley and Sons, Inc., pp. 493–506.

Houtekamer, P. L. (1995). The construction of optimal perturbations. *Mon. Wea. Rev.*, **123**, 2888–2898.

Houtekamer, P. L. and L. Lefaivre (1997). Using ensemble forecasts for model verification. *Mon. Wea. Rev.*, **125**, 2416–2426.

Houtekamer, P. L. and H. L. Mitchell (2001). A sequential ensemble Kalman filter for atmospheric data assimilation. *Mon. Wea. Rev.*, **129**, 123–137.

Houtekamer, P. L., L. Lefaivre, J. Derome, H. Ritchie, and H. L. Mitchell (1996). A system simulation approach to ensemble prediction. *Mon. Wea. Rev.*, **124**, 1225–1242.

Houze, R. A., Jr. (1982). Cloud clusters and large-scale vertical motions in the tropics. *J. Meteor. Soc. Japan*, **60**, 396–410.

(1993). *Cloud Dynamics*. Academic Press.

(1997). Stratiform precipitation in regions of convection: a meteorological paradox? *Bull. Amer. Meteor. Soc.*, **78**, 2179–2196.

Hoxit, L. R. (1975). Diurnal variations in planetary boundary-layer winds over land. *Bound.-Layer Meteor.*, **8**, 21–38.

Hunke, E. C. and J. K. Dukowicz (1997). An elastic-viscous-plastic model for sea ice dynamics. *J. Phys. Oceanogr.*, **27**, 1849–1867.

Huete, A. R., H. Q. Liu, K. Batchily, and W. van Leeuwen (1996). A comparison of vegetation indices over a global set of TM images for EOS-MODIS. *Remote Sens. Environ.*, **59**, 440–451.

Iacobellis, S. F. and R. C. J. Somerville (2000). Implications for microphysics for cloud–radiation parameterizations: lessons from TOGA COARE. *J. Atmos. Sci.*, **57**, 161–183.

Ineson, S. and M. K. Davey (1997). Interannual climate simulation and predictability in a coupled TOGA GCM. *Mon. Wea. Rev.*, **125**, 721–741.

Iwasaki, T., S. Yamada, and K. Tada (1989). A parameterization scheme of orographic gravity-wave drag with two different vertical partitionings. Part I: Impacts on medium-range forecasts. *J. Meteor. Soc. Japan*, **67**, 11–27.

Jacks, E., J. B. Bower, V. J. Dagostaro, *et al.* (1990). New NGM-based MOS guidance for maximum/minimum temperature, probability of precipitation, cloud amount, and surface wind. *Wea. Forecasting*, **5**, 128–138.

Jacquemin, B. and J. Noilhan (1990). Sensitivity study and validation of a land surface parameterization using the HAPEX-MOBILHY data set. *Bound.-Layer Meteor.*, **52**, 93–134.

Jang, K.-I., X. Zou, M. S. F. V. De Pondeca, *et al.* (2003). Incorporating TOMS ozone measurements into the prediction of the Washington, D.C., winter storm during 24–25 January 2000. *J. Appl. Meteor.*, **42**, 797–812.

Janjic, Z. I. (1994). The step-mountain Eta coordinate model: further developments of the convection, viscous sublayer, and turbulence closure schemes. *Mon. Wea. Rev.*, **122**, 927–945.

Jarvis, P. G. (1976). The interpretation of the variations in leaf water potential and stomatal conductance found in canopies in the field. *Philos. Trans. Roy. Soc. London B*, **273**, 593–610.

Johansen, O. (1975). Thermal conductivity of soils. Ph.D. Thesis, University of Trondheim. (Available from Universitetsbiblioteket I Trondheim, Hogskoleringen 1, 7034 Trondheim, Norway.)

Johns, R. H. and C. A. Doswell III (1992). Severe local storms forecasting. *Wea. Forecasting*, **7**, 588–612.

Johnson, D. E., P. K. Wang, and J. M. Straka (1993). Numerical simulations of the 2 August 1981 CCOPE supercell storm with and without ice microphysics. *J. Appl. Meteor.*, **32**, 745–759.

Johnson, R. H. (1984). Partitioning tropical heat and moisture budgets into cumulus and mesoscale components: implications for cumulus parameterization. *Mon. Wea. Rev.*, **112**, 1590–1601.

Johnson, R. H. and G. S. Young (1983). Heat and moisture budgets of tropical mesoscale anvil clouds. *J. Atmos. Sci.*, **40**, 2138–2147.

Johnson, R. H., T. M. Rickenbach, S. A. Rutledge, P. E. Ciesielski, and W. H. Schubert (1999). Trimodal characteristics of tropical convection. *J. Climate*, **12**, 2397–2418.

Jordon, C. L. (1964). On the influence of tropical cyclones on the sea surface temperature field. In *Proc. Symp. Tropical Meteorology*, Rotorua, New Zealand, November 1963. New Zealand Meteorological Service, pp. 614–622.

Joseph, J. H., W. J. Wiscombe, and J. A. Weinman (1976). The delta-Eddington approximation for radiative flux transfer. *J. Atmos. Sci.*, **33**, 2452–2459.

Juang, H.-M. H., C.-T. Lee, Y. Zhang, *et al.* (2005). Applying horizontal diffusion on pressure surface to mesoscale models on terrain-following coordinates. *Mon. Wea. Rev.*, **133**, 1384–1402.

Kaimal, J. C. and J. C. Wyngaard (1990). The Kansas and Minnesota experiments. *Bound.-Layer Meteor.*, **50**, 31–47.

Kaimal, J. C. and J. J. Finnigan (1994). *Atmospheric Boundary Layer Flows*. Oxford University Press.

Kain, J. S. (2004). The Kain–Fritsch convective parameterization: an update. *J. Appl. Meteor.*, **43**, 170–181.

Kain, J. S. and J. M. Fritsch (1990). A one-dimensional entraining/detraining plume model and its application in convective parameterization. *J. Atmos. Sci.*, **47**, 2784–2802.

(1992). The role of the convective "trigger function" in numerical forecasts of mesoscale convective systems. *Meteor. Atmos. Phys.*, **49**, 93–106.

(1993). Convective parameterization for mesoscale models: the Kain–Fritsch scheme. In *The Representation of Cumulus Convection in Numerical Models of the*

Atmosphere, ed. K. A. Emanuel and D. J. Raymond. *Meteorology Monographs*, No. 46. American Meteorological Society, pp. 165–170.

Kain, J. S., M. E. Baldwin, and S. J. Weiss (2003). Parameterized updraft mass flux as a predictor of convective intensity. *Wea. Forecasting*, **18**, 106–116.

Kalnay, E. (2003). *Atmospheric Modeling, Data Assimilation and Predictability*. Cambridge University Press.

Kanamitsu, M., A. Kumar, H.-M. H. Juang, *et al.* (2002). NCEP dynamical seasonal forecast system 2000. *Bull. Amer. Meteor. Soc.*, **83**, 1019–1037.

Katz, R. W. and A. H. Murphy (1997). *Economic Value of Weather and Climate Forecasts*. Cambridge University Press.

Kaufman, Y. J. and T. Nakajima (1993). Effect of Amazon smoke on cloud microphysics and albedo – analysis from satellite imagery. *J. Appl. Meteor.*, **32**, 729–744.

Kershaw, R. (1995). Parametrization of momentum transport by convectively generated gravity waves. *Quart. J. Roy. Meteor. Soc.*, **121**, 1023–1040.

Kershaw, R. and D. Gregory (1997). Parametrization of momentum transport by convection. I: Theory and cloud modeling results. *Quart. J. Roy. Meteor. Soc.*, **123**, 1133–1151.

Kessler, E. (1969). *On the Distribution and Continuity of Water Substance in Atmospheric Circulations. Meteorology Monographs*, No. 32. American Meteorological Society.

Keyser, D. (1986). Atmospheric fronts: an observational perspective. In *Mesoscale Meteorology and Forecasting*, ed. P. S. Ray. American Meteorological Society, pp. 216–258.

Keyser, D. and D. R. Johnson (1984). Effects of diabatic heating on the ageostrophic circulation of an upper tropospheric jet streak. *Mon. Wea. Rev.*, **112**, 1709–1724.

Kiehl, J. T. and P. R. Gent (2004). The Community Climate System Model, Version Two. *J. Clim.*, **17**, 3666–3682.

Kiehl, J. T., J. J. Hack, G. B. Bonan, *et al.* (1996). Description of the NCAR community climate model. NCAR Tech. Note NCAR/TN-420+STR.

Kim, J. and S. Verma (1991). Modeling canopy photosynthesis: scaling up from a leaf to canopy in a temperate grassland ecosystem. *Agric. For. Meteor.*, **57**, 187–208.

Kim, Y.-J. and A. Arakawa (1995). Improvement of orographic gravity wave parameterization using a mesoscale gravity wave model. *J. Atmos. Sci.*, **52**, 1875–1902.

Kim, Y.-J. and J. D. Doyle (2005). Extension of an orographic-drag parametrization scheme to incorporate orographic anisotropy and flow blocking. *Quart. J. Roy. Meteor. Soc.*, **131**, 1893–1921.

Kim, Y.-J., S. D. Eckermann, and H.-Y. Chun (2003). An overview of the past, present and future of gravity-wave drag parametrization for numerical climate and weather prediction models. *Atmos.-Ocean*, **41**, 65–98.

King, M. D., Y. J. Kaufman, D. Tanre, and T. Nakajima (1999). Remote sensing of tropospheric aerosols from space: past, present, and future. *Bull. Amer. Meteor. Soc.*, **80**, 2229–2259.

Kinzer, G. D. and R. Gunn (1951). The evaporation, temperature and thermal relaxation-time of freely falling waterdrops. *J. Atmos. Sci.*, **8**, 71–83.

Klazura, G. E. (1971). Measurements of precipitation particles in warm cumuli over southeast Texas. *J. Appl. Meteor.*, **10**, 739–750.

Knight, C. A. and N. C. Knight (2005). Very large hailstones from Aurora, Nebraska. *Bull. Amer. Meteor. Soc.*, **86**, 1773–1781.

Knight, C. A., Cooper, W. A., Breed, D. W., *et al.* (1982). Microphysics. In *Hailstorms of the Central High Plains*, vol. 1, ed. C. Knight and P. Squires. Colorado Associated University Press, pp. 151–193.

Koch, S. E. (1984). The role of an apparent mesoscale frontogenetic circulation in squall line initiation. *Mon. Wea. Rev.*, **112**, 2090–2111.

Koenig, L. R. (1971). Numerical modeling of ice deposition. *J. Atmos. Sci.*, **28**, 226–237.

Kogan, Y. L. (1991). The simulation of a convective cloud in a 3D model with explicit microphysics. Part I: Model description and sensitivity experiments. *J. Atmos. Sci.*, **48**, 1160–1189.

Koren, V., J. Schaake, K. Mitchell, *et al.* (1999). A parameterization for snowpack and frozen ground intended for NCEP weather and climate models. *J. Geophys. Res.*, **104**, 19 569–19 585.

Koster, R. D. and M. J. Suarez (1996). Energy and water balance calculations in the Mosaic LSM. NASA Tech. Memo., 104606, **9**.

  (2001). Soil moisture memory in climate models. *J. Hydrometeor.*, **2**, 558–570.

Kraus, E. B. and J. A. Businger (1994). *Atmosphere–Ocean Interaction*. Oxford University Press.

Krinner, G., N. Viovy, N. de Noblet-Ducoudré, *et al.* (2005). A dynamic global vegetation model for studies of the coupled atmosphere–biosphere system. *Global Biogeochem. Cycles*, **19**, GB1015, doi: 10.1029/2003GB002199.

Krishnamurti, T. N., M. Kanamitsu, R. Godbole, *et al.* (1976). Study of a monsoon depression (ii), dynamic structure. *J. Meteor. Soc., Japan*, **54**, 208–225.

Krishnamurti, T. N., C. M. Kishtawal, D. W. Shin, and C. Eric Williford (2000). Improving tropical precipitation forecasts from a multianalysis superensemble. *J. Climate*, **13**, 4217–4227.

Kubota, A. and M. Sugita (1994). Radiometrically determined skin temperature and scalar roughness to estimate surface heat flux. Part I: Parameterization of radiometric scalar roughness. *Bound.-Layer Meteor.*, **69**, 397–416.

Kuettner, J. P. (1959). Cloud bands in the Earth's atmosphere: observations and theory. *Tellus*, **11**, 267–294.

Kuo, H.-L. (1965). On the formation and intensification of tropical cyclones through latent heat release by cumulus convection. *J. Atmos. Sci.*, **22**, 40–63.

Kurkowski, N. P., D. J. Stensrud, and M. E. Baldwin (2003). Assessment of implementing satellite-derived land cover data in the Eta model. *Wea. Forecasting*, **18**, 404–416.

Kutchment, L. S., V. N. Demidov, and Y. G. Motovilov (1983). *River Runoff Generation*. Russian Academy of Science, Moscow (in Russian).

Lacis, A. A. and J. E. Hansen (1974). A parameterization for the absorption of solar radiation in the Earth's atmosphere. *J. Atmos. Sci.*, **31**, 118–133.

Lacis, A. A. and V. Oinas (1991). A description of the correlated k distribution method for modeling nongray gaseous absorption, thermal emission, and multiple scattering in vertically inhomogeneous atmospheres. *J. Geophys. Res.*, **96**, 9027–9063.

Laing, A. G. and J. M. Fritsch (1997). The global population of mesoscale convective complexes. *Quart. J. Roy. Meteor. Soc.*, **123**, 389–406.

Lakhtakia, M. N. and T. T. Warner (1987). A real-data numerical study of the development of precipitation along the edge of an elevated mixed layer. *Mon. Wea. Rev.*, **115**, 156–168.

Lamb, D. (2001). Rain production in convective storms. In *Severe Convective Storms*, ed. C. A. Doswell III. *Meteorology Monographs*, No. 50. American Meteorological Society, pp. 299–321.

Lamb, P. J. (1978a). Case studies of tropical Atlantic surface circulation patterns during recent sub-Saharan weather anomalies: 1967 and 1968. *Mon. Wea. Rev.*, **106**, 481–491.

(1978b). Large-scale tropical Atlantic surface circulation patterns associated with Subsaharan weather anomalies. *Tellus*, **30A**, 198–212.

Lamb, P. J. and R. A. Peppler (1992). Further case studies of tropical Atlantic surface atmospheric and oceanic patterns associated with sub-Saharan drought. *J. Climate*, **5**, 476–488.

Langmuir, I. (1948). The production of rain by chain-reaction in cumulus clouds at temperatures above freezing. *J. Meteor.*, **5**, 175–192.

Lanicci, J. M. and T. T. Warner (1991). A synoptic climatology of elevated mixed-layer inversion over the southern Great Plains in spring. Part I: Structure, dynamics, and seasonal evolution. *Wea. Forecasting*, **6**, 181–197.

Lanicci, J. M., T. N. Carlson, and T. T. Warner (1987). Sensitivity of the Great Plains severe-storm environment to soil-moisture distribution. *Mon. Wea. Rev.*, **115**, 2660–2673.

Launder, B. E., G. J. Reece, and W. Rodi (1975). Progress in the development of a Reynolds-stress turbulence closure. *J. Fluid Mech.*, **68**, 537–566.

Lee, B. D. and R. B. Wilhelmson (1997). The numerical simulation of nonsupercell tornadogenesis. Part II: Evolution of a family of tornadoes along a weak outflow boundary. *J. Atmos. Sci.*, **54**, 2387–2415.

Lee, T. J. and R. A. Pielke (1992). Estimating the soil surface specific humidity. *J. Appl. Meteor.*, **31**, 480–484.

Leftwich Jr, P. W., J. T. Schaefer, S. J. Weiss, and M. Kay (1998). Severe convective storm probabilities for local areas in watches issued by the storm prediction center. Preprints. In *19th Conf. Severe Local Storms*, Minneapolis, MN. American Meteorological Society, pp. 548–551.

Leipper, D. (1967). Observed ocean conditions and Hurricane Hilda, 1964. *J. Atmos. Sci.*, **24**, 182–196.

Leith, C. E. (1974). Theoretical skill of Monte Carlo forecasts. *Mon. Wea. Rev.*, **102**, 409–418.

Leslie, L. M. and M. S. Speer (2000). Comments on "Using ensembles for short-range forecasting." *Mon. Wea. Rev.*, **128**, 3018–3020.

Le Treut, H. and Z.-X. Li (1991). Sensitivity of an atmospheric general circulation model to prescribed SST changes: feedback effects associated with the simulation of cloud optical properties. *Climate Dyn.*, **5**, 175–187.

Leuning, R., F. M. Kelliher, D. G. G. Pury, and E.-D. Schulze (1995). Leaf nitrogen, photosynthesis, conductance and transpiration: scaling from leaves to canopy. *Plant Cell Environ.*, **18**, 1183–1200.

Levitt, J. (1974). *Introduction to Plant Physiology*. C. V. Mosby Co.

Lewis, J. M. (2005). Roots of ensemble forecasting. *Mon. Wea. Rev.*, **133**, 1865–1885.

Liang, X., E. F. Wood, and D. P. Lettenmaier (1996). Surface soil parameterization of the VIC-2L model: evaluation and modification. *Global Planet. Change*, **13**, 195–206.

Liang, X., D. P. Lettenmaier, E. F. Wood, and S. J. Burges (1994). A simple hydrologically based model of land surface water and energy fluxes for GCM. *J. Geophys. Res.*, **99**, 14 415–14 428.

Lilly, D. K. (1968). Models of cloud-topped mixed layers under a strong inversion. *Quart. J. Roy. Meteor. Soc.*, **94**, 292–309.

(1972). Wave momentum flux – a GARP problem. *Bull. Amer. Meteor. Soc.*, **53**, 17–23.

(1988). Cirrus outflow dynamics. *J. Atmos. Sci.*, **45**, 1594–1605.

Lilly, D. K. and P. J. Kennedy (1973). Observations of a stationary mountain wave and its associated momentum flux and energy dissipation. *J. Atmos. Sci.*, **30**, 1135–1152.

Lin, Y.-L., R. D. Farley, and H. D. Orville (1983). Bulk parameterization of the snow field in a cloud model. *J. Climate Appl. Meteor.*, **22**, 1065–1092.

Lindstrom, S. S. and T. E. Nordeng (1992). Parameterized slantwise convection in a numerical model. *Mon. Wea. Rev.*, **120**, 742–756.

Lindzen, R. S. (1981). Turbulence and stress due to gravity wave and tidal breakdown. *J. Geophys. Res.*, **86**, 9707–9714.

(1984). Gravity waves in the middle atmosphere. In *Dynamics of the Middle Atmosphere*, ed. J. R. Holton and T. Matsuno. Terra, pp. 3–18.

Liou, K.-N. (1980). *An Introduction to Atmospheric Radiation.* Academic Press.

(1986). Influence of cirrus clouds on weather and climate processes: a global perspective. *Mon. Wea. Rev.*, **114**, 1167–1199.

Liou, K.-N. and G. D. Wittman (1979). Parameterization of radiative properties of clouds. *J. Atmos. Sci.*, **7**, 1261–1273.

Liu, H.-C., P. K. Wang, and R. E. Schlesinger (2003). A numerical study of cirrus clouds. Part II: Effects of ambient temperature, stability, radiation, ice microphysics, and microdynamics on cirrus evolution. *J. Atmos. Sci.*, **60**, 1097–1119.

Liu, J. Y. and H. D. Orville (1969). Numerical modeling of precipitation and cloud shadow effects on mountain-induced cumuli. *J. Atmos. Sci.*, **26**, 1283–1298.

Liu, Q., J. M. Lewis, and J. M. Schneider (1992). A study of cold-air modification over the Gulf of Mexico using in situ data and mixed-layer modeling. *J. Appl. Meteor.*, **31**, 909–924.

Liu, Q. and J. Schmetz (1988). On the problem of an analytical solution to the diffusivity factor. *Beitr. Phys. Atmos.*, **61**, 23–29.

Liu, W. T., K. B. Katsaros, and J. A. Businger (1979). Bulk parameterization of the air–sea exchange of heat and water vapor including the molecular constraints at the surface. *J. Atmos. Sci.*, **36**, 1722–1735.

Lobocki, L. (1993). A procedure for the derviation of surface-layer bulk relationships from simplified second order closure models. *J. Appl. Meteor.*, **32**, 126–138.

Lohmann, U., N. McFarlane, L. Levkov, K. Abdella, and F. Albers (1999). Comparing different cloud schemes of a single column model by using mesoscale forcing and nudging technique. *J. Climate*, **12**, 438–461.

London, J., R. D. Bojkov, S. Oltmans, and J. I. Kelley (1976). Atlas of the global distribution of total ozone, July 1957–June 1967. NCAR Tech. Note 113+STR.

Lord, S. J., W. C. Chao, and A. Arakawa (1982). Interactions of a cumulus cloud ensemble with the large-scale environment. Part IV: The discrete model. *J. Atmos. Sci.*, **39**, 104–113.

Lorenc, A. C., S. P. Ballard, R. S. Bell, *et al.* (2000). The Met. Office global three-dimensional variational data assimilation scheme. *Quart. J. Roy. Meteor. Soc.*, **126**, 2991–3012.

Lorenz, E. N. (1963). Deterministic nonperiodic flow. *J. Atmos. Sci.*, **20**, 130–141.

(1969). The predictability of a flow which possesses many scales of motion. *Tellus*, **21**, 289–307.

Lott, F. and M. J. Miller (1997). A new subgrid-scale orographic drag parameterization: its formulation and testing. *Quart. J. Roy. Meteor. Soc.*, **123**, 101–127.

Louis, J. F. (1979). A parametric model of vertical eddy fluxes in the atmosphere. *Bound.-Layer Meteor.*, **17**, 187–202.

Louis, J. F., M. Tiedtke, and J. F. Geleyn (1982). A short history of the operational PBL – parameterization of ECMWF. In *Workshop on Planetary Boundary Layer Parameteriztion*, European Centre for Medium Range Weather Forecasting, Shinfield Park, Reading.

Loveland, T. R., J. W. Merchant, J. F. Brown, *et al.* (1995). Seasonal land-cover regions of the United States. *Ann. Assoc. Amer. Geographers*, **85**, 339–355.

Loveland, T. R., B. C. Reed, J. F. Brown, *et al.* (2000). Development of a global land cover characteristics database and IGBP DISCover from 1 km AVHRR data. *Int. J. Remote Sens.*, **21**, 1303–1365.

Lumley, J. L. and B. Khajeh-Nouri (1974). Computational modeling of turbulent transport. *Adv. Geophys.*, **18A**, 169–192.

Luo, H. and M. Yanai (1984). The large-scale circulation and heat sources over the Tibetan plateau and surrounding areas during the early summer of 1979. Part II: Heat and moisture budgets. *Mon. Wea. Rev.*, **112**, 966–989.

Luo, L., A. Robock, K. Y. Vinnikov, *et al.* (2003). Effects of frozen soil on soil temperature, spring infiltration, and runoff: results from the PILPS 2(d) experiment at Valdai, Russia. *J. Hydrometeor.*, **4**, 334–351.

Lynn, B. H., A. P. Khain, J. Dudhia, *et al.* (2005). Spectral (bin) microphysics coupled with a mesoscale model (MM5). Part I: Model description and first results. *Mon. Wea. Rev.*, **133**, 44–58.

Maddox, R. A. (1980). Mesoscale convective complexes. *Bull. Amer. Meteor. Soc.*, **61**, 1374–1387.

Mahfouf, J.-F. (1991). Analysis of soil moisture from near-surface parameters: a feasibility study. *J. Appl. Meteor.*, **30**, 1534–1547.

Mahfouf, J.-F., E. Richard, and P. Mascart (1987). The influence of soil and vegetation on the development of mesoscale circulations. *J. Clim. Appl. Meteor.*, **26**, 1483–1495.

Mahfouf, J.-F., A. O. Manzi, J. Noilhan, H. Giordani, and M. DéQué (1995). The land surface scheme ISBA within Météo-France climate model ARPEGE. Part I: Implementation and preliminary results. *J. Climate*, **8**, 2039–2057.

Mahrt, L. (1976). Mixed layer moisture structure. *Mon. Wea. Rev.*, **104**, 1403–1407.
  (1987). Grid-averaged surface fluxes. *Mon. Wea. Rev.*, **115**, 1550–1560.
  (1991). Boundary-layer moisture regimes. *Quart J. Roy. Meteor. Soc.*, **117**, 151–176.

Mahrt, L. and M. Ek (1984). The influence of atmospheric stability on potential evaporation. *J. Climate Appl. Meteor.*, **23**, 222–234.

Mahrt, L. and H. L. Pan (1984). A two-layer model of soil hydrology. *Bound.-Layer Meteor.*, **29**, 1–20.

Makin, V. K. and C. Mastenbroek (1996). Impact of waves on air–sea exchange of sensible heat and momentum. *Bound.-Layer Meteor.*, **79**, 279–300.

Manabe, S. (1969). The atmospheric circulation and hydrology of the Earth's surface. *Mon. Wea. Rev.*, **97**, 739–774.

Manabe, S. and R. Strickler (1964). Thermal equilibrium of the atmosphere with a convective adjustment. *J. Atmos. Sci.*, **21**, 361–385.

Manabe, S., J. Smagorinsky, and R. F. Strickler (1965). Simulated climatology of a general circulation model with a hydrologic cycle. *Mon. Wea. Rev.*, **93**, 769–798.

Mao, Q., R. T. McNider, S. F. Mueller, and H.-M. H. Juang (1999). An optimal model output calibration algorithm suitable for objective temperature forecasting. *Wea. Forecasting*, **14**, 190–202.

Mapes, B. E. (1997). Equilibrium vs. activation control of large-scale variations of tropical deep convection. In *The Physics and Parameterization of Moist Atmospheric Convection*, ed. R. K. Smith. Kluwer Academic Publishers, pp. 321–358.

Mapes, B. E. and J. Lin (2005). Doppler radar observations of mesoscale wind divergence in regions of tropical convection. *Mon. Wea. Rev.*, **133**, 1808–1824.

Mapes, B. E., P. E. Ciesielski, and R. H. Johnson (2003). Sampling errors in rawinsonde-array budgets. *J. Atmos. Sci.*, **60**, 2697–2714.

Marchuk, G., G. Mikhailov, M. Nazaraliev, *et al.* (1980). *The Monte Carlo Methods in Atmospheric Optics*. Springer-Verlag.

Margulis, S. A. and D. Entekhabi (2004). Boundary-layer entrainment estimation through assimilation of radiosonde and micrometeorological data into a mixed-layer model. *Bound.-Layer Meteor.*, **110**, 405–433.

Markowski, P. M. and J. Y. Harrington (2005). A simulation of a supercell thunderstorm with emulated radiative cooling beneath the anvil. *J. Atmos. Sci.*, **62**, 2607–2617.

Marletto, V., F. Zinoni, L. Criscuolo, *et al.* (2005). Evaluation of downscaled DEMETER multi-model ensemble seasonal hindcasts in a northern Italy location by means of a model of wheat growth and soil water balance. *Tellus*, **57A**, 488–497.

Marshak, A. and A. B. Davis (2005). *Radiative Transfer in Cloudy Atmospheres*. Springer-Verlag.

Marshall, C. H., K. C. Crawford, K. E. Mitchell, and D. J. Stensrud (2003). The impact of the land surface physics in the operational NCEP Eta model on simulating the diurnal cycle: evaluation and testing using Oklahoma Mesonet data. *Wea. Forecasting*, **18**, 748–768.

Marshall, J. S. and W. M. Palmer (1948). The distribution of raindrops with size. *J. Meteor.*, **5**, 165–166.

Marshall, T. J., J. W. Holmes, and C. W. Rose (1996). *Soil Physics*. Cambridge University Press.

Mason, B. J. (1956). On the melting of hailstones. *Quart. J. Roy. Meteor. Soc.*, **82**, 209–216.

(1971). *The Physics of Clouds*, 2nd edn. Oxford University Press.

Mass, C. F. and Y.-H. Kuo (1998). Regional real-time numerical weather prediction: current status and future potential. *Bull. Amer. Meteor. Soc.*, **79**, 253–263.

Mass, C. F., M. Albright, D. Overs, *et al.* (2003). Regional environmental prediction over the Pacific Northwest. *Bull. Amer. Meteor. Soc.*, **84**, 1353–1366.

Maykut, G. A. and N. Untersteiner (1971). Some results from a time dependent thermodynamic model of sea ice. *J. Geophys. Res.*, **76**, 1550–1575.

McCarthy, D. J., J. T. Schaefer, and M. Kay (1998). Watch verification at the storm prediction center 1970–1997. Preprints. In *19th Conf. Severe Local Storms*. American Meteorological Society, pp. 603–606.

McClain, E. P., W. G. Pichel, and C. C. Walton (1985). Comparative performance of AVHRR-based multichallen sea surface temperatures. *J. Geophys. Res.*, **90**, 11 587–11 601.

McCorcle, M. D. (1988). Simulation of surface-moisture effects on the Great Plains low-level jet. *Mon. Wea. Rev.*, **116**, 1705–1720.

McCumber, M. C. and R. A. Pielke (1981). Simulation of the effects of surface fluxes of heat and moisture in a mesoscale numerical model soil layer. *J. Geophys. Res.*, **86**, 9929–9938.

McCumber, M. C., W.-K. Tao, J. Simpson, R. Penc, and S.-T. Soong (1991). Comparison of ice-phase microphysical parameterization schemes using numerical simulations of tropical convection. *J. Appl. Meteor.*, **30**, 985–1004.

McDonald, J. E. (1958). The physics of cloud modification. *Adv. Geophys.*, **5**, 223–303.

McFarlane, N. A. (1987). The effect of orographically excited gravity wave drag on the general circulation of the lower stratosphere and troposphere. *J. Atmos. Sci.*, **44**, 1775–1800.

McPherson, R. A. and D. J. Stensrud (2005). Influences of a winter wheat belt on the evolution of the boundary layer. *Mon. Wea. Rev.*, **133**, 2178–2199.

McPherson, R. A., D. J. Stensrud, and K. C. Crawford (2004). The impact of Oklahoma's winter wheat belt on the mesoscale environment. *Mon. Wea. Rev.*, **132**, 405–421.

Meador, W. E. and W. R. Weaver (1980). Two stream approximations to radiative transfer in planetary atmospheres: a unified description of existing methods and a near improvement. *J. Atmos. Sci.*, **37**, 630–643.

Meleshko, V. P. and R. T. Wetherald (1981). The effect of a geographical cloud distribution on climate: a numerical experiment with an atmospheric general circulation model. *J. Geophys. Res.*, **86**, 11 995–12 014.

Mellor, G. L. and T. Yamada (1974). A hierarchy of turbulence closure models for planetary boundary layers. *J. Atmos. Sci.*, **31**, 1791–1806.

(1982). Development of a turbulence closure model for geophysical fluid problems. *Rev. Geophys. Space Phys.*, **20**, 851–875.

Mesinger, F., Z. I. Janjic, S. Nickovic, D. Gavrilov, and D. G. Deaven (1988). The step-mountain coordinate: model description and performance for cases of Alpine lee cyclogenesis and for a case of an Appalachian redevelopment. *Mon. Wea. Rev.*, **116**, 1493–1520.

Meyers, M. P., P. J. DeMott, and W. R. Cotton (1992). New primary ice nucleation parameterizations in an explicit cloud model. *J. Appl. Meteor.*, **31**, 708–721.

Meyers, M. P., R. L. Walko, J. Y. Harrington, and W. R. Cotton (1997). New RAMS cloud microphysics parameterization. Part II: The two-moment scheme. *Atmos. Res.*, **45**, 3–39.

Miller, D. A. and R. A. White (1998). A conterminous United States multilayer soil characteristics data set for regional climate and hydrology modeling. *Earth Interactions*, **2**. (Available online at http://EarthInteractions.org.)

Milly, P. C. D. (1997). Sensitivity of greenhouse summer dryness to changes in plant rooting characteristics. *Geophys. Res. Lett.*, **24**, 269–271.

Mitchell, H. L. and P. L. Houtekamer (2000). An adaptive ensemble Kalman filter. *Mon. Wea. Rev.*, **128**, 416–433.

Mitchell, K. E., D. Lohman, P. R. Houser, *et al.* (2004). The multi-institution North American Land Data Assimilation System (NLDAS): utilizing multiple GCIP products and partners in a continental distributed hydrological modeling system. *J. Geophys. Res.*, **109**, doi: 10.1029/2003JD0003823.

Mlawer, E. J., P. D. Brown, S. A. Clough, *et al.* (2000). Comparison of spectral direct and diffuse solar irradiance measurements and calculations for cloud-free conditions. *Geophys. Res. Lett.*, **27**, 2653–2656.

Mlawer, E. J., S. J. Taubman, P. D. Iacono, and S. A. Clough (1997). Radiative transfer for inhomogeneous atmosphere: RRTM, a validated correlated-k model for the longwave. *J. Geophys. Res.*, **102**(D14), 16 663–16 682.

Molinari, J. (1982). A method for calculating the effects of deep cumulus convection in numerical models. *Mon. Wea. Rev.*, **110**, 1527–1534.

(1985). A general form of Kuo's cumulus parameterization. *Mon. Wea. Rev.*, **113**, 1411–1416.

(1993). An overview of cumulus parameterization in mesoscale models. In *The Representation of Cumulus Convection in Numerical Models of the Atmosphere*, ed. K. A. Emanuel and D. J. Raymond. *Meteorology Monographs*, No. 46. American Meteorological Society, pp. 155–158.

Molinari, J. and M. Dudek (1992). Parameterization of convective precipitation in mesoscale models: a critical review. *Mon. Wea. Rev.*, **120**, 326–344.

Monteith, J. L. (1961). An empirical method for estimating long-wave radiation exchange in the British Isles. *Quart. J. Roy. Meteor. Soc.*, **87**, 171–179.

(1965). Evaporation and environment. *Symp. Soc. Exp. Biol.*, **19**, 205–234.

Monteith, J. L. and M. H. Unsworth (1990). *Principles of Environmental Physics*, 2nd edn. Edward Arnold.

Molteni, F., R. Buizza, T. N. Palmer, and T. Petroliagis (1996). The ECMWF ensemble prediction system: methodology and validation. *Quart. J. Roy. Meteor. Soc.*, **122**, 73–119.

Moore, K. E., D. R. Fitzjarrald, R. K. Sakai, *et al.* (1996). Seasonal variations in radiative and turbulent exchange at a deciduous forecast in Central Massachusetts. *J. Appl. Meteor.*, **35**, 122–134.

Morcrette, J.-J. (2002). The surface downward longwave radiation in the ECMWF forecast system. *J. Climate*, **15**, 1875–1892.

Morcrette, J.-J. and Y. Fouquart (1985). On systematic errors in the parametrized calculations of longwave radiation transfer. *J. Atmos. Sci.*, **43**, 321–328.

Morse, A. P., F. J. Doblas-Reyes, M. B. Hoshen, R. Hagedorn, and T. N. Palmer (2005). A forecast quality assessment of an end-to-end probabilistic multi-model seasonal forecast system using a malaria model. *Tellus*, **57A**, 464–475.

Morss, R. E., K. A. Miller, and M. S. Vasil (2005). A systematic economic approach to evaluating public investment in observations for weather forecasting. *Mon. Wea. Rev.*, **133**, 374–388.

Mullen, S. L. and D. P. Baumhefner (1994). Monte Carlo simulations of explosive cyclogenesis. *Mon. Wea. Rev.*, **122**, 1548–1567.

Müller, M. D. and D. Scherer (2005). A grid- and subgrid-scale radiation parameterization of topographic effects for mesoscale weather forecast models. *Mon. Wea. Rev.*, **133**, 1431–1442.

Munn, R. E. (1966). *Descriptive Micrometeorology*. Academic Press.

Murikami, M. (1990). Numerical modeling of dynamical and microphysical evolution of an isolated convective cloud – the 19 July 1982 CCOPE cloud. *J. Meteor. Soc. Japan*, **68**, 107–128.

Murphy, A. H. and R. L. Winkler (1979). Probabilistic temperature forecasts: the case for an operational program. *Bull. Amer. Meteor. Soc.*, **60**, 12–19.

Musil, D. J. (1970). Computer modeling of hailstone growth in feeder clouds. *J. Atmos. Sci.*, **27**, 474–482.

Nakaya, U. and T. Terada (1935). Simultaneous observations of the mass, falling velocity, and form of individual snow crystals. Hokkaido University, Ser. II, **1**, 191–201.

Ninomiya, K. (1971a). Dynamical analysis of outflow from tornado-producing thunderstorms as revealed by ATS III pictures. *J. Appl. Meteor.*, **10**, 275–294.

(1971b). Mesoscale modification of synoptic situations from thunderstorm development as revealed by ATS III and aerological data. *J. Appl. Meteor.*, **10**, 1103–1121.

Niyogi, D. S. and S. Raman (1997). Comparison of stomatal resistance simulated by four different schemes using FIFE observations. *J. Appl. Meteor.*, **36**, 903–917.

Niyogi, D. S., S. Raman, and K. Alapaty (1998). Comparison of four different stomatal resistance schemes using FIFE data. Part II: Analysis of terrestrial biospheric–atmospheric interactions. *J. Appl. Meteor.*, **37**, 1301–1320.

Niyogi, D. S., H. Chang, V. K. Saxena *et al.* (2004). Direct observations of the effects of aerosol loading on net ecosystem CO_2 exchanges over different landscapes. *Geophys. Res. Lett.*, **31**, L20506, doi: 10.1029/2004GL020915.

Noilhan, J. and S. Planton (1989). A simple parameterization of land-surface processes for meteorological models. *Mon. Wea. Rev.*, **117**, 536–549.

Nordeng, T. E. (1987). The effect of vertical and slantwise convection on the simulation of polar lows. *Tellus*, **39A**, 354–375.

(1993). Parameterization of slantwise convection in numerical weather prediction models. In *The Representation of Cumulus Convection in Numerical Models of the Atmosphere*, ed. K. A. Emanuel and D. J. Raymond. *Meteorology Monographs*, No. 46. American Meteorological Society, pp. 195–202.

Nowlin, W. D., Jr. and C. A. Parker (1974). Effects of a cold-air outbreak on shelf waters of the Gulf of Mexico. *J. Phys. Oceanogr.*, **4**, 467–486.

Oglesby, R. J. and D. J. Erickson III (1989). Soil moisture and the persistence of North American drought. *J. Climate*, **2**, 1362–1380.

Olson, D. A., N. W. Junker, and B. Korty (1995). Evaluation of 33 years of quantitative precipitation forecasting at the NMC. *Wea. Forecasting*, **10**, 498–511.

Ookouchi, Y., M. Segal, M. C. Kessler, and R. A. Pielke (1984). Evaluation of soil moisture effects on generation and modification of mesoscale circulations. *Mon. Wea. Rev.*, **112**, 2281–2292.

Oost, W. A., G. J. Komen, C. M. J. Jacobs, and C. van Oort (2002). New evidence for a relation between wind stress and wave age from measurements during ASGAMAGE. *Bound.-Layer Meteor.*, **103**, 409–438.

Otte, T. L., G. Pouliot, J. E. Pleim, *et al.* (2005). Linking the Eta model with the community multiscale air quality (CMAQ) modeling system to build a national air quality forecasting system. *Wea. Forecasting*, **20**, 367–384.

Ovtchinnikov, M. and Y. L. Kogan (2000). An investigation of ice production mechanisms in small cumuliform clouds using a 3D model with explicit microphysics. Part I: Model description. *J. Atmos. Sci.*, **57**, 2989–3003.

Palmer, T. N., G. J. Shutts, and R. Swinbank (1986). Alleviation of a systematic westerly bias in general circulation and numerical weather prediction models through an orographic gravity wave drag parametrization. *Quart. J. Roy. Meteor. Soc.*, **112**, 1001–1039.

Palmer, T. N., A. Alessandri, U. Andersen, *et al.* (2004). Development of a European multimodel ensemble system for seasonal-to-interannual prediction (DEMETER). *Bull. Amer. Meteor. Soc.*, **85**, 853–872.

Pan, D.-M. and D. A. Randall (1998). A cumulus parametrization with a prognostic closure. *Quart. J. Roy. Meteor. Soc.*, **124**, 949–981.

Pan, H. L. and L. Mahrt (1987). Interaction between soil hydrology and boundary-layer development. *Bound.-Layer Meteor.*, **38**, 185–202.

Pan, Y., A. D. McGuire, J. M. Melillo, *et al.* (2002). A biogeochemistry-based dynamic vegetation model and its application along a moisture gradient in the continental United States. *J. Vegetation Sci.*, **13**, 369–382.

Panofsky, H. A. and J. A. Dutton (1984). *Atmospheric Turbulence: Models and Methods for Engineering Applications.* John Wiley and Sons.

Paulson, C. A. (1970). The mathematical representation of wind speed and temperature profiles in an unstable atmospheric surface layer. *J. Appl. Meteor.*, **9**, 857–861.

Peixoto, J. P. and M. A. Kettani (1973). The control of the water cycle. *Sci. Amer.*, **228**, 46–61.

Peixoto, J. P. and A. H. Oort (1992). *Physics of Climate.* American Institute of Physics.

Peláez, D. V. and R. M. Bóo (1987). Plant water potential for shrubs in Argentina. *J. Range Manag.*, **40**, 6–9.

Peltier, W. R. and T. L. Clark (1979). The evolution and stability of finite-amplitude mountain waves. Part II: Surface-wave drag and severe downslope windstorms. *J. Atmos. Sci.*, **36**, 1498–1529.

Penman, H. L. (1948). Natural evaporation from open water, bare soil, and grass. *Proc. Roy. Soc. London*, **A193**, 120–195.

Persson, A. (2005a). Early operational numerical weather prediction outside the USA: an historical introduction. Part I: Internationalism and engineering, NWP in Sweden, 1952–69. *Meteorol. Appl.*, **12**, 135–159.

(2005b). Early operational numerical weather prediction outside the USA: an historical introduction. Part II: Twenty countries around the world. *Meteorol. Appl.*, **12**, 269–289.

(2005c). Early operational numerical weather prediction outside the USA: an historical introduction. Part III: Endurance and mathematics – British NWP, 1948–1965. *Meteorol. Appl.*, **12**, 381–413.

Peters-Lidard, C. D., E. Blackburn, X. Liang, and E. F. Wood (1998). The effect of soil thermal conductivity parameterization on surface energy fluxes and temperatures. *J. Atmos. Sci.*, **55**, 1209–1224.

Philander, S. G. H. (1989). *El Niño, La Niña, and the Southern Oscillation.* Academic Press.

Philipona, R. (2002). Underestimation of solar global and diffuse radiation measured at the Earth's surface. *J. Geophys. Res.*, **107**(D22), doi: 10.1019/2002JD002396.

Phillips, S. P. (1984). Analytical surface pressure and drag for linear hydrostatic flow over three-dimensional elliptical mountains. *J. Atmos. Sci.*, **41**, 1073–1084.

Pielke, R. A., G. A. Dalu, J. S. Snook, T. J. Lee, and T. G. F. Kittel (1991). Nonlinear influence of mesoscale land use on weather and climate. *J. Climate*, **4**, 1053–1069.

Pierrehumbert, R. T. (1986). An essay on the parameterization of orographic gravity-wave drag. In *Proc. Seminar/Workshop on Observation, Theory and Modeling of Orographic Effects*, vol. 1, September, Shinfield Park, Reading, ECMWF, pp. 251–282.

Pincus, R. and S. A. Ackermann (2003). Radiation in the atmosphere: foundations. In *Handbook of Weather, Climate, and Water. Dynamics, Climate, Physical Meteorology, Weather Systems, and Measurements*, ed. T. D. Potter and B. R. Colman. Wiley-Interscience, pp. 301–342.

Pleim, J. E. and J. S. Chang (1992). A non-local closure model for vertical mixing in the convective boundary layer. *Atmos. Environ.*, **26A**, 965–981.

Powers, J. G. and M. T. Stoelinga (2000). A coupled air–sea mesoscale model: experiments in atmospheric sensitivity to marine roughness. *Mon. Wea. Rev.*, **128**, 208–228.

Pressman, D. Y. (1994). Numerical model of hydrothermal processes in soil as part of the scheme of mesoscale forecasting. *Meteorol. Gidrol*, **11**, 62–75 (in Russian).

Priestley, C. H. B. (1954). Convection from a large horizontal surface. *Australian J. Phys.*, **6**, 279–290.

Priestley, C. H. B. and R. J. Taylor (1972). On the assessment of surface heat flux and evaporation using large-scale parameters. *Mon. Wea. Rev.*, **100**, 81–92.

Pruppacher, H. R. and J. D. Klett (2000). *Microphysics of Clouds and Precipitation*, 2nd edn. Kluwer Academic Publishers.

Pudykiewicz, R., R. Benoit, and J. Mailhot (1992). Inclusion and verification of a predictive cloud-water scheme in a regional numerical weather prediction model. *Mon. Wea. Rev.*, **120**, 612–626.

Rabier, F., H. Järvinen, E. Klinker, J. F. Mahfouf, and A. Simmons (2000). The ECMWF operational implementation of four-dimensional variational assimilation. Part I: Experimental results with simplified physics. *Quart. J. Roy. Meteor. Soc.*, **126**, 1143–1170.

Rabin, R. M., S. Stadler, P. J. Wetzel, D. J. Stensrud, and M. Gregory (1990). Observed effects of landscape variability on convective clouds. *Bull. Amer. Meteor. Soc.*, **71**, 272–280.

Ramanathan, V., E. J. Pitcher, R. C. Malone, and M. L. Blackman (1983). The response of a spectral general circulation model to refinements in radiative processes. *J. Atmos. Sci.*, **40**, 605–630.

Randall, D. A., J. A. Abeles, and T. G. Corsetti (1985). Seasonal simulations of the planetary boundary layer and boundary-layer stratocumulus clouds with a general circulation model. *J. Atmos. Sci.*, **42**, 641–675.

Randall, D. A., M. Khairoutdinov, A. Arakawa, and W. Grabowski (2003). Breaking the cloud parameterization deadlock. *Bull. Amer. Meteor. Soc.*, **84**, 1547–1564.

Rasmussen, E. (1985). A case study of a polar low development over the Barents Sea. *Tellus*, **37A**, 407–418.

Rasmussen, E. N. (1967). Atmospheric water vapor transport and the water balance of North America. Part I: Characteristics of the water vapor flux field. *Mon. Wea. Rev.*, **95**, 403–426.

Rasmussen, R., M. Politovich, J. Marwitz, *et al.* (1992). Winter Icing and Storms Project (WISP). *Bull. Amer. Meteor. Soc.*, **73**, 951–974.

Rauber, R. M. (2003). Microphysical processes in the atmosphere, ch. 18. In *Handbook of Weather, Climate, and Water. Dynamics, Climate, Physical Meteorology, Weather Systems, and Measurements*, ed. T. D. Potter and B. R. Colman. Wiley, pp. 255–299.

Rawls, W. J., D. L. Brakensiek, and K. E. Saxton (1982). Estimation of soil water properties. *Trans. Amer. Soc. Agric. Eng.*, **25**, 1316–1320.

Raymond, D. J. (1995). Regulation of moist convection over the west Pacific warm pool. *J. Atmos. Sci.*, **52**, 3945–3959.

Raymond, D. J. and A. M. Blyth (1986). A stochastic model for nonprecipitating cumulus clouds. *J. Atmos. Sci.*, **43**, 2708–2718.

Raymond, D. J. and K. A. Emanuel (1993). The Kuo cumulus parameterization. In *The Representation of Cumulus Convection in Numerical Models of the Atmosphere*, ed. K. A. Emanuel and D. J. Raymond. *Meteorology Monographs*, No. 46. American Meteorological Society, pp. 145–147.

Raymond, T. M. and S. N. Pandis (2002). Cloud activation of single-component organic aerosol particles. *J. Geophys. Res.*, **107**, doi: 10.1029/2002JD002159.

Raymond, W. H. and R. B. Stull (1990). Application of transilient turbulence theory to mesoscale numerical weather forecasting. *Mon. Wea. Rev.*, **118**, 2471–2499.

Reed, R. J. and E. E. Recker (1971). Structure and properties of synoptic-scale wave disturbances in the equatorial western Pacific. *J. Atmos. Sci.*, **28** 1117–1133.

Reisner, J., R. M. Rasmussen, and R. T. Bruintjes (1998). Explicit forecasting of supercooled liquid water in winter storms using the MM5 mesoscale model. *Quart. J. Royal Meteor. Soc.*, **124**, 1071–1107.

Reynolds, C. A., T. J. Jackson, and W. J. Rawls (2000). Estimating soil water-holding capacities by linking the Food and Agriculture Organization soil map of the world with global pedon databases and continuous pedotransfer functions. *Water Resources Res.*, **36**, 3653–3662.

Reynolds, D. (2003). Value-added quantitative precipitation forecasts: how valuable is the forecaster? *Bull. Amer. Meteor. Soc.*, **84**, 876–878.

Reynolds, R. W. and T. M. Smith (1994). Improved global sea surface temperature analyses using optimum interpolation. *J. Climate*, **7**, 929–948.

Richards, L. A. (1931). Capillary conduction of liquids through porous mediums. *Physics*, **1**, 318–333.

Robock, A., L. Luo, E. F. Wood, *et al.* (2003). Evaluation of the North American Land Data Assimilation System over the southern Great Plains during the warm season. *J. Geophys. Res.*, **108**, doi: 10.1029/2002JD003245.

Rodell, M., P. R. Houser, U. Jambor, *et al.* (2004). The global land data assimilation system. *Bull. Amer. Meteor. Soc.*, **85**, 381–394.

Rodgers, C. D. (1967). The use of emissivity in atmospheric radiation calculations. *Quart. J. Roy. Meteor. Soc.*, **93**, 43–54.

Rodgers, C. D. and C. D. Walshaw (1966). The computation of infrared cooling rate in planetary atmospheres. *Quart. J. Roy. Meteor. Soc.*, **92**, 67–92.

Roebber, P. J., D. M. Schultz, B. A. Colle, and D. J. Stensrud (2004). The risks and rewards of high-resolution and ensemble numerical weather prediction. *Wea. Forecasting*, **19**, 936–949.

Roeger, C., R. Stull, D. McClung, *et al.* (2003). Verification of mesoscale numerical weather forecasts in mountainous terrain for application to avalanche prediction. *Wea. Forecasting*, **18**, 1140–1160.

Rogers, R. R. (1976). *A Short Course in Cloud Physics*, 2nd edn. Pergamon.

Rogers, R. R. and M. K. Yau (1989). *A Short Course in Cloud Physics*, 3rd edn. Butterworth-Heinemann.

Ronda, R. J., B. J. J. M. van den Hurt, and A. A. M. Holtslag (2002). Spatial heterogeneity of the soil moisture content and its impact on surface flux densities and near-surface meteorology. *J. Hydrometeor.*, **3**, 556–570.

Ross, G. H. (1989). Model output statistics – an updateable scheme. Preprints. In *11th Conf. on Probability and Statistics in Atmospheric Sciences*, Monterey, CA. American Meteorological Society, pp. 93–97.

Rotta, J. C. (1951). Statistische theorie nichthomogener turbulenz. *Zeitschrift Phys.*, **129**, 547–572.

Rutledge, S. A. and P. V. Hobbs (1983). The mesoscale and microscale structure and organization of clouds and precipitation in midlatitude cyclones. Part VIII: A model for the "seeder-feeder" process in warm-frontal rainbands. *J. Atmos. Sci.*, **40**, 1185–1206.

(1984). The mesoscale and microscale structure and organization of clouds and precipitation in midlatitude cyclones. XII: A diagnostic modeling study of precipitation development in narrow cold-frontal rainbands. *J. Atmos. Sci.*, **41**, 2949–2972.

Rutter, A. J., K. A. Kershaw, P. C. Robins, and A. J. Morton (1971). A predictive model of rainfall interception in forests. I. Derivation of the model from observations in a plantation of Corsican pine. *Agric. Meteor.*, **9**, 367–384.

Rutter, A. J., A. J. Morton, and P. C. Robins (1975). A predictive model of rainfall interception in forests. II. Generalization of the model and comparison with observations in some coniferous and hardwood stands. *J. Appl. Ecol.*, **12**, 367–380.

Sanders, F. and J. R. Gyakum (1980). Synoptic-dynamic climatology of the "bomb." *Mon. Wea. Rev.*, **108**, 1589–1606.

Santanello, J. A., Jr. and T. N. Carlson (2001). Mesoscale simulation of rapid soil drying and its implication for predicting daytime temperature. *J. Hydrometeor.*, **2**, 71–88.

Sarachik, E. S. (2003). The ocean in climate, ch. 10. In *Handbook of Weather, Climate, and Water. Dynamics, Climate, Physical Meteorology, Weather Systems, and Measurements*, ed. T. D. Potter and B. R. Colman. Wiley-Interscience, pp. 129–133.

Sasamori, T. (1968). The radiative cooling calculation for application to general circulation experiments. *J. Appl. Meteor.*, **7**, 721–729.

Satheesh, S. K., V. Ramanathan, X. Li-Jones, *et al.* (1999). A model for the natural and anthropogenic aerosols over the tropical Indian Ocean derived from Indian Ocean Experiment data. *J. Geophys. Res.*, **104**, 27 421–27 440.

Sato, N., P. J. Sellers, D. A. Randall, *et al.* (1989). Effects of implementing the simple biosphere model in a general circulation model. *J. Atmos. Sci.*, **46**, 2757–2782.

Savijärvi, H. (1990). Fast radiation parameterization schemes for mesoscale and short-range forecast models. *J. Appl. Meteor.*, **29**, 437–447.

Savijärvi, H. and P. Räisänen (1998). Long-wave optical properties of water clouds and rain. *Tellus*, **50A**, 1–11.

Schaake, J. C., V. I. Koren, Q.-Y. Duan, K. Mitchell, and F. Chen (1996). Simple water balance model for estimating runoff at different spatial and temporal scales. *J. Geophys. Res.*, **101**, 7461–7475.

Schaake, J. C. *et al.* (2004). An intercomparison of soil moisture fields in the North American Land Data Assimilation System (NLDAS). *J. Geophys. Res.*, **109**, doi: 10.1019/2002JD003309.

Schlesinger, W. H. (1997). *Biogeochemistry: an Analysis of Global Change*. Academic Press.

Schneider, E. K., Z. Zhu, B. S. Giese, *et al.* (1997). Annual cycle and ENSO in a coupled ocean–atmosphere general circulation model. *Mon. Wea. Rev.*, **125**, 680–702.

Schneider, J. M. and D. K. Lilly (1999). An observational and numerical study of a sheared, convective boundary layer. Part I: Phoenix II observations, statistical description, and visualization. *J. Atmos. Sci.*, **56**, 3059–3078.

Schneider, S. H. (1972). Cloudiness as a global climate feedback mechanism: the effects on the radiation balance and surface temperature variations in cloudiness. *J. Atmos. Sci.*, **29**, 1413–1422.

Schreiner, A. J., D. A. Unger, W. P. Menzel, *et al.* (1993). A comparison of ground and satellite observations of cloud cover. *Bull. Amer. Meteor. Soc.*, **74**, 1851–1861.

Schultz, D. M., W. E. Bracken, L. F. Bosart, *et al.* (1997). The 1993 Superstorm cold surge: frontal structure, gap flow, and tropical impact. *Mon. Wea. Rev.*, **125**, 5–39; Corrigendum, **125**, 662.

Schultz, D. M., D. S. Arndt, D. J. Stensrud, and J. W. Hanna (2004). Snowbands during the cold-air outbreak of 23 January 2003. *Mon. Wea. Rev.*, **132**, 827–842.

Schultz, P. (1995). An explicit cloud physics parameterization for operational numerical weather prediction. *Mon. Wea. Rev.*, **123**, 3331–3343.

Schwarzkopf, M. D. and S. B. Fels (1991). The simplified exchange model revisited: an accurate, rapid method for computation of infrared cooling rates and fluxes. *J. Geophys. Res.*, **96**, 9075–9096.

Scinocca, J. F. and N. A. McFarlane (2000). The parametrization of drag induced by stratified flow over anisotropic orography. *Quart. J. Roy. Meteor. Soc.*, **126**, 2353–2393.

Scorer, R. S. (1949). Theory of waves in the lee of mountains. *Quart. J. Roy. Meteor. Soc.*, **75**, 41–56.

Segal, M. and R. W. Arritt (1992). Nonclassical mesoscale circulations caused by surface sensible heat-flux gradients. *Bull. Amer. Meteor. Soc.*, **73**, 1593–1604.

Segal, M., R. Avissar, M. C. McCumber, and R. A. Pielke (1988). Evaluation of vegetation effects on the generation and modification of mesoscale circulations. *J. Atmos. Sci.*, **45**, 2268–2292.

Segal, M., J. F. W. Purdom, J. L. Song, R. A. Pielke, and Y. Mahrer (1986). Evaluation of cloud shading effects on the generation and modification of mesoscale circulations. *Mon. Wea. Rev.*, **114**, 1201–1212.

Segal, M., W. E. Schreiber, G. Kallos, *et al.* (1989). The impact of crop areas in northeast Colorado on midsummer mesoscale thermal circulations. *Mon. Wea. Rev.*, **117**, 809–825.

Segal, M., W. L. Physick, J. E. Heim, and R. W. Arritt (1993). The enhancement of cold-front temperature contrast by differential cloud cover. *Mon. Wea. Rev.*, **121**, 867–873.

Segele, Z. T., D. J. Stensrud, I. C. Ratcliffe, and G. M. Henebry (2005). Influence of a hailstreak on boundary layer evolution. *Mon. Wea. Rev.*, **133**, 942–960.

Seguin, B. and N. Gignoux (1974). Etude experimentale de l'influence d'un reseau de brise-vent sur le profil vertical de vitesse du vent (Experimental study of the effects of a network of windbreaks on the vertical profile of windspeed). *Agric. For. Meteor.*, **13**, 15–23.

Sela, J. G. (1980). Spectral modeling at the National Meteorological Center. *Mon. Wea. Rev.*, **108**, 1279–1292.

Sellers, P. J. (1985). Canopy reflectance, photosynthesis and transpiration. *Int. J. Remote Sens.*, **6**, 1335–1372.

  (1987). Canopy reflectance, photosynthesis, and transpiration. II. The role of biophysics in the linearity of their interdepencence. *Rem. Sens. Env.*, **21**, 143–183.

Sellers, P. J., Y. Mintz, Y. C. Sud, and A. Dalcher (1986). A simple biosphere model (SiB) for use within general circulation models. *J. Atmos. Sci.*, **43**, 505–531.

Shettle, E. P. and J. A. Weinman (1970). The transfer of solar irradiance through inhomogeneous turbid atmospheres evaluated by Eddington's approximation. *J. Atmos. Sci.*, **27**, 1048–1055.

Shukla, J. and Y. Sud (1981). Effect of cloud–radiation feedback on the climate of a general circulation model. *J. Atmos. Sci.*, **38**, 2337–2353.

Shulman, M. L., M. C. Jacobson, R. J. Charlson, R. E. Synovec, and T. E. Young (1997). Dissolution behavior and surface tension effects of organic compounds in nucleating cloud droplets. *Geophys. Res. Lett.*, **23**, 277–280.

Shuman, F. G. (1989). History of numerical weather prediction at the National Meteorological Center. *Wea. Forecasting*, **4**, 286–296.

Shuman, F. G. and J. B. Hovermale (1968). An operational six-layer primitive equation model. *J. Appl. Meteor.*, **7**, 525–547.

Siebesma, A. P. and J. W. M. Cuijpers (1995). Evaluation of parametric assumptions for shallow cumulus convection. *J. Atmos. Sci.*, **52**, 650–666.

Simpson, J. and V. Wiggert (1969). Models of precipitating cumulus towers. *Mon. Wea. Rev.*, **97**, 471–489.

Sitch, S., B. Smith, I. C. Prentice, *et al.* (2003). Evaluation of ecosystem dynamics, plant geography and terrestrial carbon cycling in the LPJ dynamic global vegetation model. *Global Change Biol.*, **9**, 161–185.

Slater, A. G., C. A. Schlosser, C. E. Desborough, *et al.* (2001). The representation of snow in land surface schemes: results from PILPS 2(d). *J. Hydrometeor.*, **2**, 7–25.

Slingo, J. M. (1980). A cloud parameterization scheme derived from GATE data for use with a numerical model. *Quart. J. Roy. Meteor. Soc.*, **106**, 747–770.

(1987). The development and verification of a cloud prediction scheme for the ECMWF model. *Quart. J. Roy. Meteor. Soc.*, **113**, 899–927.

Smagorinsky, J. (1960). On the dynamical prediction of large-scale condensation by numerical methods. *Geophys. Monogr.*, No. 5. American Geophysical Union, pp. 71–78.

Smagorinsky, J., S. Manabe, and J. L. Holloway, Jr. (1965). Numerical results from a nine-level general circulation model of the atmosphere. *Mon. Wea. Rev.*, **93**, 727–768.

Smirnova, T. G., J. M. Brown, and S. G. Benjamin (1997). Performance of different soil model configurations in simulating ground surface temperature and surface fluxes. *Mon. Wea. Rev.*, **125**, 1870–1884.

Smirnova, T. G., J. M. Brown, S. G. Benjamin, and D. Kim (2000). Parameterization of cold-season processes in the MAPS land-surface scheme. *J. Geophys. Res.*, **105**, 4077–4086.

Smith, E. A., M. M.-K. Wai, H. J. Cooper, M. T. Rubes, and A. Hsu (1994). Linking boundary-layer circulations and surface processes during FIFE 89. Part I: Observational analysis. *J. Atmos. Sci.*, **51**, 1497–1529.

Smith, P. L. (2003). Raindrop size distributions: exponential or gamma – does the difference matter? *J. Appl. Meteor.*, **42**, 1031–1034.

Smith, R. B. (1979). The influence of mountains on the atmosphere. *Adv. Geophys.*, **33**, 87–230.

(1990). A scheme for predicting layer clouds and their water-content in a general-circulation model. *Quart. J. Roy. Meteor. Soc.*, **116**, 435–460.

Smith, S. D. (1988). Coefficients for seas surface wind stress, heat flux, and wind profiles as a function of wind speed and temperature. *J. Geophys. Res.*, **93**, 15 467–15 472.

Snyder, C. and F. Zhang (2003). Assimilation of simulated Doppler radar observations with an ensemble Kalman filter. *Mon. Wea. Rev.*, **131**, 1663–1677.

Spahn, J. F. (1976). The airborne hail disdrometer: an analysis of its 1975 performance. Rep. 76–13, *Inst. Atmos. Sci., South Dakota School of Mines and Technology*, Rapid City, SD.

Stainforth, D. A., T. Aina, C. Christensen, *et al.* (2005). Uncertainty in predictions of the climate response to rising levels of greenhouse gases. *Nature*, **433**, 403–406.

Staley, D. O. and G. M. Jurica (1970). Flux emissivity tables for water vapor, carbon dioxide and ozone. *J. Appl. Meteor.*, **9**, 365–372.

Stensrud, D. J. (1993). Elevated residual layers and their influence on surface boundary-layer evolution. *J. Atmos. Sci.*, **50**, 2284–2293.

(1996). Importance of low-level jets to climate: a review. *J. Climate*, **9**, 1698–1711.

(1996). Effects of a persistent, midlatitude mesoscale region of convection on the large-scale environment during the warm season. *J. Atmos. Sci.*, **53**, 3503–3527.

Stensrud, D. J. and J. L. Anderson (2001). Is midlatitude convection an active or a passive player in producing global circulation patterns? *J. Climate*, **14**, 2222–2237.

Stensrud, D. J. and J. M. Fritsch (1994). Mesoscale convective systems in weakly forced large-scale environments. Part III: Numerical simulations and implications for operational forecasting. *Mon. Wea. Rev.*, **112**, 2084–2104.

Stensrud, D. J. and S. J. Weiss (2002). Mesoscale model ensemble forecasts of the 3 May 1999 tornado outbreak. *Wea. Forecasting*, **17**, 526–543.

Stensrud, D. J. and N. Yussouf (2003). Short-range ensemble predictions of 2-m temperature and dewpoint temperature over New England. *Mon. Wea. Rev.*, **131**, 2510–2524.

(2005). Bias-corrected short-range ensemble forecasts of near surface variables. *Meteor. Appl.*, **12**, 217–230.

Stensrud, D. J., H. E. Brooks, J. Du, M. S. Tracton, and E. Rogers (1999). Using ensembles for short-range forecasting. *Mon. Wea. Rev.*, **127**, 433–446.

(2000a). Reply to comments on "Using ensembles for short-range forecasting." *Mon. Wea. Rev.*, **128**, 3021–3023.

Stensrud, D. J., J.-W. Bao, and T. T. Warner (2000b). Using initial condition and model physics perturbations in short-range ensembles of mesoscale convective systems. *Mon. Wea. Rev.*, **128**, 2077–2107.

Stensrud, D. J., N. Yussouf, M. E. Baldwin, *et al.* (2006). The New England High-Resolution Temperature Program (NEHRTP). *Bull. Amer. Meteor. Soc.*, **87**, 491–498.

Stephens, G. L. (1978a). Radiative properties of extended water clouds. Part I: Theory. *J. Atmos. Sci.*, **35**, 2111–2122.

(1978b). Radiative properties of extended water clouds. Part II: Parameterization schemes. *J. Atmos. Sci.*, **35**, 2123–2132.

(1984). The parameterization of radiation for numerical weather prediction and climate models. *Mon. Wea. Rev.*, **112**, 826–867.

(2005). Cloud feedbacks in the climate system: a critical review. *J. Climate*, **18**, 237–273.

Stephens, G. L. and S.-C. Tsay (1990). On the cloud absorption anomaly. *Quart. J. Roy. Meteor. Soc.*, **116**, 671–704.

Stephens, G. L., S.-C. Tsay, P. W. Stackhouse Jr., and P. J. Flatau (1990). The relevance of the microphysical and radiative properties of cirrus clouds to climatic feedback. *J. Atmos. Sci.*, **47**, 1742–1753.

Stephens, G. L., N. B. Wood, and P. M. Gabriel (2004). Assessment of the parameterization of subgrid-scale cloud effects on radiative transfer. Part I: Vertical overlap. *J. Atmos. Sci.*, **61**, 715–732.

Stern, W. F., R. T. Pierrehumbert, J. Sirutis, J. Ploshay, and K. Miyakoda (1987). Recent development in the GFDL extended-range forecasting system. *J. Meteor. Soc. Japan* (special volume WMO/IUGG NWP Smp.), 359–363.

Stevens, B., C.-H. Moeng, A. S. Ackerman, *et al.* (2005). Evaluation of large-eddy simulations via observations of nocturnal marine stratocumulus. *Mon. Wea. Rev.*, **133**, 1443–1462.

Stockdale, T. N., D. L. T. Anderson, J. O. S. Alves, and M. A. Balmaseda (1998). Global seasonal rainfall forecasts using a coupled ocean–atmosphere model. *Nature*, **392**, 370–373.

Stoelinga, M. T., P. V. Hobbs, C. V. Mass, *et al.* (2003). Improvement of Microphysical Parameterization through Observational Verification Experiment (IMPROVE). *Bull. Amer. Meteor. Soc.*, **84**, 1807–1826.

Stokes, G. M. and S. E. Schwartz (1994). The Atmospheric Radiation Measurement (ARM) program: programmatic background and design of the cloud and radiation test bed. *Bull. Amer. Meteor. Soc.*, **75**, 1201–1221.

Straka, J. M. and E. R. Mansell (2005). A bulk microphysics parameterization with multiple ice precipitation categories. *J. Appl. Meteor.*, **44**, 445–466.

Straka, J. M. and E. N. Rasmussen (1997). Toward improving microphysical parameterizations of conversion processes. *J. Appl. Meteor.*, **36**, 896–902.

Stull, R. B. (1976). The energetics of entrainment across a density interface. *J. Atmos. Sci.*, **33**, 1260–1278.

(1984). Transilient turbulence theory. Part I: The concept of eddy mixing across small distances. *J. Atmos. Sci.*, **41**, 3351–3367.

(1988). *An Introduction to Boundary Layer Meteorology*. Kluwer.

(1991). Static stability – an update. *Bull. Amer. Meteor. Soc.*, **72**, 1521–1529.

(1993). Review of non-local mixing in turbulent atmospheres: transilient turbulence theory. *Bound.-Layer Meteor.*, **62**, 21–96.

Stumpf, H. G. (1975). Satellite detection of upwelling in the Gulf of Tehuantepec, Mexico. *J. Phys. Oceanogr.*, **5**, 383–388.

Su, H. B., K. T. Paw U, and R. Shaw (1996). Development of a coupled leaf and canopy model for the simulation of plant-atmosphere interaction. *J. Appl. Meteor.*, **35**, 734–748.

Sublette, M. S. and G. S. Young (1996). Warm-season effects of the Gulf Stream on mesoscale characteristics of the atmospheric boundary layer. *Mon. Wea. Rev.*, **124**, 653–667.

Sundqvist, H. (1988). Parameterization of condensation and associated clouds for weather prediction and general circulation simulation. In *Physically-Based Modelling and Simulation of Climate and Climate Change*, ed. M. E. Schlesinger. Reidel, pp. 433–461.

Sundqvist, H., E. Berge, and J. E. Kristjansson (1989). Condensation and cloud parameterization studies with a mesoscale numerical weather prediction model. *Mon. Wea. Rev.*, **117**, 1641–1657.

Swann, H. (1998). Sensitivity to the representation of precipitating ice in CRM simulations of deep convection. *Atmos. Res.*, **47–48**, 415–435.

Sweet, W., R. Felt, J. Kerling, and P. La Violette (1981). Air–sea interaction effects in the lower troposphere across the north wall of the Gulf Stream. *Mon. Wea. Rev.*, **109**, 1042–1052.

Swinbank, W. C. (1968). A comparison between predictions of dimensional analysis for the constant-flux layer and observations in unstable conditions. *Quart. J. Roy. Meteor. Soc.*, **94**, 460–467.

Taiz, L. and E. Zeiger (2002). *Plant Physiology*. Sinauer Associates, Inc.

Takara, E. E. and R. G. Ellingson (2000). Broken cloud field longwave scattering effects. *J. Atmos. Sci.*, **57**, 1298–1310.

Tao, W.-K., J. R. Scala, B. Ferrier, and J. Simpson (1995). The effect of melting processes on the development of a tropical and a midlatitude squall line. *J. Atmos. Sci.*, **52**, 1934–1948.

Taylor, P. K. and M. A. Yelland (2001). The dependence of sea surface roughness on the height and steepness of the waves. *J. Phys. Oceanogr.*, **31**, 572–590.

Teixeira, J. and T. F. Hogan (2002). Boundary layer clouds in a global atmospheric model: simple cloud cover parameterizations. *J. Climate*, **15**, 1261–1275.

Teixeira, J., J. P. Ferreira, P. M. A. Miranda, *et al.* (2004). A new mixing-length formulation for the parameterization of dry convection: implementation and evaluation in a mesoscale model. *Mon. Wea. Rev.*, **132**, 2698–2707.

Telford, J. W. (1955). A new aspect of coalescence theory. *J. Meteor.*, **12**, 436–444.
   (1975). Turbulence, entrainment and mixing in cloud dynamics. *Pure Appl. Geophys.*, **113**, 1067–1084.

Thiebaux, J., E. Rogers, W. Wang, and B. Katz (2003). A new high-resolution blended real-time global sea surface temperature analysis. *Bull. Amer. Meteor. Soc.*, **84**, 645–656.

Thompkins, A. M. (2002). A prognostic parameterization for the subgrid-scale variability of water vapor and clouds in large-scale models and its use to diagnose cloud cover. *J. Atmos. Sci.*, **59**, 1917–1942.

Thompson, G., R. M. Rasmussen, and K. Manning (2004). Explicit forecasts of winter precipitation using an improved bulk microphysics scheme. Part I: Description and sensitivity analysis. *Mon. Wea. Rev.*, **132**, 519–542.

Thompson, P. D. (1957). Uncertainty of the initial state as a factor in the predictability of large scale atmospheric flow patterns. *Tellus*, **9**, 275–295.

Tian, L. and J. A. Curry (1989). Cloud overlap statistics. *J. Geophys. Res.*, **94**, 9925–9935.

Tibaldi, S. (1986). Envelope orography and the maintenance of quasi-stationary waves in the ECMWF model. *Adv. Geophys.*, **29**, 339–374.

Tiedtke, M. (1989). A comprehensive mass flux scheme for cumulus parameterization in large-scale models. *Mon. Wea. Rev.*, **117**, 1779–1800.
   (1993). Representation of clouds in large-scale models. *Mon. Wea. Rev.*, **121**, 3040–3061.

Tokay, A. and D. A. Short (1996). Evidence from tropical raindrop spectra of the origin of rain from stratiform versus convective clouds. *J. Appl. Meteor.*, **35**, 355–371.

Tompkins, A. M. (2001). Organization of tropical convection in low vertical wind shears: the role of water vapor. *J. Atmos. Sci.*, **58**, 529–545.

Toth, Z. and E. Kalnay (1993). Ensemble forecasting at NMC: the generation of perturbations. *Bull. Amer. Meteor. Soc.*, **74**, 2317–2330.
   (1997). Ensemble forecasting at NCEP and the breeding method. *Mon. Wea. Rev.*, **125**, 3297–3319.

Trenberth, K. E. and C. J. Guillemot (1996). Physical processes involved in the 1988 drought and 1993 floods in North America. *J. Climate*, **9**, 1288–1298.

Tribbia, J. J. (1991). The rudimentary theory of atmospheric teleconnections associated with ENSO. In *Teleconnections Linking Worldwide Climate Anomalies*, ed. M. H. Glantz, R. W. Katz, and N. Nicholls. Cambridge University Press, pp. 285–308.

Tribbia, J. J. and D. P. Baumhefner (2004). Scale interactions and atmospheric predictability: an updated perspective. *Mon. Wea. Rev.*, **132**, 703–713.

Trier, S. B., C. A. Davis, and J. D. Tuttle (2000). Long-lived mesoconvective vortices and their environment. Part I: Observations from the central United States during the 1998 warm season. *Mon. Wea. Rev.*, **128**, 3376–3395.

Tripoli, G. J. and W. R. Cotton (1980). A numerical investigation of several factors contributing to the observed variable intensity of deep convection over south Florida. *J. Appl. Meteor.*, **19**, 1037–1063.

Troen, I. and L. Mahrt (1986). A simple model of the atmospheric boundary layer: sensitivity to surface evaporation. *Bound.-Layer Meteor.*, **37**, 129–148.

Tucker, C. J., J. A. Gatlin, and S. R. Schneider (1984). Monitoring vegetation in the Nile delta with NOAA-6 and NOAA-7 AVHRR imagery. *Photo. Eng. Remote Sens.*, **50**, 53–61.

Twomey, S. (1976). Computations of the absorption of solar radiation by clouds. *J. Atmos. Sci.*, **33**, 1087–1091.

(1977). The influence of pollution on the shortwave albedo of clouds. *J. Atmos. Sci.*, **34**, 1149–1152.

Unsworth, M. H. and J. L. Monteith (1975). Long-wave radiation at the ground. I. Angular distribution of incoming radiation. *Quart. J. Roy. Meteor. Soc.*, **101**, 13–24.

Van de Hulst, H. C. (1945). Theory of absorption lines in the atmosphere of the Earth. *Ann. Astrophys.*, **8**, 21–34.

Viterbo, P. and A. C. M. Beljaars (1995). An improved land surface parameterization scheme in the ECMWF model and its validation. *J. Climate*, **8**, 2716–2748.

Walcek, C. J. (1994). Cloud cover and its relationship to relative humidity during a springtime midlatitude cyclone. *Mon. Wea. Rev.*, **122**, 1021–1035.

Walko, R., W. R. Cotton, M. P. Meyers, and J. Y. Harrington (1995). New RAMS cloud microphysics parameterization. Part I: The single-moment scheme. *Atmos. Res.*, **38**, 29–62.

Wallace, J. M. and P. V. Hobbs (1977). *Atmospheric Science: an Introductory Survey.* Academic Press.

Wallace, J. M., S. Tibaldi, and A. J. Simmons (1983). Reduction of systematic forecast errors in the ECMWF model through the introduction of an envelope orography. *Quart. J. Roy. Meteor. Soc.*, **109**, 683–717.

Walters, M. K. (2000). Comments on "The differentiation between grid spacing and resolution and their application to numerical modeling." *Bull. Amer. Meteor. Soc.*, **81**, 2475–2477.

Walton, C. C. (1988). Nonlinear multichallen algorithms for estimating sea surface temperature with AVHRR satellite data. *J. Appl. Meteor.*, **27**, 115–124.

Wandishin, M. S., S. L. Mullen, D. J. Stensrud, and H. E. Brooks (2001). Evaluation of a short-range multimodel ensemble system. *Mon. Wea. Rev.*, **129**, 729–747.

Wang, W. and N. L. Seaman (1997). A comparison study of convective parameterization schemes in a mesoscale model. *Mon. Wea. Rev.*, **125**, 252–278.

Wang, Y.-P. and P. J. Jarvis (1990). Description and validation of an array model – MAESTRO. *Agric. For. Meteor.*, **51**, 257–280.

Warner, T. T. and N. L. Seaman (1990). A real-time, mesoscale numerical weather prediction system used for research, teaching, and public service at The Pennsylvania State University. *Bull. Amer. Meteor. Soc.*, **71**, 792–805.

Webb, R. S., C. E. Rosenzweig, and E. R. Levine (1993). Specifying land surface characteristics in general circulation models: soil profile data and derived water-holding capacities. *Global Biogeochemical Cycles*, **7**, 97–108.

Weber, S. L., H. von Storch, P. Viterbo, and L. Zambresky (1993). Coupling an ocean wave model to an atmospheric general circulation model. *Climate Dyn.*, **9**, 53–61.

Webster, P. J. and G. L. Stephens (1983). Cloud–radiation interaction and the climate problem. In *The Global Climate*, ed. J. T. Houghton. Cambridge University Press, pp. 63–78.

Weidinger, T., J. Pinto, and L. Horváth (2000). Effects of uncertainties in universal functions, roughness length, and displacement height on the calibration of surface layer fluxes. *Meteor. Zeit.*, **9**, 139–154.

Weisman, M. L., W. C. Skamarock, and J. B. Klemp (1997). The resolution dependence of explicitly modeled convective systems. *Mon. Wea. Rev.*, **125**, 527–548.

Welch, R. M., S. K. Cox, and J. M. Davis (1980). *Solar Radiation and Clouds. Meteorology Monographs*, No. 39. American Meteorological Society.

Wetzel, P. and J.-T. Chang (1987). Concerning the relationship between evapotranspiration and soil moisture. *J. Clim. Appl. Meteor.*, **26**, 18–27.

Wichmann, M. and E. Schaller (1986). On the determination of the closure parameters in higher-order closure models. *Bound.-Layer Meteor.*, **37**, 323–341.

Williams, R. J., K. Broersma, and A. L. Van Ryswyk (1978). Equilibrium and actual evapotranspiration from a very dry vegetated surface. *J. Appl. Meteor.*, **17**, 1827–1832.

Wilson, D. R. and S. P. Ballard (1999). A microphysically based precipitation scheme for the UK Meteorological Office Unified Model. *Quart. J. Roy. Meteor. Soc.*, **125**, 1607–1636.

Wilson, L. J. and M. Vallée (2002). The Canadian updateable model output statistics (UMOS) system: design and development tests. *Wea. Forecasting*, **17**, 206–222.

Wolf, B. J. and D. R. Johnson (1995a). The mesoscale forcing of a midlatitude upper-tropospheric jet streak by a simulated convective system. Part I: Mass circulation and ageostrophic processes. *Mon. Wea. Rev.*, **123**, 1059–1087.

(1995b). The mesoscale forcing of a midlatitude upper-tropospheric jet streak by a simulated convective system. Part II: Kinetic energy and resolution analysis. *Mon. Wea. Rev.*, **123**, 1088–1111.

Woodcock, F. and C. Engel (2005). Operational consensus forecasts. *Wea. Forecasting*, **20**, 101–111.

Wu, M. L. (1980). The exchange of infrared radiative energy in the troposphere. *J. Geophys. Res.*, **85**, 4084–4090.

Wyngaard, J. C. (1985). Structure of the planetary boundary layer and implications for its modeling. *J. Clim. Appl. Meteor.*, **24**, 1131–1142.

Xu, M., J.-W. Bao, T. T. Warner, and D. J. Stensrud (2001a). Effect of time step size in MM5 simulations of a mesoscale convective system. *Mon. Wea. Rev.*, **129**, 502–516.

Xu, M., D. J. Stensrud, J.-W. Bao, and T. T. Warner (2001b). Applications of the adjoint technique to short-range ensemble forecasting of mesoscale convective systems. *Mon. Wea. Rev.*, **129**, 1395–1418.

Xue, H. and J. M. Bane, Jr. (1997). A numerical investigation of the Gulf Stream and its meanders in response to cold air outbreaks. *J. Phys. Oceanogr.*, **27**, 2606–2629.

Xue, H., J. M. Bane, Jr., and L. M. Goodman (1995). Modification of the Gulf Stream through strong air–sea interactions in winter: observations and numerical simulations. *J. Phys. Oceanogr.*, **25**, 533–557.

Xue, H., Z. Pan, and J. M. Bane, Jr. (2000). A 2D coupled atmosphere–ocean model study of air–sea interactions during a cold air outbreak over the Gulf Stream. *Mon. Wea. Rev.*, **128**, 973–996.

Xue, Y., M. J. Fennessy, and P. J. Sellers (1996). Impact of vegetation properties on U.S. summer weather prediction. *J. Geophys. Res.*, **101**, 7419–7430.

Yaglom, J. M. (1977). Comments on wind and temperature flux-profile relationships. *Bound.-Layer Meteor.*, **11**, 89–102.

Yamada, T. and G. Mellor (1975). A simulation of the Wangara boundary layer data. *J. Atmos. Sci.*, **32**, 2309–2329.

Yanai, M. and R. H. Johnson (1993). Impacts of cumulus convection on thermodynamic fields. In *The Representation of Cumulus Convection in Numerical Models of the Atmosphere*, ed. K. A. Emanuel and D. J. Raymond. *Meteorology Monographs*, No. 46. American Meteorological Society, pp. 39–62.

Yanai, M., S. Esbensen, and J. Chu (1973). Determination of bulk properties of tropical cloud clusters from large-scale heat and moisture budgets. *J. Atmos. Sci.*, **30**, 611–627.

Yanai, M., J.-H. Chu, T. E. Stark, and T. Nitta (1976). Response of deep and shallow tropical maritime cumuli to large-scale processes. *J. Atmos. Sci.*, **33**, 976–991.

Yin, Z. and T. H. L. Williams (1997). Obtaining spatial and temporal vegetation data from Landsat MSS and AVHRR/NOAA satellite images for a hydrological model. *Photogr. Eng. Remote Sens.*, **63**, 69–77.

Young, G. S. (1988). Turbulence structure of the convective boundary layer. Part II: Phoenix 78 aircraft observations of thermals and their environments. *J. Atmos. Sci.*, **45**, 727–735.

Yussouf, N., D. J. Stensrud, and S. Lakshmivarahan (2004). Cluster analysis of multimodel ensemble data over New England. *Mon. Wea. Rev.*, **132**, 2452–2462.

Zamora, R. J., S. Solomon, E. G. Dutton, *et al.* (2003). Comparing MM5 radiative fluxes with observations gathered during the 1995 and 1999 Nashville southern oxidant studies. *J. Geophys. Res.*, **108**, doi: 10.1029/2002JD002122.

Zamora, R. J., E. G. Dutton, M. Trainer, *et al.* (2005). The accuracy of solar irradiance calculations used in mesoscale numerical weather prediction. *Mon. Wea. Rev.*, **133**, 783–792.

Zängl, G. (2002). An improved method for computing horizontal diffusion in a sigma-coordinate model and its application to simulations over mountainous topography. *Mon. Wea. Rev.*, **130**, 1423–1432.

Zeman, O. (1981). Progress in the modeling of planetary boundary layers. *Ann. Rev. Fluid Mech.*, **13**, 253–272.

Zeng, X. (2001). Global vegetation root distribution for land modeling. *J. Hydrometeor.*, **2**, 525–530.

Zeng, X. and R. E. Dickinson (1998). Effect of surface sublayer on surface skin temperature and fluxes. *J. Climate*, **11**, 537–550.

Zeng, X., M. Zhao, and R. E. Dickinson (1998). Comparison of bulk aerodynamic algorithms for the computation of sea surface fluxes using the TOGA COARE data. *J. Climate*, **11**, 2628–2644.

Zhang, D.-L. and R. A. Anthes (1982). A high-resolution model of the planetary boundary layer – sensitivity tests and comparisons with SESAME-79 data. *J. Appl. Meteor.*, **21**, 1594–1609.

Zhang, D.-L. and R. Harvey (1995). Enhancement of extratropical cyclogenesis by a mesoscale convective system. *J. Atmos. Sci.*, **52**, 1107–1127.

Zhang, D.-L. and W.-Z. Zheng (2004). Diurnal cycles of surface winds and temperatures as simulated by five boundary layer parameterizations. *J. Appl. Meteor.*, **43**, 157–169.

Zhang, D.-L., E.-Y. Hsie, and M. W. Moncrieff (1988). A comparison of explicit and implicit predictions of convective and stratiform precipitating weather systems with a meso-β scale numerical model. *Quart. J. Roy. Meteor. Soc.*, **114**, 31–60.

Zhang, G. J. and H.-R. Cho (1991). Parameterization of the vertical transport of momentum by cumulus clouds. Part II: Application. *J. Atmos. Sci.*, **48**, 2448–2457.

Zhang, H. and C. S. Frederiksen (2003). Local and nonlocal impacts of soil moisture initialization on AGCM seasonal forecasts: a model sensitivity study. *J. Climate*, **16**, 2117–2137.

Zhao, Q. and F. H. Carr (1997). A prognostic cloud scheme for operational NWP models. *Mon. Wea. Rev.*, **125**, 1931–1953.

Zhou, Y. P. and R. D. Cess (2000). Validation of longwave atmospheric radiation models using atmospheric radiation measurement data. *J. Geophys. Res.*, **105**, 29 703–29 716.

Zhu, Y. and R. E. Newell (1998). A proposed algorithm for moisture fluxes from atmospheric rivers. *Mon. Wea. Rev.*, **126**, 725–735.

Ziegler, C. L. (1985). Retrieval of thermal and microphysical variables in observed convective storms. Part I: Model development and preliminary testing. *J. Atmos. Sci.*, **42**, 1487–1509.

(1988). Retrieval of thermal and microphysical variables in observed convective storms. Part II: Sensitivity of cloud processes to variation of the microphysical parameterization. *J. Atmos. Sci.*, **45**, 1072–1090.

Ziegler, C. L., P. S. Ray, and N. C. Knight (1983). Hail growth in an Oklahoma multicell storm. *J. Atmos. Sci.*, **40**, 1768–1791.

Ziegler, C. L., T. J. Lee, and R. A. Pielke, Sr. (1997). Convective initiation at the dryline: a modeling study. *Mon. Wea. Rev.*, **125**, 1001–1026.

Ziehmann, C. (2000). Comparison of a single-model EPS with a multi-model ensemble consisting of a few operational models. *Tellus*, **52A**, 280–299.

Zilitinkevich, S. (1995). Non-local turbulent transport: pollution dispersion aspects of coherent structure of convective flows. In *Air Pollution Theory and Simulation, Air Pollution III*, vol. I, ed. H. Power, N. Moussiopoulos, and C. A. Brebbia. Computational Mechanics Publications, pp. 53–60.

Zilitinkevich, S. and D. V. Chalikov (1968). On the determination of the universal wind and temperature profiles in the surface layer of the atmosphere. *Izv. Acad. Sci. U.S.S.R., Atmos. Oceanic Phys.*, **4**, 294–302.

Zilitinkevich, S., A. A. Grachev, and C. W. Fairall (2001). Scaling reasoning and field data on the sea surface roughness lengths for scalars. *J. Atmos. Sci.*, **58**, 320–325.

Zobler, L. (1986). A world soil file for global climate modeling. NASA Tech. Memo. 87802.

Županski, D. and F. Mesinger (1995). Four-dimensional variational assimilation of precipitation data. *Mon. Wea. Rev.*, **123**, 1112–1127.

Županski, M., D. Županski, T. Vukicevic, K. Eis, and T. Vonder Haar (2005). CIRA/CSU four-dimensional variational data assimilation system. *Mon. Wea. Rev.*, **133**(4), 829–843.

Index